AF677921

MICROWAVE INTEGRATED CIRCUITS

MICROWAVE INTEGRATED CIRCUITS

Edited by

K. C. Gupta
Indian Institute of Technology, Kanpur

Amarjit Singh
Central Electronics Engineering Research Institute, Pilani

A HALSTED PRESS BOOK

JOHN WILEY & SONS, INC.
New York • London • Sydney • Toronto

Published in the Western Hemisphere
by Halsted Press, a Division of
John Wiley & Sons, Inc., New York

Library of Congress Cataloging in Publication Data

Gupta, K C

Microwave integrated circuits.

"A Halsted Press book."
Bibliography : p.
1. Microwave circuits. 2. Integrated circuits.
I. Singh, Amarjit 1924—joint author. II Title.
TK7876.G8 621.381'73 74-8772

ISBN 0-470-33640-4

Copyright © 1974, Wiley Eastern Private Limited
New Delhi

Printed in India

PREFACE

Microwave integrated circuit (MIC) techniques represent an extension of integrated circuit technology to microwave frequencies. MICs are now becoming available to microwave systems designers as an alternative to conventional waveguide or coaxial circuits. The interest in MICs arises mainly from their potential advantages in respect of their small size, low weight, high reliability and reproducibility. When produced in large numbers they also have an advantage of lower cost.

MICs combined with current developments in microwave solid-state devices are likely to bring about a radical change in microwave technology just as digital and linear ICs have brought for low frequency circuit applications. Cost reduction and other advantages will lead to opening of many new applications for non-defence usages and the expansion of present microwave markets.

Technology of MICs is interdisciplinary in nature. It draws on a variety of areas like semiconductor devices, integrated circuit technology, microwave circuits and microwave system design. In a technology of this type boundaries between the roles of component designer, circuit designer and system designers get blurred and the MIC designer is called upon to have some expertise in all of these areas.

This monograph presents some basic material useful for understanding and design of microwave integrated circuits. Transmission structures, passive components, lumped elements and non-reciprocal devices are discussed in Chapters 2 to 7. Chapters 8 and 9 describe microwave semiconductor devices used in MICs. Circuits and technological aspects of MICs are treated in Chapters 10 to 13. Typical system applications are discussed in Chapter 14. Chapter 15 is a classified bibliography of research papers on MICs published during the period 1967 to 1973. Subject and author indices to the bibliography are included.

Preparation of this monograph started with an Intensive Course on this subject organised by Indian Institute of Technology, Kanpur (IIT/K) in October, 1972. This was one of a series of courses IIT/K organises for informing the people working in industry and research institutes about the latest trends in certain specialised areas. For this course, lecture notes were prepared and distributed to the participants. As such material has not been available at one place elsewhere, there was an extensive demand for these notes. It was, therefore, considered desirable to revise the lecture material and to publish it as an edited monograph. The present volume is the outcome of this endeavour.

The material for this volume has been contributed by twelve persons, seven from IIT/K, four from Central Electronic Engineering Research Institute, Pilani and one from University of Waterloo, Waterloo (Canada). In a text of this nature, certain diversity of approach is inevitable even after editing. The editors hope that the monograph would serve its basic purpose inspite of this. It is expected that the monograph will be useful for post-graduate students, research scholars, working engineers and teachers interested in microwave circuits.

Publication of the monograph has been made possible by the facilities extended by Advanced Centre for Electronic Systems, IIT/K and Department of Electrical Engineering, IIT/K. Financial help by Advanced Centre for Electronic Systems for the purpose is gratefully acknowledged. We are thankful to Dr. B. Prasada and Dr. K.R. Sarma of I.I.T. Kanpur for supporting this project. Assistance of Mr. Ramesh Garg in editing the typescript is thankfully appreciated. We acknowledge with thanks the assistance rendered by Mr. C.M. Abraham in typing the manuscript and Mr. J.C. Verma in preparing the figures.

Kanpur, India
May 18, 1974

K.C. GUPTA
AMARJIT SINGH

CONTENTS

CONTRIBUTORS

R.N. Agnihotri	Central Electronics Engineering Research Institute, Pilani (Rajasthan)
Amarjit Singh	Central Electronics Engineering Research Institute, Pilani (Rajasthan)
K.C. Gupta	Department of Electrical Engineering and Advanced Centre for Electronic Systems, Indian Institute of Technology, Kanpur(U.P.)
M.M. Hasan	Department of Electrical Engineering and Advanced Centre for Electronic Systems, Indian Institute of Technology, Kanpur(U.P.)
G.T. Mirchandani	Central Electronics Engineering Research Institute, Pilani (Rajasthan)
S.K. Mullick	Department of Electrical Engineering and Advanced Centre for Electronic Systems, Indian Institute of Technology,Kanpur(U.P.)
Nagendra Singh	Department of Electrical Engineering and Advanced Centre for Electronic Systems, Indian Institute of Technology,Kanpur(U.P.)
Ramesh Garg	Department of Physics, Indian Institute of Technology, Kanpur (U.P.)
D.J. Roulston	Department of Electrical Engineering, University of Waterloo, Waterloo (Ontario), Canada.
R. Sharan	Department of Electrical Engineering and Advanced Centre for Electronic Systems, Indian Institute of Technology,Kanpur(U.P.)
B.G. Sharma	Central Electronics Engineering Research Institute, Pilani (Rajasthan)
P.K. Wahi	Advanced Centre for Electronic Systems, Indian Institute of Technology,Kanpur(U.P.)

1

MICROWAVE INTEGRATED CIRCUITS, AN OVERVIEW

by

Amarjit Singh

Chapter One

1.1 INTRODUCTION

The original development of microwave systems involved the introduction of new microwave tubes, together with advancement in circuits with distributed constants. Lumped-constant elements became practically obsolete for use at microwave frequencies, and circuitry took the shape of 'plumbing' using coaxial transmission lines, waveguides and cavity resonators, etc. This situation remained unchanged for about two decades, and it still is basically the same at the high power end of microwave systems. However, at the lower power end the look of microwave sub-assemblies is getting radically changed with the arrival of microwave integrated circuits.

In the evolution of technology one sees continuous interaction between techniques and needs. At times, techniques developed in one field may lead to better fulfilment of needs in another field. Often new developments take place where two new areas overlap. The development of microwave integrated circuits (MICs) is an example of this phenomenon.

In the sixties two types of development converged to give rise to the evolution of microwave integrated circuits. These two areas are, development of semiconductor devices operating at microwave frequencies, and rapid strides made by integrated circuit technology at the lower frequencies.

Common to both the streams of development was the introduction of photo-lithographic techniques into device fabrication. Localisation of various processes such as diffusion, was controlled by photographic masks. These consist of repetitive patterns of microscopic dimensions.

Besides this, device technology also drew upon techniques of epitaxial growth, diffusion of impurities at controlled rates, deposition of thin metallic films, etc. All this had a profound influence on the cost structure of devices and their ability to go to higher frequencies. It became easy to produce devices repetitively in large numbers thus bringing the prices down. Also, microscopic dimensions could be achieved reproducibly. This included microscopic interconnections between devices. It thus became possible to have entire circuits on chips no larger than the size of a pin-head, referred to as integrated circuits.

Besides improving the capabilities of transistors to operate at high frequencies, a number of new semi-conductor devices were introduced for operation in the microwave range. PIN junction diodes were introduced for controlling the amplitude and switching of microwave signals. New categories of devices were developed which were based on phenomena taking place in the bulk of the semiconductor, as distinct from junctions. These included avalanche transit time diodes (IMPATTS), and transferred electron devices (TEDs). These could generate and amplify at power levels and frequencies into the millimeter wave range, far beyond the capabilities of transistors. A better understanding of the phenomena at junctions and in the bulk material, is leading to still newer devices.

There is a growing trend towards the use of microwave integrated circuits. It would be of interest at this stage to take note of broad economic and technical factors behind this trend.

1.2 RATIONALE FOR MICROWAVE INTEGRATED CIRCUITS

For any technology to make rapid progress, it must provide sizeable benefits to the users in return for the investment in research, design and development. Such advantages exist in considerable measure in the case of microwave integrated circuits. Most of these are common to those which resulted in a similar phenomenal growth of integrated circuits at lower frequencies. These features are (a) reduction in volume and weight (b) increase in reliability; and (c) reduction in cost when a standardised module is required in substantial numbers.

One application which has this requirement of standardised modules in large numbers is phased array radar [1,2,3]. This type of radar has received considerable attention in recent times. As the name implies the antenna consists of an array of individual elements, radiating coherently, with controlled phase differences among them. The relative phases control the nature and direction of the radiated beam. These phase differences can be controlled by digital computers whose programming can be such that the radar set can perform a variety of functions at the same time. For instance, it may track certain targets while also scanning others. The microwave source could be a high power tube or a large number of semiconductor sources. When solid state devices are used the meantime between failures is increased. Also, failure of a few of them does not shut off the system completely. There is a graceful degradation of the performance rather than abrupt stoppage. Phased array radar using MICs therefore, has advantages both for large and complex ground based systems with multiple functions, and for air-borne systems ensuring light weight and high reliability.

Although the development of microwave integrated circuits was primed mainly by requirements of phased array radar, MICs are now finding substantial use in the field of microwave communications also.

Projections of requirement abroad have been given in a number of publications[4]. In the case of the Indian industry, there are certain projections for microwave equipment given in the report of the Bhabha Committee "Electronics in India". The total projected installed capacity in 1975 is given as Rs.570 million based on a linear expansion assumption and Rs. 1100 million on the basis of a geometric expansion. One can work out more upto date estimates from a summary of the Indian industry and projections which have been given [5]. Estimates on this basis come closer to the linear expansion estimate of the Bhabha Committee, namely, approximately Rs. 600 million per year installed capacity in 1975.

The components cost in professional equipment of this nature is usually about 1/3rd of the total cost, giving a figure of approximately Rs 200 million per year. Since tubes will continue to be used especially at the high power ends, it may be estimated that half of this component cost would lie in subassemblies involving solid-state devices, giving a figure of approximately Rs 100 million per year. One could assume that about half of this could be put in the form of microwave integrated circuit modules giving a broad estimate of the order of Rs 50 million per year for the year 1975. This is only an indicating figure intended to give what the approximate size of the MIC industry could be in the year 1975.

Use of MICs in the front ends of augmented TV receivers for direct reception of broadcast from satellites will mean a sizeable new requirement. Microwave integrated circuits would be a natural choice in such a case, since standardised modules will be required in large numbers and cost as well as reliability would be prime factors for consideration. The Department of Atomic Energy (Government of India) and the National Aeronautics and Space Administration (NASA) of U. S.A. are cooperating in an experiment on the satellite ATS-F to be launched in the year 1975. There are approximately 500,000 villages in India which would be candidates for coverage through direct broadcasts. In different schemes, different proportions of this number are being considered. One of the recent studies takes a figure of 150,000 villages for direct reception. However, one can see that the outlay for microwave integrated circuits would be quite large.

1.3 TYPES OF MICs AND THEIR TECHNOLOGY

As in the case of conventional integrated circuits, MICs can be broadly divided into two categories namely, hybrid and monolithic. Hybrid circuits can in turn be divided into thick and thin films circuits. In the case of monolithic technology various active devices and their interconnection patterns are formed simultaneously in single crystal substrates of semiconductor material. In hybrid technology the active devices are in the form of chips which are added to a substrate that carries the rest of the circuit.

The major thrust in development of MICs so far has been in hybrid technology for several reasons. One of these is that microwave active devices constitute a large variety, such as transistors, mixer diodes, varactors, PIN diodes, IMPATT diodes, TEDs, etc. It is not easy to standardise the process specifications so that several kinds of devices are simultaneously produced in an optimum way. Further, in MICs the interconnections between the active components have relatively large dimensions, especially where distributed elements are used. This necessitates large areas of substrates which in the case of monolithic technology turn out to be quite expensive.

Hybrid technology itself can be divided into two parts, namely, thin film and thick film technology. Thick films are usually deposited as conductive, resistive or insulating layers, through a screen printing process. Thin films are generally deposited in vacuum by evaporation or sputtering. The thickness can be further augmented by electro-plating. The technology and relative advantages and disadvantages at different frequency ranges will be discussed in detail later.

For passive components, distributed elements are preferred at the higher frequencies whereas lumped constant elements can be used right upto X-band. The limitations on the use of lumped constants come from the increased losses and excessive reduction in the size of the elements at higher frequencies. However, they do offer advantages of economy in those frequency ranges where they can be used.

Among distributed elements, microstrip is the technique commonly employed for MICs. Microstrip lends iteself well to fabrication by photo-lithographic and screen printing techniques used in MICs, the delineation of the circuit being in one plane. It can be used for passive components as well as for interconnections. Other planar guides which have been used, though not as commonly as microstrip, are slot line and coplanar waveguide. In their case, shunt mounting of devices in the plane of the circuit is also possible. However, slot line and coplanar waveguide are not as good as microstrip, as regards, size, weight and circuit losses.

Monolithic integrated circuits become more attractive in the millimeter wave region, where dimensions and sizes of substrates required become smaller. Here, two types of substrates are normally considered, namely, silicon and gallium arsenide. The latter, though more expensive has the advantage that some of the newer active devices such as TEDs can be grown right into the substrates.

1.4 SCOPE OF MONOGRAPH

The subsequent chapters give detailed discussions of the different areas.

Chapters 2 to 7 deal with passive circuits and transmission lines for MICs. In particular, the second chapter introduces the basic design considerations for microstrip circuits. Design of impedance transformers, filters and bias decoupling circuits is discussed. Some useful data relating to microstrips is presented in the form of graphs.

The third chapter presents a theoretical analysis of microstrip, the most commonly used transmission line in MICs. Wheeler's evaluation of the characteristic impedance for transverse-electro-magnetic (TEM) mode is presented. It uses conformal transformations and a concept of effective dielectric constant. A numerical method for more accurate calculations for microstrip is outlined. Wave analysis for a non-TEM hybrid mode on microstrip line is also presented.

The two other types of transmission lines for possible use in MICs, namely, 'slot line' and 'coplanar waveguide' are discussed in chapter 4.

Chapter 5 contains an analysis of a coupled pair of microstrip lines and discusses the design of directional couplers in microstrip form.

Use of lumped elements (capacitors, inductors, etc.) in MICs is discussed in chapter 6. Design considerations for capacitors and inductors are discussed with special reference to the quality factor, Q. Fabrication techniques are described. Several circuits using lumped elements are given as illustrations. These include an S-band low pass filter, a branch line coupler, an impedance transformer at C-band and a transistor amplifier at S-band. A comparison of performance obtained with lumped elements versus that with distributed element circuits is also included.

Chapter 7 discusses the theory and design of nonreciprocal components for use in MICs. Two methods of obtaining such components are discussed. In one, a ferrimagnetic disc is used as an insert in a dielectric substrate. In the other, the entire substrate is made of a ferrimagnetic material. Design of microstrip circulators,isolators and phase shifters is discussed.

Chapters 8 and 9 deal with the theory and applications of active semiconductor devices with special reference to MICs. Junction devices like microwave transistors, microwave FETs Schottky barrier diodes, varactors and PIN diodes are treated in Chapter 8. Bulk effect and avalanche devices are taken up in Chapter 9. This chapter also covers some recent devices like baritts and double drift avalanche diodes.

Design considerations and practical examples for several microstrip circuits are given in chapter 10. These include high power circuits like a transistor oscillator, a step recovery diode frequency multiplier, an avalanche diode oscillator and a PIN diode switch. Among low power circuits which have been described are: a Schottky diode balanced mixer, a parametric amplifier and a PIN diode limiter. Design for microstrip versions of PIN diode phase shifters is also discussed.

Technology of hybrid MICs is the subject matter of chapter 11. Requirements of materials for dielectric substrate is outlined. Thick and thin film techniques for fabrication of passive components are discussed. Techniques for mounting active and discrete components on substrates are also described.

Chapter 12 describes the main features of monolithic MICs. Properties of microstrips on semiconductor substrates are presented. Possibility of using microstrips on silicon-silicon dioxide (or GaAs-silicon dioxide) substrate for monolithic MICs is discussed. Monolithic MICs reported in literature are described.

Chapter 13 contains a discussion of technology for monolithic MICs. This includes formation of silicon dioxide and silicon nitride layers and growing of epitaxial layers. Various techniques for p-n junction fabrication, namely, diffusion and ion implantation are described.

Chapter 14 discusses the use of MICs in phased array radar systems and in satellite television systems.

The last chapter is a bibliography on MICs covering the period from 1967 to 1973. Indices are provided for getting cross-references to different topics and authors.

References

1. Proc. IEEE, Vol.56, Nov. 1968. A special issue on Electronic Scanning.

2. Vincent, B.T.,and M.E. Wallace, Microwave Integrated Circuits for phased array applications, Microwave Journal, Vol.12, pp. 53-59, Sept. 1969.

3. Microwave Journal, Vol.14, Jan. 1971. This issue has two articles dealing with phased array antennas, as well as editorial comment on Phased Array Radar - Perspective, and notes/ views regarding applications in different configurations of radar using phased arrays.

4. Webster, W.M., Integrated Microwave modules, a prospectus, IEEE Trans. Vol. ED-15, pp. 446-450, July 1968.

5. Seshagiri, N., Statistical outline of the Indian Electronics Industry, Jour. Asia Electronics Union, Vol.5, pp. 36-41, Jan. 1971.

Some special issues regarding MICs and related areas

1. Microwave integrated circuits, IEEE Trans. on Electron Devices, Vol.ED-15, July 1968.

2. Electronic scanning, Proc. IEEE, Vol. 56, Nov. 1968.

3. Solid state microwave circuits, IEEE Jour. of Solid State Circuits, Vol. SC-5, Dec. 1970.

4. Microwave integrated circuits, IEEE Trans. on Microwave Theory and Techniques, Vol. MTT-19, July 1971.

5. Microwave semiconductors, Proc. IEEE, Vol. 59, Aug.1971.

6. Thick and thin films for electronic applications, Proc. IEEE Vol. 59, Oct. 1971.

7. Solid state microwave circuits, IEEE Jour. of Solid State Circuits, Vol. SC-7, Feb. 1972.

2

MICROSTRIP CIRCUIT DESIGN, BASIC CONSIDERATIONS

by

D.J. Roulston

Chapter Two

2.1 INTRODUCTION

One of the basic problems in the design of any active circuit consists in impedance transformation and/or filtering. There are many ways of achieving the same end result with microstrip circuits. We shall discuss briefly a few of the commonly used methods in the following sections. Illustrations of some of the methods will be given in subsequent chapters including specifically chapter 10.

Some general comments can be made concerning the choice of circuit. Tolerances should not be too critical for two reasons: firstly, the active device parameters will usually vary over a fairly wide range (± 20% is typical) from one device to another, not to mention variations as a function of temperature; secondly there is a limit set by the technology on the precision of reproduceability of the strip itself. It goes without saying that the less critical a circuit is to parameter variations, the more likely it is to operate under a wide range of ambient conditions (temperature, vibrations, shock, etc.) hence the higher will be its reliability-an important aspect of MIcs. In general the simpler the geometry of the circuit, the more likely it is to perform according to theoretical predictions. It is also easier to make intelligent adjustments to simpler circuits at the design and development stage. Power loss and power dissipation are additional parameters which are usually determined both by the microstrip line and by the active devices. Wide bandwidth is also frequently a requirement which helps to reduce criticalness of circuit operation.

2.2 IMPEDANCE TRANSFORMERS

2.2.1 The λ/4 transformer

This is the simplest type of impedance transformer. For a quarter wave section of line of characteristic impedance Z_o, the following relation applies:

$$Z_{in} Z_L = Z_o^2 \qquad \text{... (2.1)}$$

Since the line impedance is a direct function of the width, spacing and dielectric constant it is possible to obtain moderate impedance transformations by this method. For larger impedance ratios, two or more quarter wave sections may be cascaded. A double section gives for example, the following relation:

$$Z_{in} = (Z_{o1}/Z_{o2})^2 Z_L \qquad \text{... (2.2)}$$

where Z_{o1} and Z_{o2} are the two transformer impedances and Z_L is the load impedance. Fig. 2.1 illustrates these two methods. Note that in order to maintain TEM propagation, on which the designs are based, and in order to have reasonable dimensions, there is an upper and lower limit to the widths (and hence the Z_o values) of each transformer.

Fig.2.2 gives the characteristic impedance of the line for various values of w/d, where w is the width of the line and d the spacing between the line and the ground plane. Note that

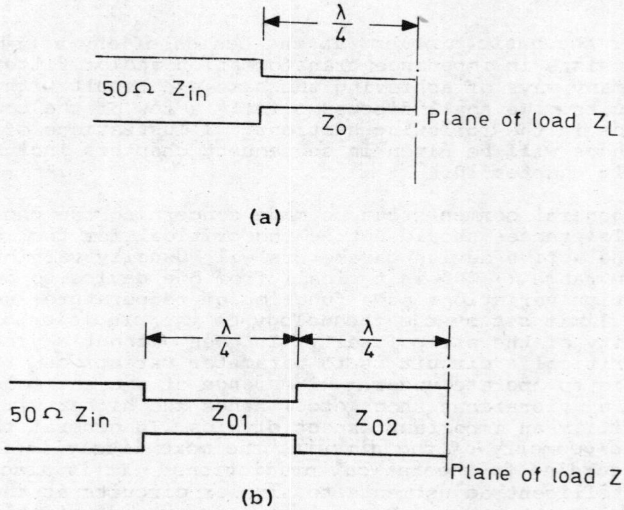

Fig. 2.1 (a) Single quarter wave transformer
(b) Double quarter wave transformer

Fig. 2.3 should be used to determine the effective values of ε for wavelength calculations (see also [4]). Impedances and wavelength calculations would be discussed in detail in chapter 3.

This simple transformer suffers from the drawback that the obtainable impedance ratio is limited by the ratio of the widths of the adjacent lines; propagation of higher order modes limits the maximum width to $\lambda/4$ whilst power loss (and possibly dissipation) limits the minimum width. The presence of discontinuities adds to the complication of exact design of this type of impedance transformer. Finally, it should be noted that any alteration in the centre frequency is nearly impossible once a transformer of this type has been constructed; this is not the case with other type as will be seen below.

2.2.2 Double stub impedance transformers

Using conventional transmission line properties, we can always obtain impedance transformations by placing appropriate values of shunt admittance at suitable positions on a line of given characteristic impedance. The simplest such impedance matching device is shown in Fig.2.4 for the case of a reverse biased diode to be matched to the 50 ohm line impedance. The determination of stub spacing and stub lengths using the Smith chart has been discussed in several text books on transmission lines.

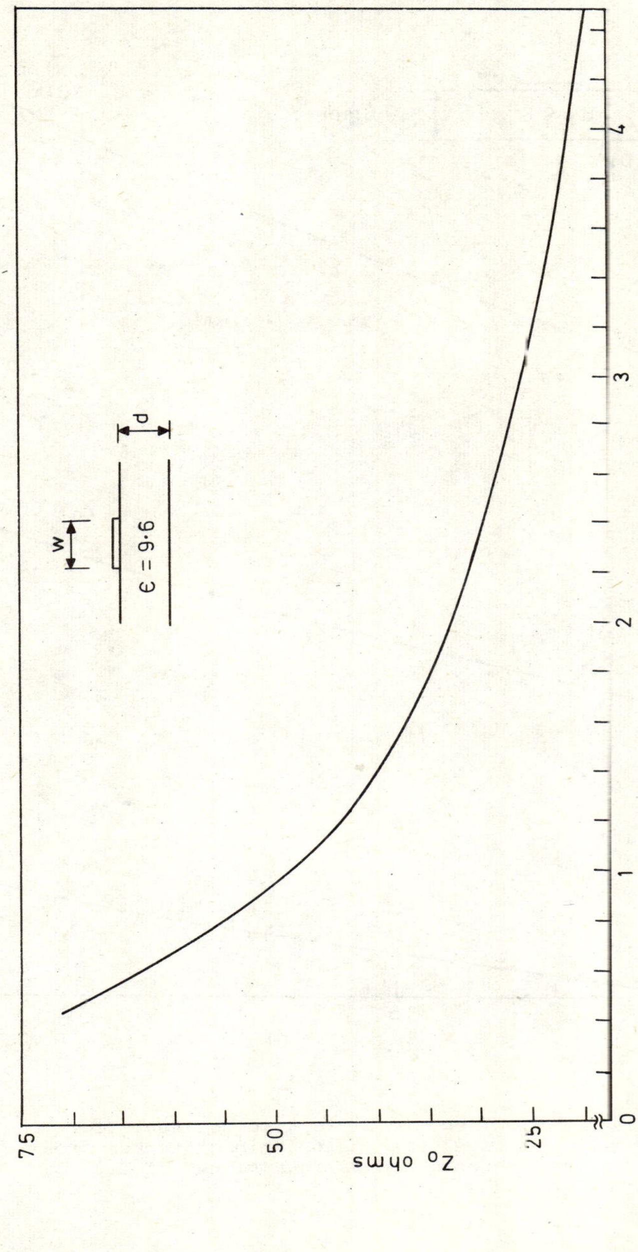

Fig. 2.2 Z_o versus w/d for single microstrip line

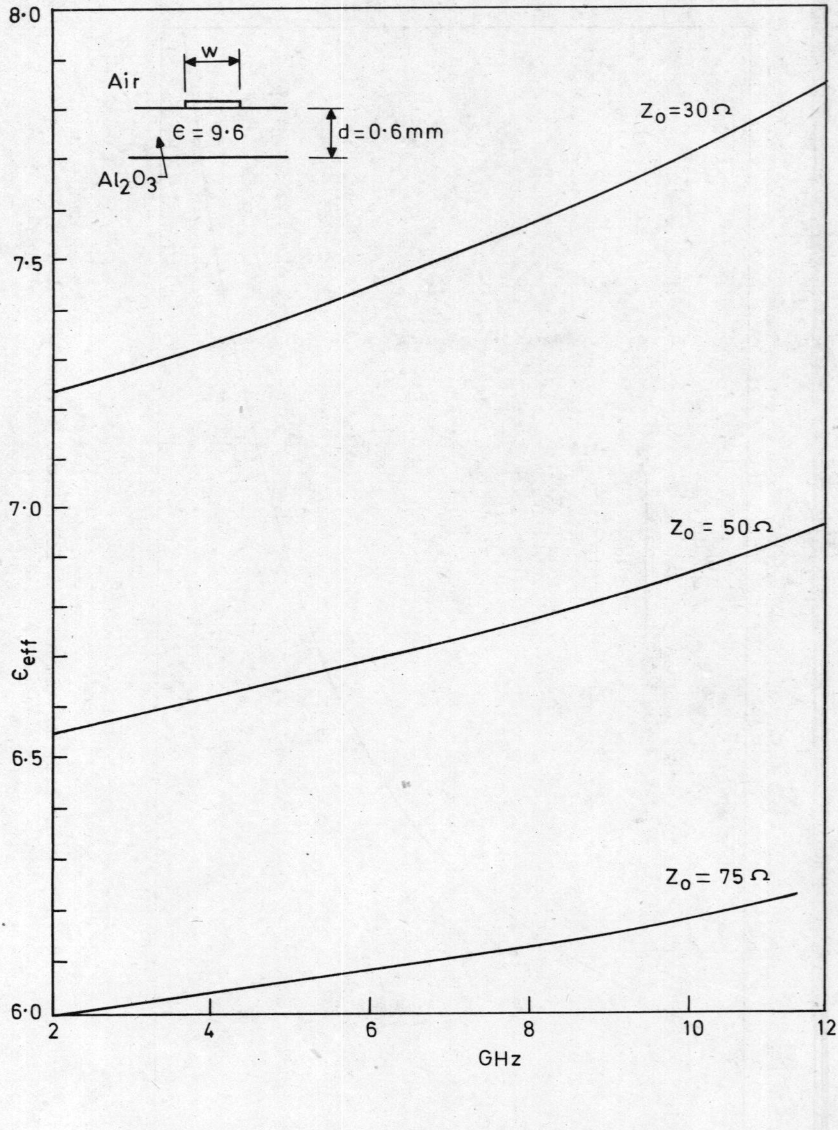

Fig. 2.3 Effective dielectric constant (for wavelength calculations) versus frequency for various values of Z_o

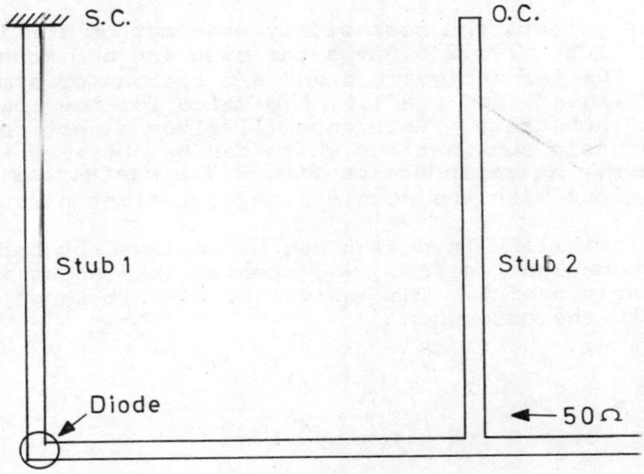

Fig. 2.4 Example of double stub microstrip impedance matching

It is worth noting that this method does allow a certain amount of final adjustment after the circuit has been fabricated. Reduction of the stub lengths will alter both the reactive and resistive components and can be accomplished by scraping and etching techniques.

A very simple technique for adding shunt capacitance to a completed circuit consists in placing a small piece of dielectric, with metalization on top, on the line at the required position. This technique is illustrated in Fig.2.5.

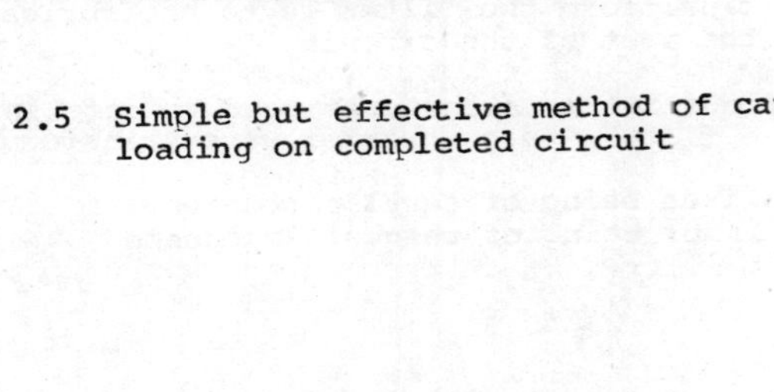

Fig. 2.5 Simple but effective method of capacitive loading on completed circuit

2.2.3 Coupled line impedance transformers

This is perhaps the most widely used method for impedance matching in MICs. Fig.2.6 gives the even and odd mode impedances Z_{oe} and Z_{oo} for various w/d and s/d ratios for symmetrical coupled lines*, for $\varepsilon = 9.6$, the value for the most commonly used ceramic substrate. Reference [1] gives a very complete list of the basic combinations which can be obtained with such coupled lines. We reproduce on Fig. 2.7 a useful case of $\lambda/4$ coupled sections with the double stub equivalent circuit.

More flexibility in design can be achieved by using non symmetrical coupled lines**. The special case shown in Fig. 2.8 is particularly useful. The equivalent circuit shown in Fig.2.8 holds good in the case when

$$Z^a_{oe} + Z^a_{oo} = 2Z_A \qquad \text{... (2.3)}$$

Z_s and turns ratio N are given by:

$$Z_s = Z_A\,(N^2 - 1)/N^2 + Z^b_{oo} - Z^a_{oo} \qquad (2.4)$$

$$N = 2Z_A/[Z^a_{oe} - Z^a_{oo}] \qquad \text{... (2.5)}$$

It is interesting to note that in the second case, final adjustment of the resonant frequency can be achieved by removal of the metalization at one end of one of the lines. Use of non-symmetrical coupled lines also allows the designer to choose dimensions so as to reduce power loss and dissipation. Unfortunately due to multiple combinations of the four characteristic impedances, it is not possible to give design curves similar to those given in Fig.2.6. It is not too difficult, however, to compute the different parameters on the computer (e.g.[2]).

2.3 FILTERS

Any of the above impedance transformers will, of course, act as a filter. It is often convenient, however, to separate the two functions at the design stage; the category 'filters' is then concerned with a combination of coupled resonators giving the desired characteristics such as band-pass, low-pass,etc. with, normally 50 ohm input and output impedance. In designing MICs it is often convenient to fabricate and measure the filter separately during the development stage of a complex MIC even if in the final form the filter is to be fabricated simultaneously with the rest of the circuit.

Of the various filter structures used to date in conventional stripline filters (e.g. quarter wave transformers,interd-

* two coupled lines being of equal impedance
** two coupled lines being of unequal impedance

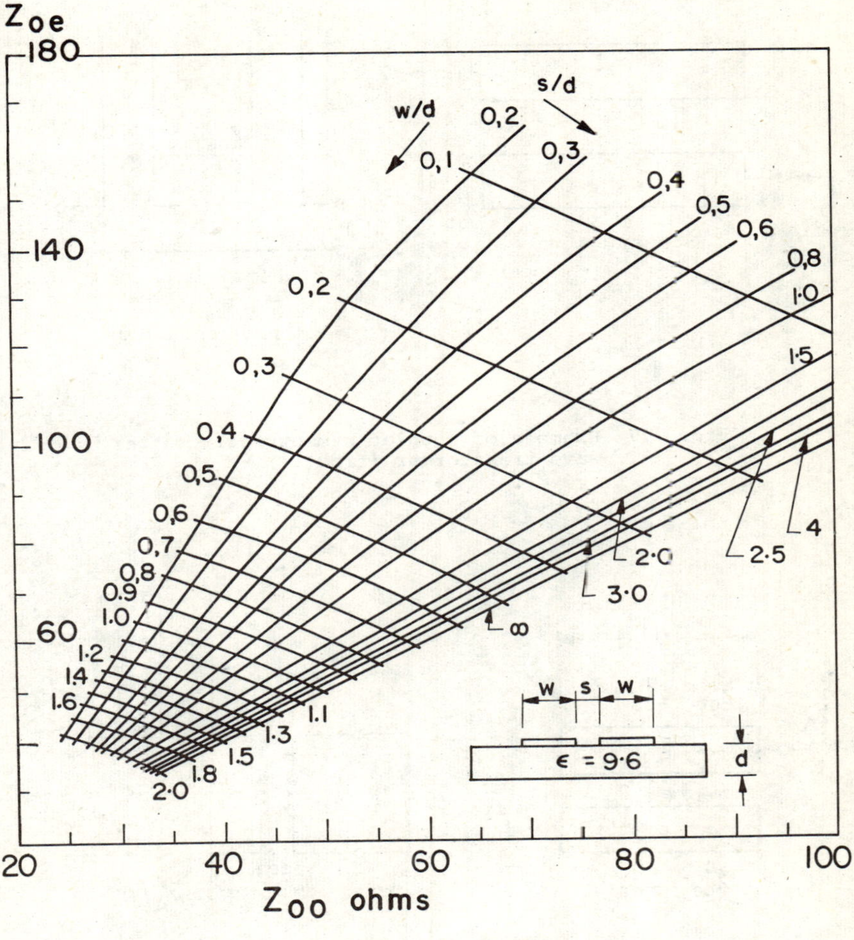

Fig. 2.6 Even (Z_{oe}) and odd (Z_{oo}) mode impedances for various spacings and dielectric heights for symmetric parallel coupled lines

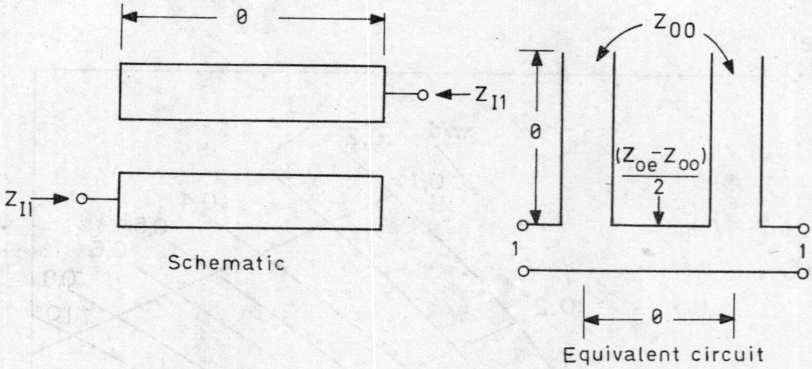

Fig. 2.7 Example of coupled symmetrical quarter wave transformer (from[1])

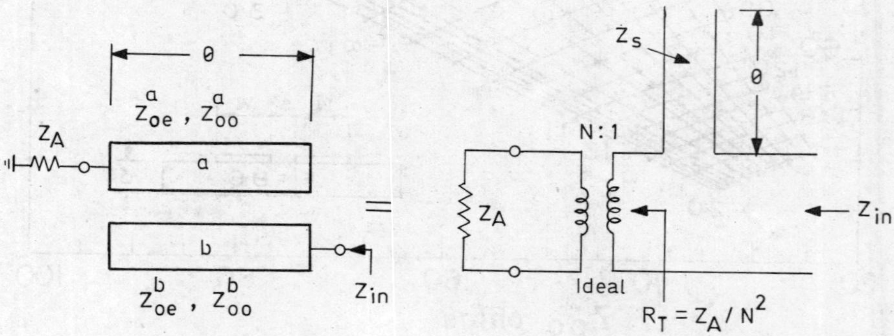

Fig. 2.8 Example of non symmetrical coupled quarter wave transformer (from[1])

igital coupled quarter wave sections, comb-line filters) it is properly the parallel coupled resonators with the configuration shown in Fig.2.9 which has found the most widespread use in MIC applications. Although not as compact as the interdigital structure, it is not too unwieldy and has the advantage, from a design point of view, that coupling between non overlaping sections can be neglected with negligible error in nearly all cases. The filter so designed will therefore give a performance which is very close to that which is predicted. The basic design equations for this filter are discussed in [1]. The strip dimensions can be determined from Fig.2.6. It is outside the scope of this monograph to include the very extensive amount of data contained in [1] relating the prototype filter parameters to the desired response characteristic.

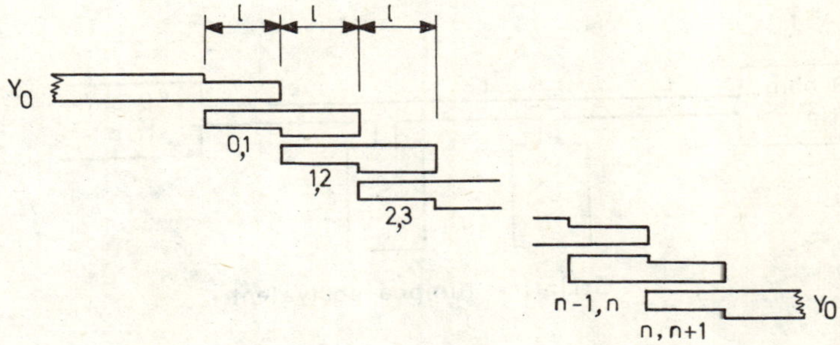

Fig. 2.9 A parallel coupled quarter wave resonator filter (from[1])

2.4 LUMPED CONSTANT MICROSTRIP CIRCUITS

It is worth noting that short sections of line can be used to approximate inductance or capacitance. Using the fact that the reactance X_L of a length of line of impedance Z_o shorted at a distance 'ℓ' is given by:

$$X_L = Z_o \tan(\omega\ell/v) \tag{2.6}$$

where v is the velocity of propagation along the line.

The admittance of a similar open circuited length of line is given by:

$$Y_C = Y_o \tan(\omega\ell/v) \tag{2.7}$$

It is apparent that for short lengths of line we can approximate an inductance L by: (for shorted line)

$$L \simeq Z_o\ell/v \tag{2.8}$$

and a capacitance C by: (for open line)

$$C \simeq Y_o \ell / v \tag{2.9}$$

A typical series L, shunt C, filter can thus be built as shown in Fig.2.10 using this technique. This topic is considered in more detail in Chapter 6.

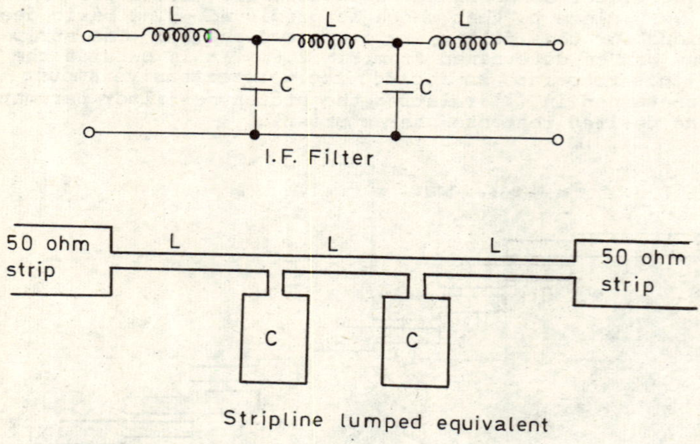

Fig. 2.10 Example of series L, shunt C, filter

2.5 BIAS DECOUPLING CIRCUITS

Two common techniques exist for introducing bias to the active element of MICs. These are illustrated in Fig.2.11.

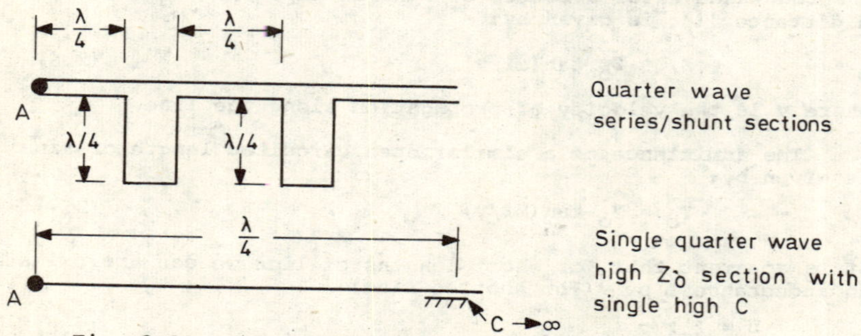

Fig. 2.11 Two methods for applying bias to an active device

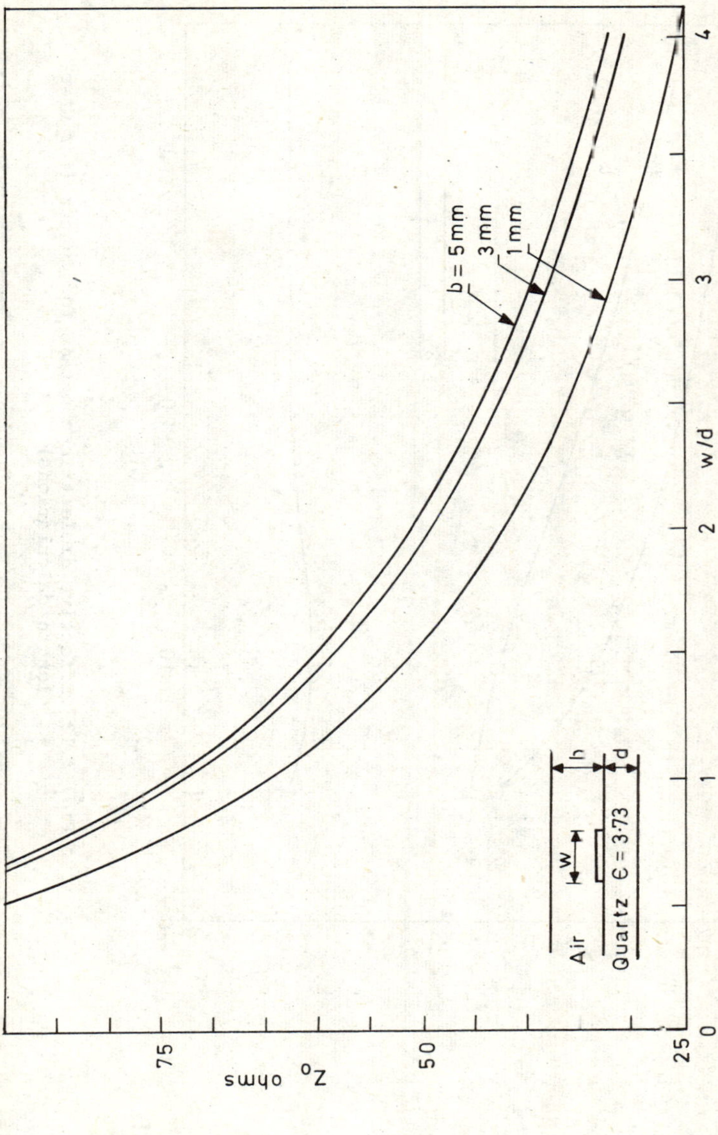

Fig. 2.12 Z_o versus w/d for a single stripline on quartz with top plane included

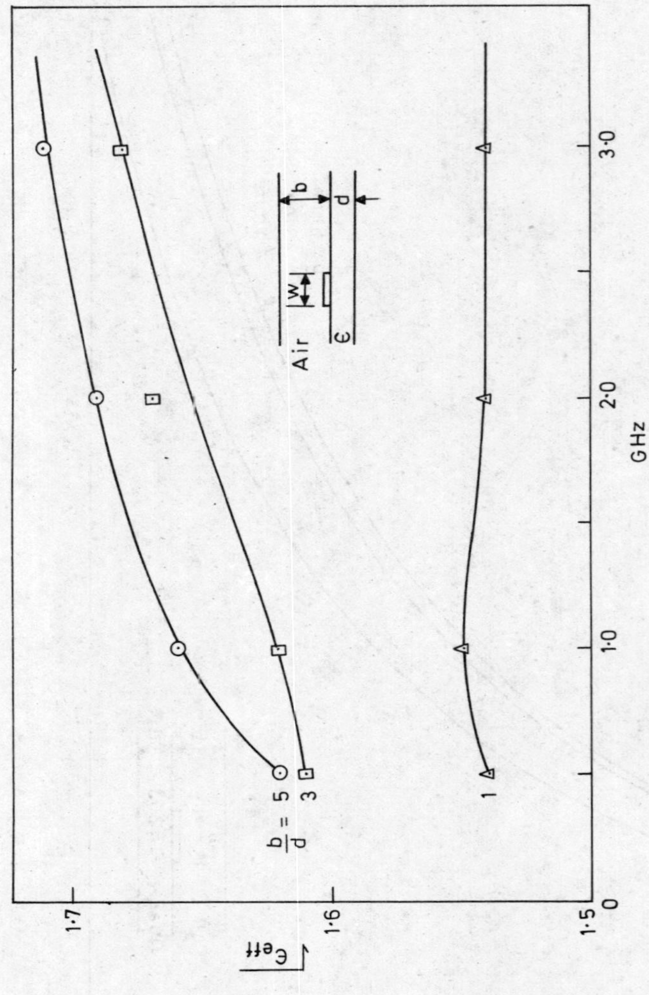

Fig. 2.13 Effective dielectric constant for quartz (for wave-length calculations)

The first method uses two or more series/shunt $\lambda/4$ sections. This gives a high impedance at point 'A' and has the advantage of being easily fabricated. Note the low impedance shunt and high impedances series sections to give a wide bandwidth. The second method is more difficult (i.e.costly) to fabricate and consists of a very high impedance $\lambda/4$ thin wire with a very large capacitance at the d.c. input end. This capacitance and wire have to be added after fabrication of the rest of the circuit.

2.6 USE OF QUARTZ ABOVE X-BAND

At frequencies higher than about 12 GHz, dimensions become so small using a ceramic dielectric that circuits become impractical. A convenient dielectric at these frequencies is quartz. Fig.2.12 gives the characteristic impedance as a function of w/d. Note that the spacing between the microstrip and the metal cover, b (i.e. the height of the air dielectric) is taken into account in these curves. Fig. 2.13 gives the effective dielectric constant for wavelength calculations.

2.7 APPLICATION TO PRACTICAL CIRCUITS

In chapter 10 the above discussed design concepts will be applied to a wide range of practical (and currently fabricated) hybrid MICs. The importance of some of the impedance adaptation methods and charts will then become clear.

References

1. Matthaei, G.L. et al, Microwave filters, impedance matching networks and coupling structures, McGraw-Hill, 1964.

2. Wheeler, H.A., Transmission line properties of parallel wide strips by a conformal mapping approximation, IEEE Trans.Microwave Theo.Tech., Vol.MTT-12, pp.280-289, May 1964.

3. Obregon, J. and P. Zamansky, Calcul des lignes microstrips et application aux oscillateurs a avalanche, Revue technique Thomson-CSF, Vol.3, pp. 5-18, March 1971.

4. Zamansky, P., Private communication.

5. Jain, O.P. et al., Dispersive behaviour of microstrip transmission lines, Proc. European Microwave Conf. 1971, C3/5:1 - 5:4

6. Obregon, J., private communication.

7. Weir Ather, B., Microstrip can reduce multiplier size, Electronic Design Vol.2, p.49, Jan. 21, 1971.

3

ANALYSES OF MICROSTRIP LINE

by

K.C. Gupta* and Ramesh Garg**

* Sections 3.1 to 3.3, 3.5 and 3.6
** Section 3.4

Chapter Three

3.1 INTRODUCTION

For transmission lines to be used in MICs it is desirable that the characteristics of the line (impedance etc.) should depend upon the dimensions in one plane only. This facilitates their fabrication by photoetching of metallised substrates. Three geometries that meet this requirement are shown in Fig.3.1 (a,b and c). These are known as microstrip, slot line and coplanar waveguide respectively. Out of these three, microstrip line is the one most commonly used. This is analysed in the present chapter. Slot line and coplanar waveguide are described in Chapter 4.

For using any transmission line in circuits it is necessary to know the following three parameters for the line: characteristic impedance, guide wavelength and attenuation constant. These parameters for microstrip line would be discussed in subsequent sections.

As shown in Fig. 3.1(a), the microstrip structure contains a ground plane, a plane dielectric slab and a conducting strip on the top surface of the dielectric. The microstrip line is fabricated by photoetching one side of a metallised dielectric substrate*. The ground plane and the strip constitute the two conductors of a two-wire transmission line. In case the dielectric slab were considered to be replaced by air, the mode of propagation along the microstrip line would be identical to that for a wave on a low frequency two-wire or coaxial transmission line. The dominant mode in a line consisting of two-wires in air or in any single dielectric, is the transverse electromagnetic (TEM) mode. For TEM mode both the electric field and the magnetic field lines lie in a plane transverse to the direction of propagation, and the field distribution may be obtained by a solution for static fields in the transverse plane. In such a case the characteristic impedance of the transmission line may be obtained from the electrostatic capacitance per unit length of the line by using the following simple relation.

$$Z_o = 1/v_p C \tag{3.1}$$

where Z_o = characteristic impedance of the line

v_p = phase velocity for the wave

C = capacitance per unit length of the line

In the case of microstrip line we have two dielectric media, namely the dielectric substrate and air. As will be shown in Sec. 3.4 of this chapter, in such a case of mixed dielectric a pure TEM wave cannot exist. However the deviation from TEM mode is small and in sections 3.2 and 3.3 a TEM mode will be assumed for microstrip analysis.

*Technology of microstrip lines would be discussed in detail in Chapter 11.

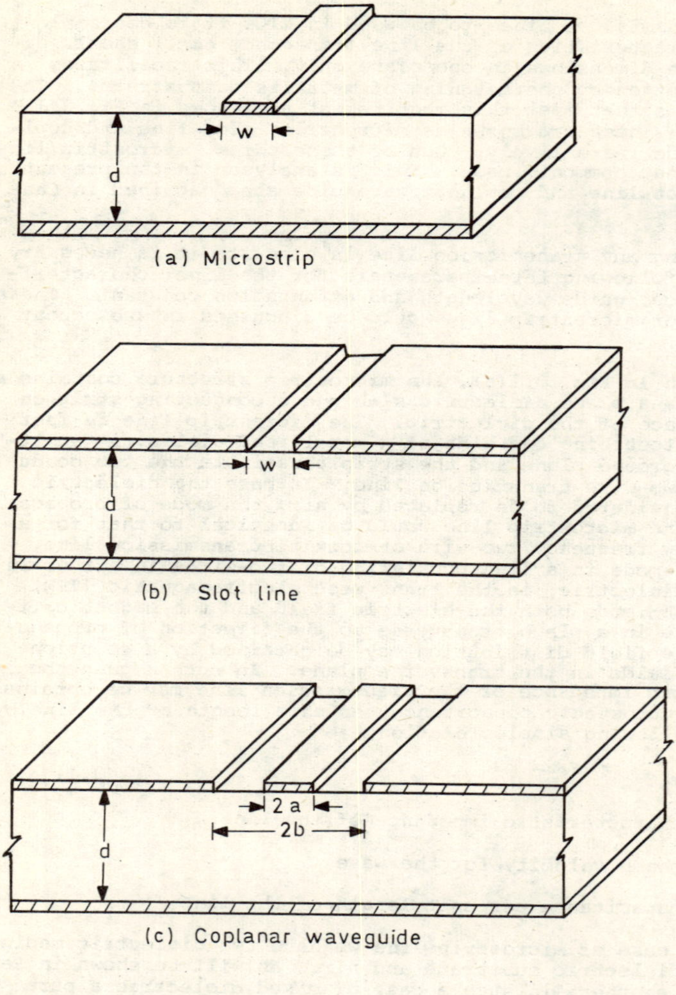

Fig. 3.1 Transmission lines for MICs

The characteristic parameters of microstrip line (Fig.3.1a) depend on strip width W, thickness d and the dielectric constant ε_r of the substrate material. For a given thickness and dielectric constant of the substrate, strip width W controls the characteristics of the line. This feature makes microstrip line very well suited for microwave integrated circuits. Approximate electric and magentic field distribution for the microstrip line is shown in Figs.3.2 (a and b). As suggested by the electric field distribution shown in Fig.3.2a, most of the energy carried by the microstrip propagates in the dielectric just below the strip conductor. The rest of it is in a fringe field which extends in the air space above the dielectric.

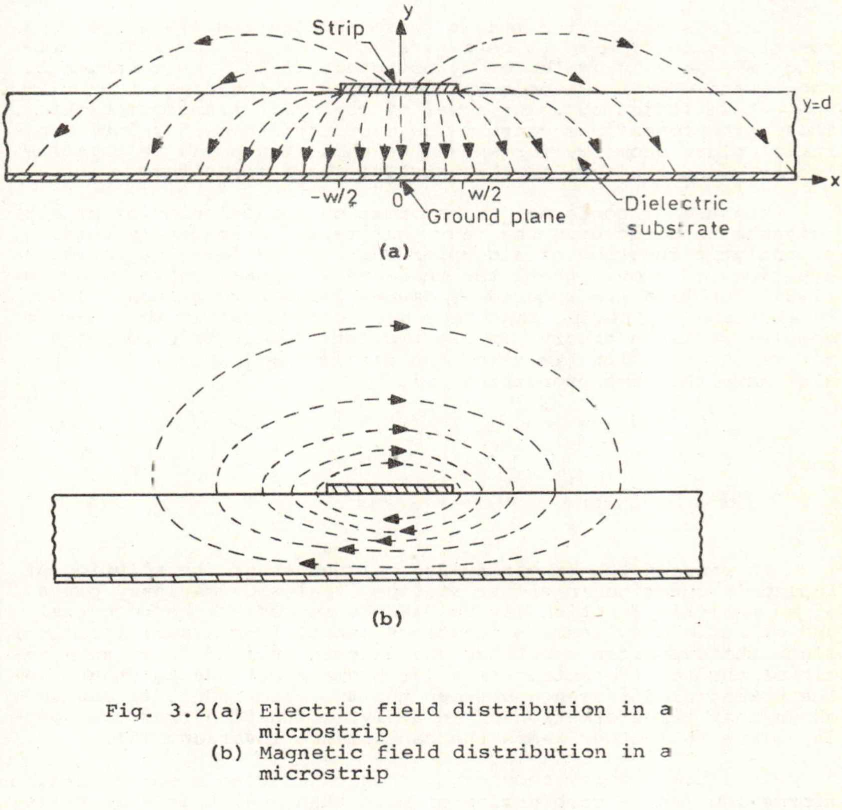

Fig. 3.2(a) Electric field distribution in a microstrip
(b) Magnetic field distribution in a microstrip

There are three methods that are commonly used for analysing microstrip lines:

1. Method of conformal transformation[1,2]

2. Numerical analysis using relaxation technique[3]

3. Hybrid mode analysis[4]

The first two methods assume the mode of propagation to be a pure TEM mode, whereas the wave analysis method is more rigorous and considers a hybrid (non-TEM) mode. These three methods are discussed in the next three sections of this chapter.

3.2 METHOD OF CONFORMAL TRANSFORMATION

3.2.1 Conformal Transformation

In this method the mode of wave propagation along the microstrip is considered to be a pure TEM mode and the characteristic impedance is evaluated by using eq. (3.1). Calculation of capacitance for the geometry encountered in the present problem may be simplified by using a set of conformal transformations. These transformations convert the microstrip geometry into a parallel plate geometry for which the capacitance may be calculated by using the formula for parallel plate capacitor.

The use of conformal transformation for calculation of capacitance is based upon the fact that real and imaginary parts of an analytic function of a complex variable satisfy the Laplace's equation. Not only that, the real and imaginary parts of an analytic function are related by Cauchy-Reimann relations. This is expressed by saying that they are 'conjugate' in the sense of complex variable theory. In electrostatics, electric potential $V(x,y)$ and the flux $\phi(x,y)$ of the displacement vector ($D = \varepsilon E$), also have the same properties, viz.,

$$\nabla^2 V(x,y) = 0, \quad \nabla^2 \phi(x,y) = 0 \tag{3.2}$$

and

$$\partial V/\partial x = \partial \phi/\partial y \text{ and } - \partial V/\partial y = \partial \phi/\partial x \tag{3.3}$$

In view of the linearity and uniqueness of the solution of Laplace's equation in electrostatics, real and imaginary parts of an analytic function may be used to represent the potential and electric displacement functions (apart from a scalar factor). Since the capacitance between the electrodes is defined as a ratio of the total electric flux (from one electrode to other) to the potential difference between the two electrodes, it can be shown that transformation of an analytic function from one complex plane to another keeps the capacitance invariant [5].

Thus the problem reduces to that of finding a suitable transformation (or a combination of more than one) that will transform microstrip geometry into a parallel plate geometry for which the capacitance may be easily calculated.

3.2.2 Transformation used for microstrip analysis

A conformal transformation suited for analysing microstrip geometry has been proposed by Wheeler[1]. In reference[1] he deals with transmission line properties of parallel wide strips. Microstrip geometry can be obtained by inserting a plane of symmetry half way between the two strips as shown in Fig.3.3a. However the effect of dielectric has not been taken into account. This has been dealt with in a subsequent paper by Wheeler[2], where transmission line properties of parallel strips separated by a dielectric sheet are discussed.

Wheeler[1] has analysed the parallel strips geometry by using Schwarz-Christoffel transformation. This transformation[5], in its general form, is defined by the following differential equations.

$$dz/dt = f'(t) = A/[(t-r_1)^{\alpha_1} (t-r_2)^{\alpha_2} \ldots\ldots (t-r_k)^{\alpha_k}] \qquad (3.4)$$

where z (= x + jy) and t(= r + js) are complex variables. A is a complex constant and r_1, r_2 etc. are real finite numbers. Equation (3.4) may be solved by a contour integral in the complex plane. This equation is used to transform a k-sided polygon in z-plane to the real axis in t-plane. The interior of the polygon is transformed into the upper half t-plane. α's represent the exterior angles (normalized with respect to π) at various corners of the polygon. $\Sigma\alpha_k = 2$ for a closed polygon. For transforming a polygon in z-plane into the real axis in t-plane one has to select values of r's. Segments corresponding to the various sides of the polygon are separated by points r_1, r_2....etc. on real axis in t-plane. It emerges that [5] we can choose arbitrarily the values of three r's: two at finite distance from the origin and one at infinite.

Eq.(3.4) forms the first part of the transformation used to solve the electrostatic field problem with the boundaries specified by a polygon. Another transformation is necessary to convert the geometry in t-plane to the geometry of a parallel plate capacitor. This is achieved by another Schwarz-Christoffel transformation from t-plane to w-plane specified by

$$dw/dt = B/[(t-r_m)\ (t-r_n)] \qquad (3.5)$$

This transformation contains only two singular points corresponding to $t = r_m$ and $t = r_n$ which correspond to points $Re(w) = \pm\infty$ in the w-plane. Since the angles at both of these points are equal to π radians, corresponding α's are unity. In w-plane, equipotential lines are parallel to the real axis (u-axis) and flux lines are parallel to the imaginary axis (v-axis).

In order to use Schwarz-Christoffel transformation for microstrip analysis it is required to represent the geometry in terms of a polygon. Wheeler has selected the polygon in such a manner that the desired transformation is obtained in terms of fu-

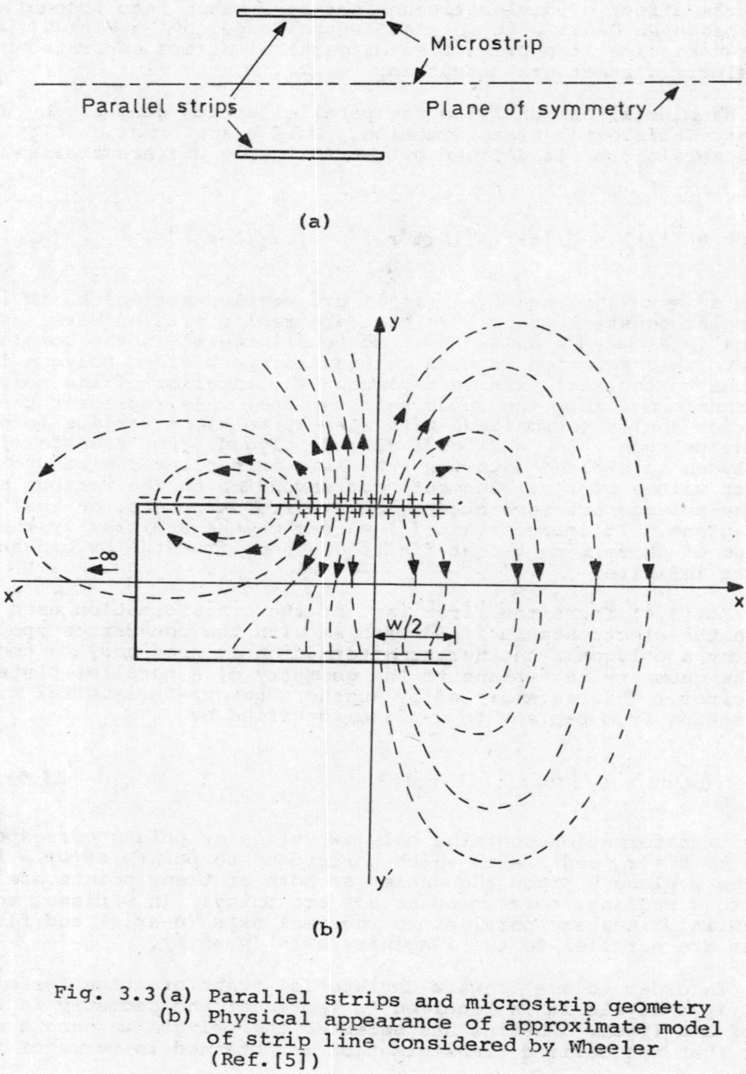

Fig. 3.3(a) Parallel strips and microstrip geometry
(b) Physical appearance of approximate model of strip line considered by Wheeler (Ref.[5])

nctions that can be easily evaluated. This is achieved when the angles describing the polygon are only $\pm\pi$ radians, i.e. values of α's in Eq.(3.4) are ± 1 only. This simplifies the integration of Eq.(3.4). When angles of $\pi/2$ or $3\pi/2$ are included (i.e. α's are 1/2 or 3/2), integration of Eq. (3.4) is not very straight forward. Because of the presence of square root sign in the denominator on right hand side, the integral cannot be evaluated explicitly but is expressed in terms of elliptic integrals. For any other angle the integration cannot be handled easily.

Because of symmetry considerations, only one quarter of the parallel strip geometry is considered. As it is desirable to include only angles of $\pm\pi$, the half strip (in one quarter cross-section) is extended to $-\infty$ as shown in Fig. 3.3b. An insulating slit is located at the points I on the upper conductor. This ensures that the field line at the origin is vertical. The contour of the polygon in z-plane is shown in Fig.3.4. This polygon lacks a vertical plane of symmetry. Wheeler[1] succeeded in re-establishing the vertical symmetry plane by adopting a double layer representation of the model as shown in Fig.3.5. Now the polygon is symmetrical about y-axis. The quantity W/2 is now dictated by the symmetry and becomes one of the initial data. In the absence of this symmetry, it becomes difficult to determine equivalent value of W/2 after transformation. However, now the strip is described twice by the transformation. We need only consider each half transformation to lie on a different layer of the Reimann surface, the two layers intersecting on the y-axis. For the physical solution, we use only half of each layer, the half containing the strip.

Using the method outlined above, Wheeler[1] has obtained the following transformation for the microstrip geometry:

z-plane to t-plane transformation:

$$z = j\pi + t - 2\tanh^{-1}(t/\nu)$$

where

$$\nu = 1 + \sqrt{(1 + \mu^2)}$$

μ is given by

$$\pi W/2d = \mu - \sinh^{-1}\mu$$

t-plane to w-plane transformation:

$$t = \nu' \tanh(W/2)$$

where

$$\nu' = \nu(1 - 4\nu\, e^{-2\nu})$$

Characteristic impedance

The line capacitance for one quarter of a profile is equal to that for complete profile and may be expressed in terms of m

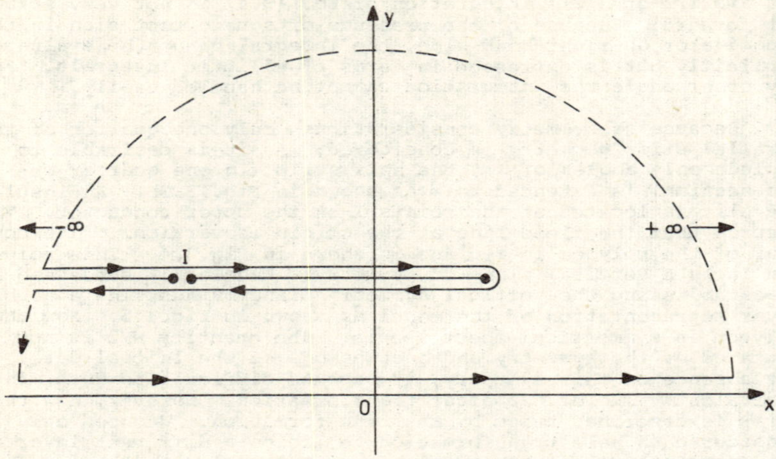

Fig. 3.4 Contour of the polygon in z-plane (Ref.[5])

as[5]

$$C/(\varepsilon_o W/2d) = \{m + 1 + \ln [2\pi(m/\pi + 0.94)]\}/m \qquad (3.6)$$

where $m = \pi W/2d$

Hence the characteristic impedance Z_o of the parallel strip line of Fig.3.3a may be written as[5]

$$Z_o/(\eta 2d/W) = m/\{m + 1 + \ln [2\pi\ (m/\pi + 0.94)]\} \qquad (3.7)$$

where $\eta = \sqrt{\mu_o/\varepsilon_o}$

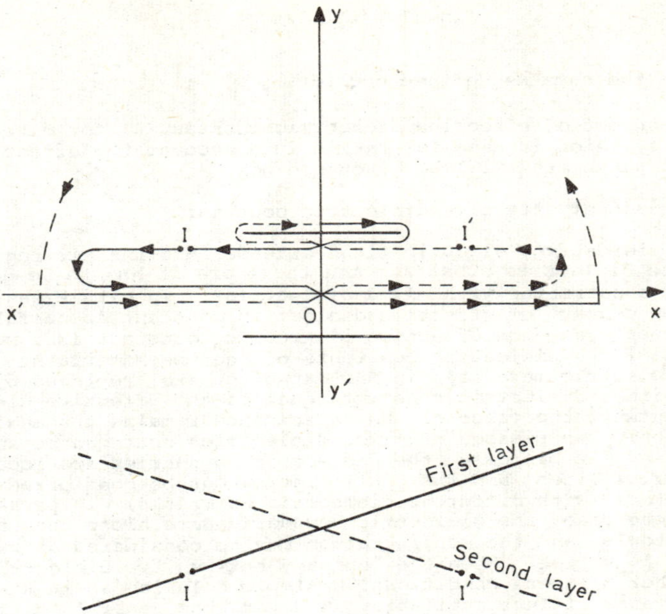

Fig. 3.5 The "double layer" representation of the theoretical contour of Fig.3.4 (Ref.[5])

Characteristic impedance for the case of an 'air' microstrip line, i.e. without a dielectric substrate being present between the ground plane and the strip, is half of Z_o in eqn.(3.7). Wheeler[2] has derived the expression for the characteristic impedance of a microstrip line when the space between the microstrip and the ground plane is filled with a dielectric of relative dielectric constant ε_r. The impedance of wide strip line (W>>d) is given as

$$Z_o = (\eta/\sqrt{\varepsilon_r})(d/W)[1+(d/\pi W)\{2\ln 4+(1+1/\varepsilon_r)\ln[\pi e(W/2d+0.94)/2] + (1/\varepsilon_r-1/\varepsilon_r^2)\ln(e\pi^2/16)\}]^{-1}$$

For the narrow strip case (W<<d) the impedance is

$$Z_o = (\eta/2\pi)\sqrt{(2/[\varepsilon_r+1])}\ \left[\ln(8d/W)+(W/2d)^2/8-0.5[(\varepsilon_r-1)/(\varepsilon_r+1)]\ \{\ln\pi/2+1/\varepsilon_r \ln 4/\pi\}\right]$$

where η is the same as defined earlier.

The concept of effective dielectric constant[2] for mixed dielectrics, which is used for taking into account the effect of dielectric substrate, will be discussed now.

3.2.3 Concept of effective dielectric constant

In a microstrip, electric field extends into the air region outside the dielectric substrate and therefore it has to be considered as a capacitor with mixed dielectrics. In situations like this the concept of effective dielectric constant is useful. In the present case the effective dielectric constant is a weighted mean of the dielectric constants of the two materials. If the two dielectric materials in the capacitor are replaced by a material with a dielectric constant equal to the effective dielectric constant, the value of the capacitance remains unchanged. There are only two classes of mixed dielectrics that can be analysed easily. In one case, the dielectric boundaries are located along the flux lines, and the configuration can be considered as two capacitors (with different dielectric materials) in parallel. In the second case, the dielectric boundaries are along the potential contours, and the configuration may be considered as two capacitors in series. When the boundary between two dielectrics meets neither of these conditions, a simple solution as in the above two cases, is not available. However, the general configuration may be interpretted as a combination of the two specific cases mentioned above.

Another definition, that is useful especially for capacitors that are partially filled with a dielectric (and remaining part filled with air) is that of effective filling fraction. The effective filling fraction (q) is related to the effective dielectric constant (ε_{reff}) by the following relation

$$\varepsilon_{reff} = (1-q) + q\ \varepsilon_r = 1 + q\ (\varepsilon_r - 1) \qquad (3.8)$$

In the special case when the boundary between the dielectric and the air region is along a flux line (parallel combination case) q is identical with the actual filling fraction i.e. fraction of volume filled with dielectric material. However, in general, the volume occupied by dielectric can be divided into an equivalent parallel volume v_p and an equivalent series volume v_s. If v_d is the volume occupied by the dielectric out of a total volume v, the actual filling fraction q' is given by

$$q' = V_d/V \qquad (3.9)$$

Parallel part of the filling fraction is

$$q'' = (V_d - V_s)/V = q' - V_s/V \qquad (3.10)$$

It can be shown that for small values of v_s series volume is only $1/\varepsilon_r$ times as effective as v_p. The effective filling fraction is therefore given by

$$q = q'' + (q'-q'')/\varepsilon_r = q'/\varepsilon_r + (1-1/\varepsilon_r)q'' \tag{3.11}$$

Effective dielectric constant for microstrip

The computation of the effective dielectric constant is carried out differently for the case of wide strips and narrow strips.

The derivation for the case of wide strip is based on conformal transformation discussed earlier. The concept of effective dielectric constant, developed above, is applied after transformation to the w-plane. The expression for dielectric filling fraction q thus obtained for wide strip case is as follows [25]

$$q\nu = \nu - \ln\,[(\nu+\mu)/(\nu-\mu)] + 0.732\,\{\ln\,[(\nu+\mu)/(\nu-\mu)] - \cosh^{-1}\,(0.358\nu + 0.953)\}/\varepsilon_r + (\varepsilon_r-1)\,[0.386-0.5/(\nu-1)]/\varepsilon_r \tag{3.12}$$

Effective dielectric constant may be obtained by using this expression for q in eqn.(3.8).

In the case of narrow strips ($W/d \ll 1$) free space formulas have been used. When the strip is very narrow the flux outside the dielectric and inside the dielectric is about the same and the effective dielectric constant may be written as

$$\varepsilon_{reff} = (\varepsilon_r + 1)/2 \tag{3.13}$$

Wheeler[2] has derived free space impedance for narrow strips to a second approximation and this has been used in evaluating the effective dielectric constant. Effective dielectric constant thus obtained is given by:

$$\varepsilon_{reff} = [(\varepsilon_r+1)/2]\{1+[(\varepsilon_r-1)/(\varepsilon_r+1)][(\ln\pi/2 + 1/\varepsilon_r\,\ln 4/\pi)/\,\ln 8d/W\,]\} \tag{3.14}$$

3.2.4 Discussion of results

The analysis presented in the previous subsections has been used to calculate the parameters of a microstrip line on alumina substrate[6]. These results are presented in Figs.3.6 and 3.7. In Fig.3.6, characteristic impedance for the microstrip line is plotted against W/d, the ratio of width of the line to thickness of dielectric substrate. We note that Z_o decreases as the width of the line increases, increase in the line width increases the

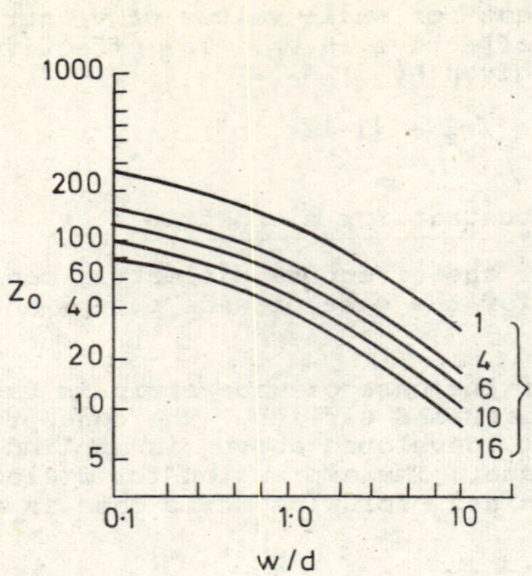

Fig. 3.6 Impedance of microstrip as a function of W/d (Ref.[6])

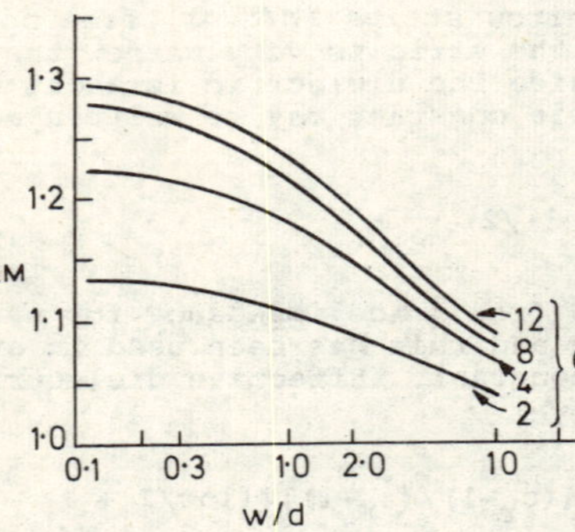

Fig. 3.7 Normalised guide wavelength of microstrip as a function of W/d (Ref.[6])

capacitance per unit length which decreases the characteristic impedance. For the same reason, lower Z_o is obtained for higher values of dielectric constant of the substrate material.

In Fig.3.7 λ/λ_{TEM} is plotted versus W/d. λ_{TEM} is the guide wavelength in a TEM mode line completely filled with dielectric. We have

$$\lambda/\lambda_{TEM} = \sqrt{\varepsilon_r/\varepsilon_{reff}} \tag{3.15}$$

We note that the factor λ/λ_{TEM} decreases as the width of the microstrip is increased and tends to unity for large width. This behaviour is because of the fact that for larger widths more and more of the power flowing along the microstrip is contained in the dielectric below the strip and the effective filling fraction becomes higher. Also from Fig.3.7, we find that λ/λ_{TEM} is larger for higher values of dielectric constant of the substrate, i.e. for higher dielectric constants $\varepsilon_r/\varepsilon_{reff}$ is larger.

3.3 NUMERICAL METHOD FOR ANALYSIS OF MICROSTRIP LINES.

Electrostatic capacitance for microstrip calculation may also be obtained by solving Laplace's equation numerically.

Since the microstrip structure is an open structure, the electric field region is essentially semi-infinite. Although the Laplace equation can be solved numerically for such a semi-infinite region, it is more convenient to consider the microstrip line to be enclosed in a box as shown in Fig.3.8. Since a similar enclosing structure is invariably used in most of the microstrip circuits, the configuration shown in Fig.3.8 is quite realistic. In fact, it is an advantage of the numerical method that the effect of the enclosing box is taken into account. This fac-

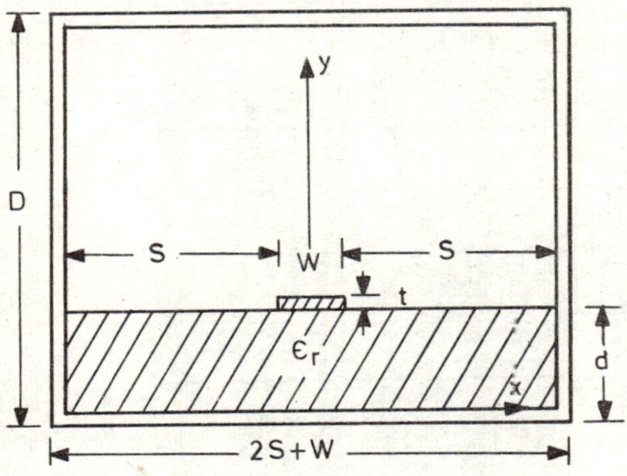

Fig. 3.8 Enclosed Microstrip

tor was not considered in the conformal transformation method discussed earlier.

3.3.1 Relaxation method

Solution of partial differential equations by finite difference methods has been described in the literature[7]. A very commonly used method for solving Laplace's equation numerically is known as Relaxation technique. For using this method,the total electric field region (inside of the box in Fig.3.8) is divided into small squares by means of rectangular grids. Points of intersection of these grids are referred to as net points. For illustrating the relaxation method, let us consider the net point with adjacent points 1 ,2,3 and 4 shown in Fig.3.9a. Let the potential at point 0 with coordinates (x,y) be V_a. We expand the potential in Taylor's series about 0 to find the potential at the the other points. At 1,

$$V_1 = V(x+h,y) = V_o+h(\partial V/\partial x)_o + 0.5h^2(\partial^2V/\partial x^2)_o + \ldots \quad (3.16)$$

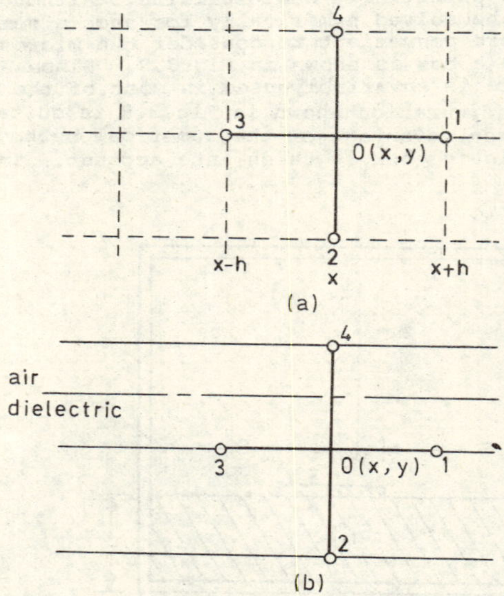

Fig. 3.9 Net points for Finite difference method
(a) A mesh point O(x,y) away from the dielectric-air boundary
(b) A mesh-point O(x,y) near the dielectric-air boundary

and at 3,

$$V_3 = V(x-h,y) = V_0 - h\,(\partial V/\partial x)_0 + 0.5h^2(\partial^2 V/\partial x^2)_0 - \ldots \qquad (3.17)$$

Note the sign reversal for all odd powers of h in this equation. Adding these two equations, we have

$$V_1 + V_3 = 2V_0 + h^2(\partial^2 V/\partial x^2)_0 + \text{terms in } h^4 \text{ and higher.}$$

In a similar way, calculating potentials at 2 and 4 and adding we get

$$V_1 + V_2 + V_3 + V_4 = 4V_0 + h^2\,(\partial^2 V/\partial x^2 + \partial^2 V/\partial y^2)_0 + \text{terms in } h^4 \text{ and in higher powers of } h.$$

The quantity in the parenthesis above is the Laplacian of V. In charge free space it is zero. Since h is small, terms in h^4 and higher order terms may be ignored. Therefore we have

$$V_1 + V_2 + V_3 + V_4 = 4V_0 \qquad (3.18)$$

This relation is an important property of the solution of Laplace's equation and holds good for every point in the region where $\nabla^2 V = 0$. However, for the points at dielectric-air boundary a modification in eqn.(3.18) would be nedded. This is discussed in Sec.3.3.2.

The relaxation method consists of a systematic iterative approach. It starts with an assumed distribution of the net point potentials at every unknown point. This is essentially an intelligent guess. Next,eqn.(3.15) is used to know how bad the assumed value is at any given point. The amount by which the assumed value is different from the computed value is called the residual. Numerically, the residual at point 0 (for points not on dielectric-air boundary) may be written as:

$$R_0 = V_0 - (V_1 + V_2 + V_3 + V_4)/4 \qquad (3.19)$$

Relaxation is the process of changing the assumed value of potential at each net point in a systematic manner to obtain smaller residual values at each point. When this is done repeatedly, the residuals become vanishingly small and the net point potentials give the correct potential distribution. The rate at which the potentials are changed or corrected is determined by the residual calculated at that point. This process is best illustrated by the formula

$$\text{new } V_0 = \text{old } V_0 - \alpha R_0$$

where R_0 is residual at A and α is the rate of correction. Of course, changing the potential at point 0 will also affect the residuals of the neighbouring points, making the process complicated. Indeed, the process requires many repeated corrections, but with an appropriate selection of α, the process does converge, giving the potentials at each net point. These values for potential satisfy both the Laplace equation and the boundary conditions.

3.3.2 Modification for dielectric-air boundary

When a net point next to the dielectric air boundary is considred, the relationship given by eqn.(3.18) needs to be modified. Considering the point 0 shown in Fig.3.9(b), we note that the field component, given by $\partial V/\partial y$, near 0 is different for the dielectric and air regions. Values of $\partial V/\partial y$ in the dielectric and the air regions are related by

$$(\partial V/\partial y)_a = \varepsilon_r(\partial V/\partial y)_d$$

where subscripts a and d denote the values in air and dielectric respectively. Potentials at 2 (inside the dielectric) and at 4 may be written as

$$V_2 = V(x,y-h) = V_o-h(\partial V/\partial y)_o + 0.5h^2\ (\partial^2V/\partial y^2)_o- \ldots$$

$$V_4 = V(x,y+h) = V_o+h(\partial V/\partial y)_o + 0.5h^2\ (\partial^2V/\partial y^2)_o+ \ldots$$

From these relations, we may write

$$V_1 + \varepsilon_r V_2 + V_3 + V_4 = (3 + \varepsilon_r) V_o \tag{3.20}$$

which should be used in place of eqn.(3.18). Using eqn.(3.20) the residual for a point near dielectric-air boundary is obtained as

$$R_o = (3+\varepsilon_r)V_o/4 - (V_1+\varepsilon_rV_2+V_3+V_4)/4 \tag{3.21}$$

which should be used in place of eqn.(3.19)

3.3.3 Calculation of characteristic impedance

The numerical method described above yields the values of the electric potential at the net points. Electric field may be calculated from these values. The capacitance of the microstrip line is computed from the line integral of the electric field normal to the boundary. By Gauss's Law the charge on the enclosing box is given by the integral

$$Q = \varepsilon_o \oint_s \varepsilon_r(E_n.ds) \tag{3.22}$$

where s is the inner surface of the box. The strip could also be considered for this calculation but since the field varies rapidly near the strip, the boundary s is normally used. If the values of E_n are computed for a potential difference of unity between the strip and the box, capacitance per unit length of the microstrip line may be written as

$$C = \varepsilon_o \oint_s \varepsilon_r(E_n.ds) \tag{3.23}$$

Relative propagation constant (or the relative phase velocity) may be computed by comparing with the case where the dielectric in the microstrip is replaced by air. Let the capacitance in this case be denoted by C_a. The phase velocity for the microstrip line v_p is related to the phase velocity, for all-air line v_a by

$$v_p/v_a = \sqrt{C_a/C} \tag{3.24}$$

For the TEM mode v_a is equal to the velocity of light. Thus, v_p may be calculated using the above relation. Using these values of C and v_p in eqn.(3.1), the characteristic impedance for the microstrip line may be written as

$$Z_o = 1/(V_a\sqrt{C_a C}) \tag{3.25}$$

3.3.4 Discussion of results

Numerical results based on the method described above have been given by Stinehelfer[3]. He has calculated the phase velocity, the characteristic impedance and also the loss factor for a microstrip line enclosed in a box. A comparison of his results for Z_o with Wheeler's conformal transformation method is given in Fig.3.10. We observe that the characteristic impedance for microstrip in a box is lower than that for an open microstrip structure. The difference in the two values increases with impedance and is about 5 percent for a 50 ohm line. This is an expected result since the presence of the enclosure increases the cap-

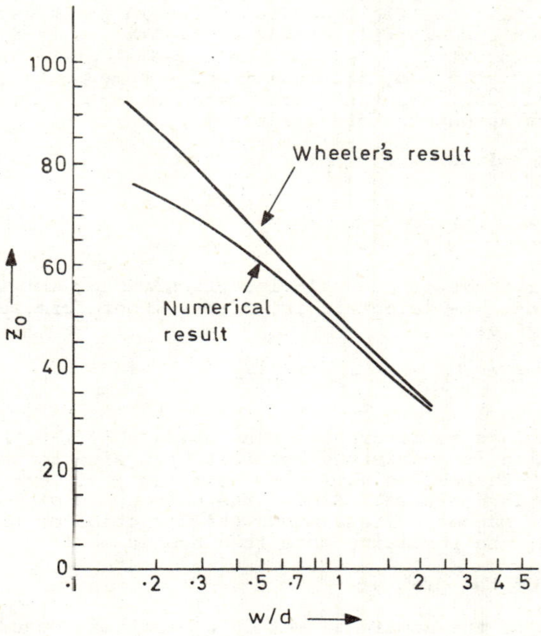

Fig. 3.10 Comparison of Z_o by Stinehelfer's results with Wheeler's results (Ref.[3]) (t = 0.005 in., 2S+W = 0.5 in.)

acitance between the two conductors and thereby decreases the characteristic impedance.

The numerical method of analysis can be extended for analysing non-TEM modes also. This is discussed in Sec.3.4.

3.4 HYBRID-MODE ANALYSIS

3.4.1 Need for hybrid mode analysis

The two methods of microstrip analysis discussed earlier are based on the assumption that microstrip lines support a TEM mode. This assumption is not truly valid and the analyses based on this assumption are inadequate for estimating the dispersion properties of the line at higher frequencies. The presence of air-dielectric interface in these lines gives rise to the propagation of hybrid modes, i.e., a superposition of TE and TM modes , and a rigorous procedure based on hybrid mode analysis of the structure is required to study high frequency performance of these lines.

Using Maxwell's equations it can be seen that hybrid modes are necessary to explain the field configuration in a microstrip line. Fig.3.2a shows the electric field configuration in a cross-section of the microstrip line. At the air-dielectric interface defined by $x>W/2$ at $y = d$, both the x and y-components of electric field exist. E_x, the tangetial component of electric field, is continuous across the air-dielectric interface.

The continuity of E_x gives

$$(\partial H_y/\partial z-\partial H_z/\partial y)_d = \varepsilon_r(\partial H_y/\partial z - \partial H_z/\partial y)_a \tag{3.26}$$

Also , the normal component of magnetic induction vector (B) is continuous across the air-dielectric interface. Therefore, eqn. (3.26) may be written as

$$\varepsilon_r \; \partial H_z/\partial y|_a - \partial H_z/\partial y|_d = (\varepsilon_r-1) \; \partial H_y/\partial z$$

since H_y is non-zero and ε_r is greater than unity the right-hand side of above equation is non-zero. For left hand side to be non-zero it is necessary that H_z is non-zero. Similarly, it can be proved that E_z also is non-zero. Since , the components of both the electric and the magnetic field along the direction of propagation are non-zero, the resulting mode is a hybrid mode.

3.4.2 Formulation of the problem

As in the case of the numerical method of analysis described in Sec.3.3, the microstrip line is enclosed in a box as shown in Fig.3.8.

The enclosed microstrip line can also be viewed as a dielectric-loaded rectangular waveguide with a microstrip line. An inhomogeneously filled waveguide, in general, supports a hybrid mode, having both the TE and TM mode components[8]. The TM and TE field configurations are derivable from scaler potentials Ψ and ϕ which are linearly related to E_z and H_z respectively.

The modes in the microstrip line may be subdivided into the $(H_z)_{odd}-(E_z)_{even}$ mode and $(H_z)_{even}-(E_z)_{odd}$ mode with the even and the odd symmetries occuring about the y-axis. The $(H_z)_{odd}-(E_z)_{even}$ mode is discussed here. A similar formulation can be written for the other mode also.

The initial steps in the formulation of the problem are as follows:

(a) The electric and magnetic fields in subregions 1 and 2 are written in terms of scalar potentials. The scalar potentials are expressed in a form (e.g. infinite series) such that the boundary conditions on the side walls are satisfied by the functions involved.

(b) The boundary conditions at the interface y=d on the strip ($|x|<W/2$) and at the interface ($|x|>W/2$) (assuming the metal strip to be infinitely thin) are applied.

(c) The resulting equations for the dispersion characteristics are solved.

TM mode field distribution may be written as[4]

$$E_{zi} = j[(k_i^2 - \beta^2)/\beta]\ \Psi_i\ e^{-j\beta z}$$

$$E_{ti}^{(e)} = \nabla_t \Psi_i\ e^{-j\beta z}$$

$$H_{ti}^{(e)} = (\omega\varepsilon_i/\beta)\ \hat{z} \times \nabla_t \Psi_i\ e^{-j\beta z}$$

TE mode field distribution may similarly be written in terms of ϕ_i.

$$H_{zi} = j[(k_i^2 - \beta^2)/\beta]\ \phi_i\ e^{-j\beta z}$$

$$E_{ti}^{(h)} = -(\omega\mu/\beta)\ \hat{z} \times \nabla_t \phi_i\ e^{-j\beta z}$$

$$H_{ti}^{(h)} = \nabla_t \phi_i\ e^{-j\beta z}$$

where β is the propagation constant in the z direction, the subscript t implies the transverse direction, $\hat{z}$ is the z-directed unit vector, and superscripts (e) and (h) are associated with E(TM) and H(TE) fields, respectively. The subscript i = 1,2 designates the region 1 or 2, and

$$k_1 = \omega\sqrt{\varepsilon_1\mu_o} = \omega\sqrt{\varepsilon_r\varepsilon_o\mu_o}$$

$$k_2 = \omega\sqrt{\varepsilon_2\mu_o} = \omega\sqrt{\varepsilon_o\mu_o}$$

where ω is the operating frequency, ε_o and μ_o are the permittivity and permeability of free space and ε_r is the relative dielectric constant of the substrate.

The scalar potentials Ψ_i and ϕ_i satisfy the two-dimensional Helmholtz equations in regions 1 and 2 as well as the requirements that the tangential electric fields vanish on the metallic periphery of the box and the tangential magnetic fields H_y and H_z vanish at the plane of symmetry x = 0. In view of the boundary conditions on the side walls it is appropriate to express the scalar potentials as

$$\Psi_1 = \sum_{n=1}^{\infty} A_n^{(e)} \sinh(\alpha_{n1}y) \cos k_n x$$

$$\Psi_2 = \sum_{n=1}^{\infty} B_n^{(e)} \sinh\{\alpha_{n2}(y-h)\} \cos k_n y$$

$$\phi_1 = \sum_{n=1}^{\infty} A_n^{(h)} \cosh(\alpha_{n1}y) \sin k_n x$$

$$\phi_2 = \sum_{n=1}^{\infty} B_n^{(h)} \cosh\{\alpha_{n2}(y-h)\} \sin k_n y$$

where

$$k_n = (n-1/2)\ \pi/L$$

$$\alpha_{n1} = \sqrt{(k_n^2 + \beta^2 - \varepsilon_r k_o^2)}$$

$$\alpha_{n2} = \sqrt{(k_n^2 + \beta^2 - k_o^2)}$$

and $k_o = \omega\sqrt{\varepsilon_o \mu_o}$ is the free space wavenumber.

Now, the boundary conditions are applied on the field components and the resulting equations are solved for the dispersion characteristics. The various techniques, for determining the dispersion characteristics of microstrip lines, differ in the application of boundary conditions and the method of solving resulting equations.

We shall review the different techniques employed by Hornsby and Gopinath [9,10] and Mitra and Itoh [4], for calculating the dispersion characteristics of microstrip line and discuss the results obtained by them.

3.4.3 Fourier analysis method [9]

The dispersion relation for the microstrip is obtained by applying the boundary conditions on the field components, and eliminating the coefficients $A_n^{(e)}$, $B_n^{(e)}$, $A_n^{(h)}$ and $B_n^{(h)}$ from the resulting equations. It is not necessary to consider all the boundary conditions. In fact, satisfaction of a particular subset of boundary conditions will ensure the satisfaction of others.

Here, a sufficient subset is: E_z, H_x, H_y and H_z are continuous for $W/2 \leq x \leq S$ with

$$E_{z1} = E_{z2} = \partial H_{z1}/\partial y = \partial H_{z2}/\partial y = 0 \quad \text{for } 0 \leq x \leq W/2$$

The various steps in the analysis are: (a) The boundary conditions are divided into three groups H_{z1}, H_{z2}, $\partial H_{z1}/\partial y$; H_{y1}, H_{y2}, $\partial H_{z2}/\partial y$; and H_{x1}, H_{x2}, E_z. (b) In a group the boundary conditions are written in a functional form such that the functions $F_i(x)$, $i = 1,2,3$ so obtained are identically zero at $y = d$ and across the width of the microstrip line $(o \leq x \leq W/2+S)$. The functions $F_i(x)$ are obtained as

$$F_1(x) = [D(x)(H_{z1}-H_{z2}) + \{1-D(x)\} \partial H_{z1}/\partial y]_{y=d}$$

$$F_2(x) = [D(x)\ (H_{y1}-H_{y2}) + \{1-D(x)\} \partial H_{z2}/\partial y]_{y=d}$$

$$F_3(x) = [D(x)\ (H_{x1}-H_{x2}) + \{1-D(x)\} E_z]_{y=d}$$

where $D(x)$ is 0 on the strip and 1 on the interface. The usefulness of this step lies in the fact that instead of having six functions, $E_x, E_y, E_z, H_x, H_y, H_z$; satisfying non-uniform boundary conditions across the width of microstrip line we now have three functions $F_i(x)$ which satisfy the identical boundary conditions (i.e. zero along $y = d$, $o \leq x \leq (W/2 + S)$). (c) The condition in step (b) is satisfied by constructing a complete set of functions, and requiring each $F_i(x)$ to be orthogonal to every member of the complete set. (d) The expressions for $F_i(x)$ are substituted in the orthogonality relation. (e) The necessary steps of differentiation and integration are performed to obtain a homogeneous set of 3J linear equations for the unknows $A_n^{(e)}$, $A_n^{(h)}$ and $B_n^{(h)}$. $B_n^{(e)}$ is related to $A_n^{(e)}$ through the continuity of E_z across the interface. Here J=n is the number of terms after which the series expression for Ψ_i and ϕ_i is terminated. (f) Because the coefficients of these equations are known functions of the unknowns α_{n1} and α_{n2}, the nonzero solution of the problem gives the value of β.

Method of calculation

For any selected fixed value of the propagation coefficient an estimate of the corresponding frequency is made, α_{n1} and α_{n2} are calculated, and the determinant is evaluated. The frequency is then varied in a systematic manner until a value is found for which the determinant is zero.

3.4.4 Singular integral equation approach[4]

Fourier analysis method suffers from the slow convergence of the series due to the singular behaviour of the fields at the edge of the strip. As a result of this the size of the matrix to be solved for the dispersion characteristic, becomes very large (of the order of 100x100). In the present method this difficulty does not exist.

The key step in this method involves the transformation of

the equations obtained by applying boundary conditions into an auxiliary set of equations that can be solved by methods used for solving singular integral equations. The final result of these steps is a determinantal equation which has the property of more rapid convergence as compared to the determinantal equation corresponding to the original set of equations.

The determinantal equation is studied for a suitable truncation size, and it is found that a 2x2 size is adequate for commonly used values of parameters. The resulting determinantal equation, which is rather simple in form, is then solved for the wavenumber β as a function of the free-space wavenumber k_o.

The computational effort in this method is at least ten times smaller than in the method discussed in the previous section. Apart from the numerical efficiency, the simplicity of the method makes it possible to solve conveniently for higher order modes that affect the high frequency behaviour of microstrip lines.

Another advantage of this method is that a user interested in numerical results only, may bypass the mathematical details completely and use the final eigenvalue equation which is easily programmable on the computer.

3.4.5 Numerical method

Hornsby and Gopinath[10] have used the finite difference method to study the lowest order hybrid mode in the boxed microstrip line. The finite difference method is applied to a uniform mesh and the resulting matrix eigenvalue problem is solved using a method developed by Martin and Wilkinson[13]. The dispersion curve has been plotted and results are compared with those obtained by Wheeler's method. The difference between the two results is of the order of 10 percent for a microstrip with $\varepsilon_r = 9.0$, $(S+W/2) = 1.75$ mm, $W = 0.5$ mm, $d = 0.5$ mm and $D = 2.0$ mm operated at a frequency higher than 20 GHz.

Corr and Davies[11] have used this method to obtain the dispersion curves for the fundamental and higher order hybrid modes in microstrip line. A graded mesh has been used to investigate realistic structures, where the enclosing box is large compared to the strip width and the substrate thickness. In order that the coefficients of the finite difference equations yield a symmetric matrix, a variational method has been used. Using this approach, dispersion curves are plotted in the form of ε_{reff} versus frequency. These results indicate that the dominant mode in the microstrip is dispersive, but in the zero frequency limit it can be approximated with the static case. Also, there is a close relation between the higher order modes, and the LSM modes in the corresponding slab loaded waveguide. A simple estimate of the frequency at which the higher order modes will propagate may be obtained through the consideration of the LSM modes.

3.4.6 Discussion of results

Table 3.1 compares the value of propagation constant obtained by Fourier analysis Method[9] and integral equation method [4] for a microstrip line with the following parameters; ε_r = 9.0, (S+W/2) = 1.75 mm, W = 0.5 mm, d = 0.5 mm and D = 2.0 mm.

Table 3.1 Propagation constant β (mm^{-1})

Frequency (GHz)	Integral equation method 2x2 matrix	Fourier Analysis method 100x100 matrix	TEM Mode analysis
10	0.53	0.55	0.53
20	1.10	1.17	1.05
30	1.71	1.77	1.55

The above results are plotted in Fig.3.11. It shows that the results obtained by the two methods compare quite well. The curve shown here is for the lowest order symmetric mode (dominant mode) which does not exhibit any low frequency cut-off. Moreover, the dominant mode is slightly dispersive. Higher order modes are not present in the frequency range considered. The two dotted lines correspond to air-filled and dielectric-filled transmission lines. The quasi-TEM solution for open microstrip line is also shown for comparison.

Figure 3.12 shows the dispersion characteristics of a microstrip line which supports higher order modes also. The number of higher order propagating modes increases with increasing frequency and/or dielectric constant. Also, with the increase in frequency the dispersion curve for the lowest order mode becomes parallel to the dotted line corresponding to the dielectric-filled TEM transmission line. Thus for high frequencies the normalised guide wavelength for the dominant mode approaches $1/\sqrt{\varepsilon_r}$, the ratio of guide wavelength in the dielectric-filled TEM transmission line to that in an air-filled line.

In microstrip lines most of the energy is concentrated in the dielectric directly below the centre strip, and in this region the longitudinal fields are relatively weak. Consequently, one would expect TEM calculations of guide wavelength, phase constant etc. to be reasonably close to the actual ones[12].

3.5 OTHER METHODS OF ANALYSIS

3.5.1 Coupled mode analysis

This method of analysis takes into account the coupling between quasi-TEM mode along the microstrip line (as discussed in sections 3.2 and 3.3) and TM_{10} surface wave mode on the dielectric substrate with metallisation on the bottom surface. This formulation has been reported by Hartwig et al.[14] and Jain et al. [15,16]. These investigators assume that a strong coupling exists between the two modes.

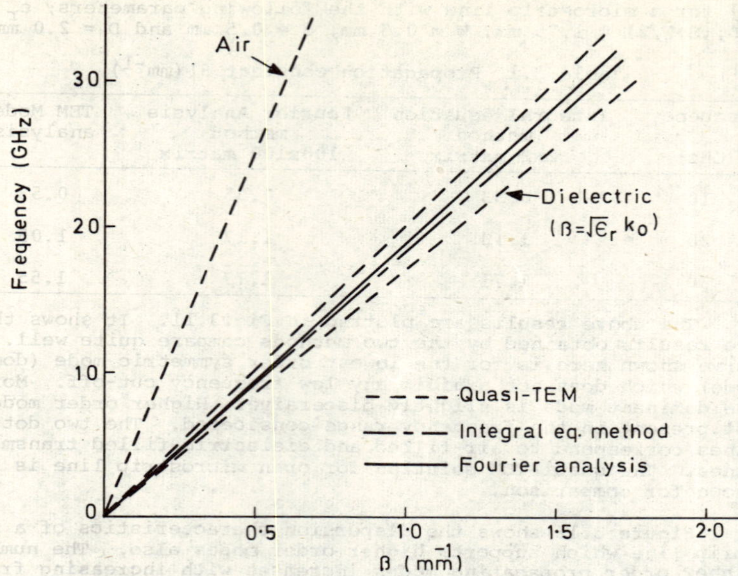

Fig. 3.11 Dispersion diagram of shielded microstrip line (Ref. [4])

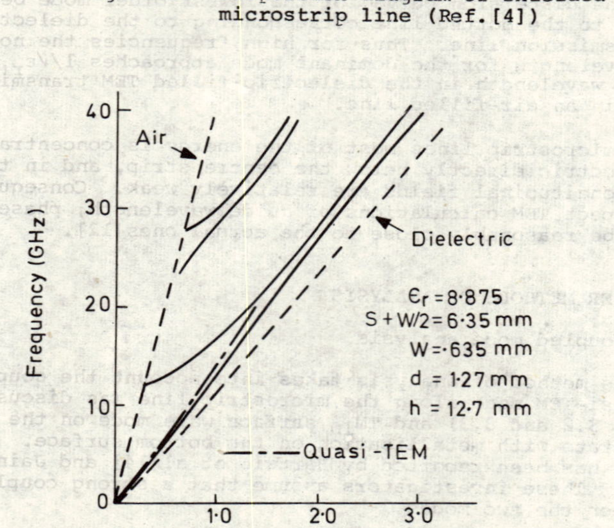

Fig. 3.12 Higher order hybrid modes in the shielded microstrip line (Ref. [4])

An upper cut off frequency f_L is defined as the frequency at which the phase velocity of the quasi-TEM mode along the microstrip line is equal to phase velocity of TM_{10} surface wave on the substrate. At frequencies below f_L, the effective dielectric constant for the coupled modes (ε_{eff} 1,2) is obtained in terms of effective dielectric constants of the uncoupled modes ε_{effo} and ε_{effTM} by using the coupled mode theory as given below [15, 16, 17]

$$\varepsilon_{eff1,2} = [(\sqrt{\varepsilon_{effo}} + \sqrt{\varepsilon_{effTM}})/2 \pm \sqrt{\{\varepsilon_{12}\varepsilon_{21} + (\sqrt{\varepsilon_{effo}} - \sqrt{\varepsilon_{effTM}})^2/4\}}]^2 \qquad (3.27)$$

where $\varepsilon_{12}/\varepsilon_m = \varepsilon_{21}/\varepsilon_m$ = coupling coefficients between the two modes and

$$\varepsilon_m = \sqrt{(\varepsilon_{effo} \cdot \varepsilon_{effTM})}$$

ε_{effo} is known from Wheeler's analysis (see eqn.(3.14) of sec. 3.2). $\varepsilon_{eff\ TM}$ is the effective relative permittivity of TM surface wave mode and may be evaluated by solving the dispersion relation graphically[18].

The difficulty in using eqn.(3.27) is mainly due to lack of explicit expression for coupling coefficients $\varepsilon_{12}/\varepsilon_m$ and $\varepsilon_{21}/\varepsilon_m$. Hartwig et al.[14] have derived two different empirical relations for ε_{12} to fit experimentally determined dispersion data for an alumina substrate (ε_r = 9.7, d = 50 x 10^{-3} in., W/d = 0.96) and for a rutile substrate (ε_r = 104, d = 50 x 10^{-3} in. and W/d = 0.6) respectively. Jain et al.[15,16] have derived an expression for ε_{12} based on qualitative arguments and experimental results. For small thickness of the strip this expression is as follows:

$$\varepsilon_{12}^2 = 0.22\ (\sqrt{\varepsilon_r}-1)^2\ (W/d)^{3/4}\ (f/f_L)^{4/3} \qquad (3.28)$$

Equation (3.28) may be used to predict accurately (within 2%) the dispersion characteristics of the microstrip lines over a wide range of relative permittivity ($2 \leq \varepsilon_r \leq 104$), substrate thickness (0.024 in $\leq d \leq$ 0.125 in) and W/d ratio ($0.9 \leq W/d \leq 6$).

Results based on equation (3.28) have been compared with experimental results[15,16]. A very good agreement, as shown in Fig.3.13, has been reported.

It has been pointed out[15] that the above described model may be used upto a frequency 0.5 f_L. The analysis and experimental results both indicate that additional coupled modes may exist at frequencies above 0.5 f_L.

3.5.2 Method of images

In addition to the conformal transformation method described in Sec.3.2, TEM properties of microstrip transmission lines may also be analysed by considering the multiple images at the dielectric air boundaries and solution of the integral equation obtained by applying Coulomb's Law. This method has been descr-

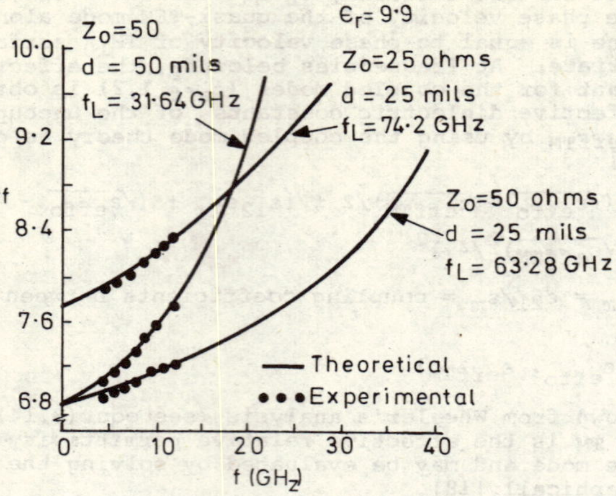

Fig. 3.13 Results obtained by coupled mode analysis (From Ref.[16])

ibed by Silvester[19], Bryant and Weiss[20] and Pal and Babu[21].

As in the case of Wheeler's conformal transformation method [2], Silvester[19] has analysed the microstrip geometry in terms of two parallel strips separated by a dielectric sheet. Each strip is considered as a summation of line charges distributed over the surface of the strip. The potential distribution resulting from a single line charge (carrying a unit charge) is calculated by using Coulomb's Law and is called two-dimensional Green's function. The total potential at any point P(x,y) may then be written in terms of the Green's function, superposing the contributions of all charges, as

$$V(x,y) = \int \sigma(Q) \quad G(P,Q) \; dQ \tag{3.29}$$

when the integration is carried out all over the conductor surface Q. Here $\sigma(Q)$ denotes the surface charge density on the conductors and is the desired unknown quantity. The problem of computing the capacitance may now be divided into two parts.

a) calculation of Green's function G(Q,Q) and

b) solution of the above integral equation

Concept of multiple images is used for calculation of the Green's function. An image coefficient K is defined to take into account the partial reflection of flux line at the interface between two dielectric media with relative permittivities given by

ε_o and ε_1. K is given by

$$K = (\varepsilon_o - \varepsilon_1)/(\varepsilon_o + \varepsilon_1)$$

Multiple images of a line charge q near a dielectric slab 2W wide are shown in Fig.3.14. G may be written as the sum of the potentials produced by various images of the line charge q.

The solution of the integral equation is carried out by numerical techniques. Considering the strip to be made up of small sub-strips, it is possible to replace the integral in eqn. (3.29) by summation and thus write a matrix equation in place of the integral equation. The matrix equation is solved for charge on each substrip by standard matrix inversion techniques.

Silvester[19] has compared the characteristic impedance computed by this method with the experimental measurements reported by Kaupp[22] and found them to be in fair agreement.

Bryant and Weiss[20] have extended this method for calculation of even and odd mode impedances of coupled microstrips. Also, results for single microstrip by this method have been compared with Wheeler's results[2]. Wheeler has described two approximations appropriate for wide and narrow strips respectively. Agreement between these two approximations is best for W/d equal to unity. The results obtained by the method of images are intermediate between the two, the disagreement being within about 1%.

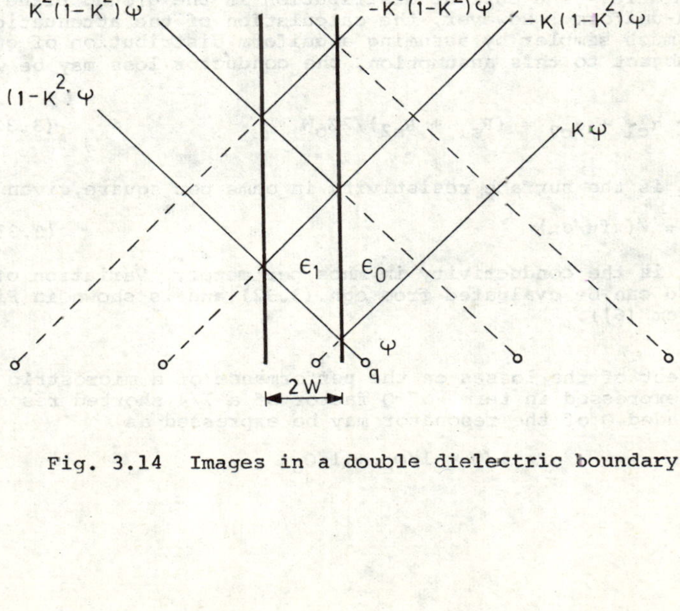

Fig. 3.14 Images in a double dielectric boundary

Pal and Babu[21] have also reported the impedance computations based on the method of images. Results have been compared with those reported by Bryant and Wiess[20]. Small disagreement between the two results is attributed to computational procedure . The charge distribution on the strip conductor has been computed by this method and found to be of parabolic shape[21].

3.6 LOSSES IN MICROSTRIPS

As in the case of other transmission lines, losses in microstrips are of two types: Conductor loss and dielectric loss. The total loss per unit length of the line may be written as

$$\alpha = \alpha_{c1} + \alpha_{c2} + \alpha_d \tag{3.30}$$

Where, α_{c1} and α_{c2} are the attenuation constants because of loss in two conductors and α_d is due to the dielectric loss in the substrate . α_c's and α_d may be expressed in terms of parameters of transmission line as

$$\alpha_{c1} = r_1/2Z_o, \ \alpha_{c2} = r_2/2Z_o \text{ and } \alpha_d = gZ_o/2 \tag{3.31}$$

where r_1, r_2 are series resistances per unit length, g is the shunt conductance per unit length and Z_o is the characteristic impedance of the line.

3.6.1 Conductor loss

Calculation of conductor loss in a microstrip becomes complicated because of the fact that the current distribution is not uniform across the width of the microstrip conductor[23]. The current densities at the edges of the strip is higher than that in the middle. The current distribution in the ground plane is also non-uniform. However, the calculation of the attenuation becomes much simpler by assuming a uniform distribution of current. Subject to this assumption, the conductor loss may be written as

$$\alpha_c = \alpha_{c1} + \alpha_{c2} = (R_{s1} + R_{s2})/2Z_oW \tag{3.32}$$

where R_s is the surface resistivity in ohms per square, given by

$$R_s = \sqrt{(\pi f \mu/\sigma_c)} \tag{3.33}$$

where σ_c is the conductivity in mhos per meter. Variation of α_c with W/d can be evaluated from eqn.(3.32) and is shown in Fig. 3.15 (from [6]).

Effect of the losses on the performance of a microstrip can also be expressed in terms of Q factor of a $\lambda/4$ shorted resonator. The unloaded Q of the resonator may be expressed as

$$1/Q = 1/Q_c + 1/Q_d$$

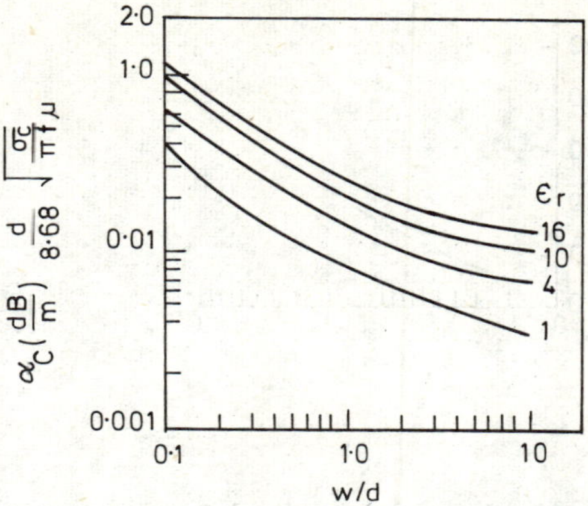

Fig. 3.15 Calculated normalised conductor loss as a function of W/d and ε_r. (Ref. [6])

where Q_c is the Q factor of the conductors and Q_d is the Q factor of the dielectric. Q_c for the case when the strip and the ground plane are fabricated from the same material may be written as

$$Q_c = \omega_o Z_o^2 C / r_s \tag{3.34}$$

where ω_o is the resonant frequency, C is the capacitance per unit length and r_s is the total series resistance of the conductors per unit length. Q_c can be expressed in normalised form as [6]

$$6Q_c \,/\, d\sqrt{f_{GHz}\sigma_c} = (\lambda_{TEM}/\lambda)\ (W/d)\ Z_o\sqrt{\varepsilon_r} \tag{3.35}$$

This variation is plotted in Fig.3.16. Even though Z_o decreases with W/d normalised Q_c increases with W/d.

Effect of nonuniform current distribution on the conductor loss is discussed in [23].

3.6.2 Dielectric loss

For a uniformly filled transmission line the dielectric loss α_d is independent of the geometry of the line and may be written as (from eqn. (3.31))

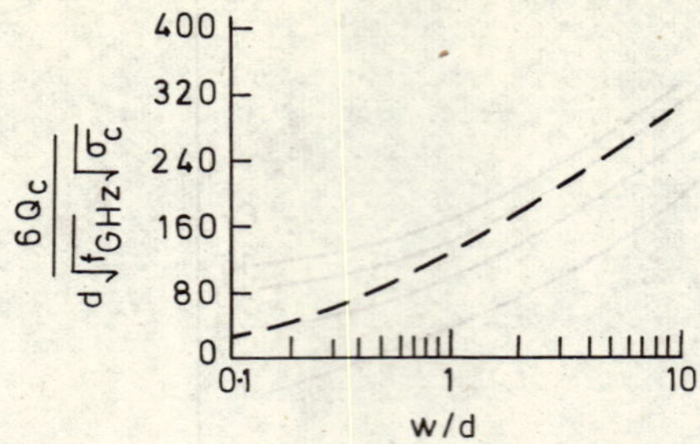

Fig. 3.16 Calculated normalised Q_c as a function of W/d (curve is independent of ε_r) (Ref.[6])

$$\alpha_d = gZ_o/2 = \omega\varepsilon''\sqrt{\mu/\varepsilon'}\ /2 = (\omega\sqrt{\mu\varepsilon}\ \tan\delta)/2 \tag{3.36}$$

where ε' is the real part of the permittivity of the dielectric and ε'' is the imaginary part of the permittivity. However, when the ε is not uniform over the cross section of the line, the above expression does not hold good. Calculation of loss for mixed dielectric case has been considered by Welch and Pratt[20]. In this analysis a 'filling factor' for the loss tangent ($\tan\delta$) is derived. When the upper dielectric (air) is assumed lossless the following expression for α'_d (attenuation constant considering mixed dielectrics) is obtained [24].

$$\alpha'_d = \omega\sqrt{\mu\varepsilon_{eff}}\ (\tan\delta)_{eff}/2$$

$$= \omega\sqrt{\mu\varepsilon_{eff}}\ (\sigma_{eff}/\omega\varepsilon_{eff})/2 \tag{3.37}$$

since $\varepsilon_{eff} = q\varepsilon + (1-q)\ \varepsilon_o$

and $\sigma_{eff} = q\sigma + (1-q)\ \sigma_o \simeq q\sigma$

We have

$$\alpha'_d = q\ [\ \varepsilon/(q\varepsilon+(1-q)\varepsilon_o)]^{1/2}\ \alpha_d$$

where q is the dielectric filling fraction (see Sec.3.2.3) and α_d is attenuation constant for a line uniformly filled with the dielectric ε.

Comparing the expression for α_d and $\alpha'd$, the loss tangent filling factor may be written as

$$q_{d\ell} = q\varepsilon_r/\varepsilon_{reff}$$

$q_{d\ell}$ is of the order of unity and approaches this value as a limit when $\varepsilon_r \to \infty$. For most practical purposes, $q_{d\ell}$ may be approximated by unity.

References

1. Wheeler, H.A., Transmission line properties of parallel wide strips by conformal mapping approximation, IEEE Trans. Microwave Theory Tech., Vol. MTT-12, 1964, pp.280-289.

2. Wheeler, H.A., Transmission line properties of parallel strips separated by a dielectric sheet. IEEE Trans. Microwave Theory Tech., Vol.MTT-13, 1965, pp.172-185.

3. Stinehelfer, H.E., An accurate calculation of uniform microstrip transmission lines, IEEE Trans. Electron Devices, Vol. ED-13, No.7, July 1968, p.501.

4. Mittra, R. and T. Itoh, A new technique for the analysis of the dispersion characteristics of microstrip lines, IEEE Trans. Microwave Theory Tech., Vol.MTT-19, 1971, p.47.

5. Grivet, P., The physics of transmission lines at high and very high frequencies, Vol.1, Academic Press 1970, p.131.

6. Caulton, M., J.J. Hughes and H. Sobol, Measurements on the properties of microstrip transmission lines for microwave integrated circuits, RCA Rev. Vol.27, Sept. 1966, pp.377-391.

7. Forsythe, C.E., and W.R. Wasow, Finite difference methods for partial differential equations, New York, John Wiley, 1960.

8. Collin, R.E., Field theory of guided waves, McGraw-Hill, 1960, p.224.

9. Hornsby, J.S. and A. Gopinath, Fourier Analysis of a dielectric loaded waveguide with a microstrip line, Electron.Lett., Vol.5, June 12, 1969, pp.265-267.

10. Hornsby, J.S. and A. Gopinath, Numerical analysis of the dielectric-loaded waveguide with a microstrip line-Finite-difference methods, IEEE Trans.Microwave Theory Tech.,Vol.MTT-17, 1969.

11. Corr, D.G. and J.B. Davies, Computer analysis of the fundamental and higher order modes in single and coupled microstrip, IEEE Trans. Microwave Theory Tech., Vol.MTT-20, 1972,p.664.

12. Wharton, R.P. and G.P. Rodrigue, A dominant mode analysis of

microstrip, IEEE Trans.Microwave Theory Tech., Vol.MTT-20, 1972, pp.552.

13. Martin, R.S. and J.H. Wilkinson, Solution of symmetric and unsymmetric band equations and the calculation of eigenvector of band matrices, Numerische Math., Vol.9, 1967, p.279.

14. Hartwig, C.P., D. Masse and R.A. Pucel., Frequency dependant behaviour of microstrip, Int. Microwave Symposium, Detroit, Mich., 20th-22nd May 1968, pp.110-116.

15. Jain, O.P. et al., Coupled-mode model of dispersion in microstrip, Elecncs Lett, 15 Jul 1971, Vol.7, pp 405-407.

16. Jain, O.P. et al., Dispersive behaviour of microstrip transmission lines, Proc. European Microwave Conf., 1971, C3/5:1-5:4.

17. Johnson, C.C., Field and Wave Electrodynamics, McGraw Hill 1965, pp 296-299.

18. Collin, R.E., Field Theory of guided waves, McGraw Hill 1960, pp 453-474.

19. Silvester, P., TEM wave properties of microstrip transmission lines, Proc. IEE, Vol. 115, p.43-48, 1968.

20. Bryant, T.G., J.A. Weiss, Parameters of microstrip transmission lines and of coupled pair of microstrip lines, IEEE Trans. Microwave Theory Tech., Vol. MTT-16, p.1021-1027, Dec. 1968.

21. Pal, S. and G.R. Babu, Characteristic impedance of microstrip transmission line, JAEU, 5, p.21-27, 1972.

22. Kaupp, H.R., Characteristics of microstrip transmission lines, IEEE Trans., Vol. EC-16, p.185-193, 1967.

23. Pucel, R.A., D.J. Masse and C.P. Hartwig, Losses in microstrip, IEEE Trans. Microwave Theory Tech., Vol.MTT-16, June 1968, pp.342-350.

24. Welch, J.D. and H.J. Pratt, Losses in microstrip transmission system for integrated microwave circuits, NEREM Rec., Vol.8, pp.100-101, 1966.

25. Sobol, H., Applications of integrated circuit technology to microwave frequencies, Proc. IEEE, Vol.59, pp. 1200-1211, Aug. 1971.

4

SLOT LINE AND COPLANAR WAVEGUIDE

by

K.C. Gupta

Chapter Four

4.1 INTRODUCTION

Apart from the microstrip line discussed in Chapter three, two other types of transmission lines have been proposed for use in MICs. These are known as slot line[1,2] (Fig.4.1) and coplanar waveguide[3] (Fig.4.2). One significant aspect in which these lines differ from the microstrip line is that for both the slot line and the coplanar waveguide, the mode of propagation is essentially non-TEM. This gives rise to two special features that are not available with the microstrip line. These are: the shunt mounting of components across the line and the availability of elliptically polarised magnetic field regions suitable for ferrite devices.

Section 4.2 and 4.3 describe salient features of slot line and coplanar waveguide respectively.

4.2 SLOT LINE

As shown in Fig.4.1 a slot line consists of a slot or gap in a conducting coating on a dielectric substrate. The technique of fabrication is identical to that for the microstrip line and is described in Chapter Eleven. However, because of the presence of a non TEM mode, hybrid mode analysis is necessary. Quasi-static methods based on the solution of Laplace's equation (similar to those described for microstrip line in Chapter Three) are not applicable in the present case.

4.2.1 Guide wavelength and field distribution

The simplest model for calculating the guide wavelength for a slot line is based on a solution by Galejs[4]. In this method,

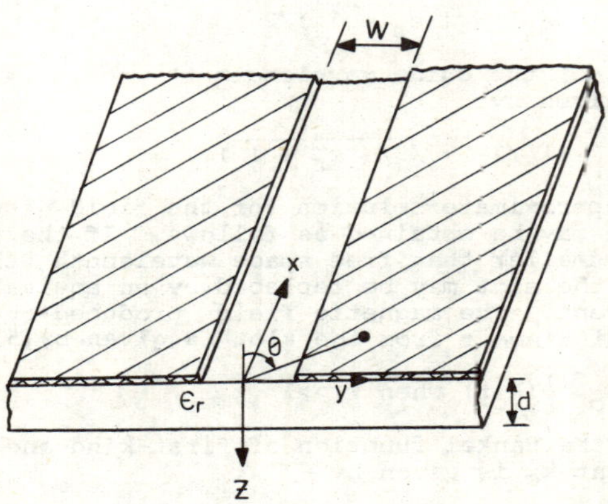

Fig. 4.1 Slot line

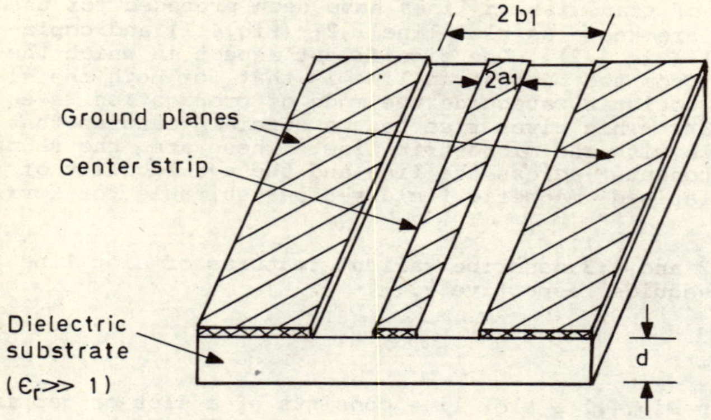

Fig. 4.2 Coplanar waveguide

the slot line is approximated by a slot in a zero thickness perfectly conducting sheet separating free space from a dielectric medium of infinite extent. For this configuration, the effective permittivity may be written as

$$\varepsilon_r' = (\varepsilon_r + 1)/2 \tag{4.1}$$

and the ratio of the guide wavelength λ' to the free space wavelength λ is given by

$$\lambda'/\lambda = \sqrt{2/(\varepsilon_r + 1)} \tag{4.2}$$

A very approximate solution for the field distribution of the slot line may be obtained as follows. If the width of the slot is much smaller than free space wavelength, the electric field across the slot may be replaced by an equivalent source of magnetic current. The magnetic field produced by such a line source at a distance r from the slot is given by [5]

$$H_x = A\, H_o^{(1)} (k_c r) \text{ when } r >> \lambda \tag{4.3}$$

$H_o^{(1)}$ denotes the Hankel function of first kind and zeroth order. The coefficient k_c is given by

$$k_c = j\ 2\pi\sqrt{[(\lambda /\lambda')^2 - 1]}\ /\lambda \tag{4.4}$$

Substituting eqn.(4.2) in eqn.(4.4) a first order approximation for k_c is obtained as

$$k_c = j\ (2\pi/\lambda)\ \sqrt{[(\varepsilon_r-1)/2]} \qquad (4.5)$$

Other field components, H_r and E_ϕ, are related to H_x and may be evaluated by using Maxwell's equations.

Equation (4.4) indicates that the argument of Hankel function in eqn.(4.3), is imaginary since (λ'/λ) is less than unity. For imaginary arguments, the value of Hankel function decreases monotonically. Thus, the field decays with distance r. Hence, λ'/λ less than unity is a sufficient condition to ensure the decay of the slot line field with radial distance. As λ'/λ is decreased, the decay becomes sharper and the field gets more tightly bound to the slot. A detailed behaviour of slot line is obtained by waveguide type analysis outlined below.

4.2.2 Waveguide analysis

Slot line configuration has been analysed in terms of waveguide modes by Cohn[1]. The key feature of this approach is the introduction of boundary walls permitting the slot line configuration to be treated as a rectangular waveguide problem rather than a problem in cylindrical coordinates. Conversion of the slot line geometry into a rectangular waveguide configuration is illustrated in Fig.4.3. Electric walls are inserted in planes perpendicular to the substrate at x = 0 and x = a = $\lambda'/2$, and also at y = ± b/2. Introduction of these walls creates the configuration of a capacitive iris in a rectangular waveguide of width b oriented in z-direction.

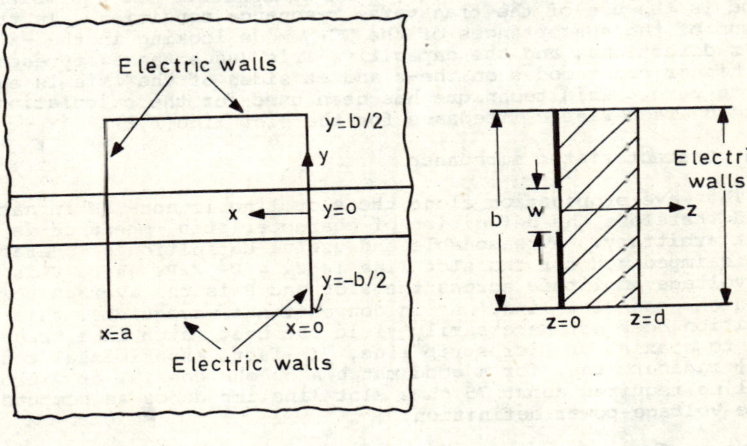

Fig. 4.3 Configuration for waveguide type analysis

In x-directions, walls are placed at a distance of $\lambda'/2$, the field distribution is identical to that obtained by the superposition of two slot line waves of equal amplitudes travelling in +x and -x directions respectively. In ±y directions, b/2 is larger than λ' and therefore the field at the walls (at $y = \pm b/2$) is very small. In such a case the value of b has negligible effect on the fields near the slot.

Field in the waveguide configuration of Fig.4.3 is analysed in terms of various waveguide modes that can exist on either side of the iris plane at $z = 0$. The solution for the field problem is obtained by matching the field components at $z = 0$.

Because of the presence of electric walls at $x = 0$ and $x = a = \lambda'/2$, all waveguide modes must have the $\lambda'/2$ variation of the slot wave in the x direction. Also, because of the symmetry of the structure, all modes must have an E field maximum at $y = 0$. Therefore, the complete set of modes satisfying the boundary conditions are $TE_{10}, TE_{12}, TE_{14}$...and TM_{12}, TM_{14} ..., that is, $TE_{1,2n}$ for $n > 0$. and $TM_{1,2n}$ for $n > 1$. For the wave along the slot line λ' is less than λ and hence $\lambda/2$ is greater than a. Therefore, the TE_{10} and all higher modes along z-direction are cut off, or nonpropagating, in air regions. In the dielectric region the TE_{10} mode is propagating and the first few higher modes may propagate or all higher modes may be cutoff, depending on the values of a and b. Since all modes are cutoff in the two air regions, the energy of the slot wave is trapped near the slot. The amplitudes of various modes in each region must be such that when the complete set of modes are superimposed, the boundary coditions in the iris plane at $z = 0$ are met and all field components on either side of the dielectric air interface at $z = d$ are matched. This method has been used to evaluate the field componenets for the slot line wave[6]. An alternative but equivalent method is the use of the transverse resonance condition. In this the sum of the susceptances of the TE_{10} mode looking in the -z and +z directions, and the capacitive-iris susceptance representing higher order modes on the-z and +z sides of the iris is equated to zero. This technique has been used for the calculation of the characteristic impedance for the slot line[1,2].

4.2.3 Characteristic impedance

The wave propagation along the slot line is non-TEM in nature and therefore the definition of characteristic impedance is somewhat arbitrary. A reasonable and useful definition of characteristic impedance for the slot line is $z_o = V^2 /2P$, where V is the peak voltage amplitude across the slot and P is the average power flow along the slot line. As in conventional waveguides, this definition does not necessarily yield the best match in a transition to coaxial or microstrip line. In fact, experimental results[2] indicate that for a good match a 50 ohm coaxial or microstrip line requires about 75 ohms slot line impedance as computed by the voltage-power definition.

In the present case, relationship between power and voltage may be obtained by considering the resonant cavity formed by inserting electrical walls at $x = 0$ and $x = a$. Energy stored in the resonant cavity may be written as:

$$W_t = (P_+ + P_-)\ (\lambda'/2v_g) \tag{4.6}$$

where P_+ and P_- denote the power carried by waves travelling in +x and -x directions respectively. v_g is the group velocity (or the velocity of energy transport) along the slot line. In low loss case

$P_+ = P_- = P$ and

$$W_t = (2\pi P/\omega)\ (v/v_g) \tag{4.7}$$

Let V_o be the maximum voltage across the slot. Since the waves in the +x and -x directions have equal voltages, $V_+ = V_-$, and the maximum voltage is $V_o = 2V_+$, Z_o may be written as

$$Z_o = V_+^2/2P_+ = \pi V_o^2 v\ /(4\omega v_g W_t) \tag{4.8}$$

W_t can also be expressed in terms of susceptance B at the port of cavity as [7]

$$W_t = (V^2/4)\ (dB/d\omega) \tag{4.9}$$

In the present case we shall assume the port to be at the iris plane z = 0. We shall set B equal to the total waveguide susceptance B_t at the plane and V equal to the slot voltage V_o at $x = \lambda/4$. These choices of B and V are consistant with the waveguide impedance and slot impedance definitions, both of which are on a voltage power basis. Therefore ,

$$Z_o = (v/v_g)\ \pi/(\omega \partial B_t/\partial\omega) \tag{4.10}$$

where $\partial B_t/\partial\omega$ is evaluated at the resonant frequency, that is when $B_t = 0$.

Detailed derivation based on this method has been given by Cohn[1]. Mariani et al.[2] have presented computations based on Cohn's results. Some typical results are given in Fig.4.4.

4.2.4 Coupling to the slot line

Because of the non-TEM nature of the wave along the slot line, coupling from a coaxial line or microstrip to the slot line is not very straight forward. A coaxial line to slot line transition is shown in Fig.4.5. A miniature, semirigid coaxial line is placed at the end of an open-circuited slot line with both the inner and the outer conductors of the coaxial line soldered to the conductive plating of the slot line. Reported experimental results[2] on this type of transition are as follows:

Coaxial line impedance	50 ohms
Theoretical Z_o for slot line	75 ohms
Frequency	3 GHz

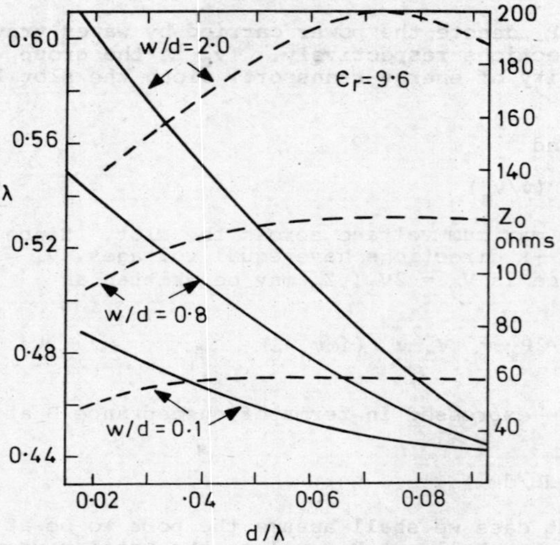

Fig. 4.4 Characteristic impedance and λ'/λ for the slot line (From Ref. [2])

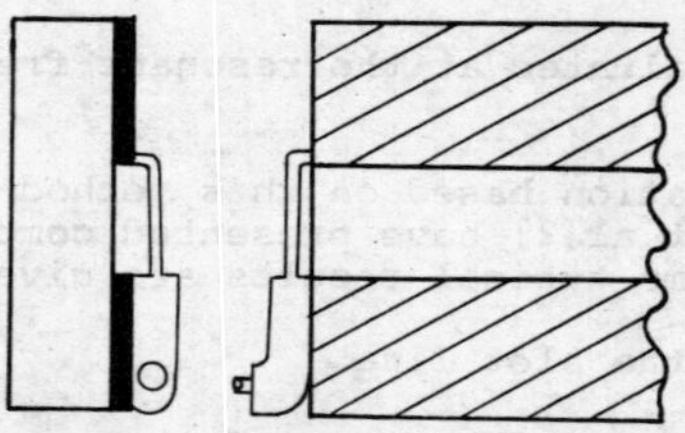

Fig. 4.5 A coaxial line to slot line transition

Bandwidth	500 MHz
VSWR	less than 1.2.

The slot width was 0.031 inch on a 0.062 inch thick Trans-Tech D-16 substrate.

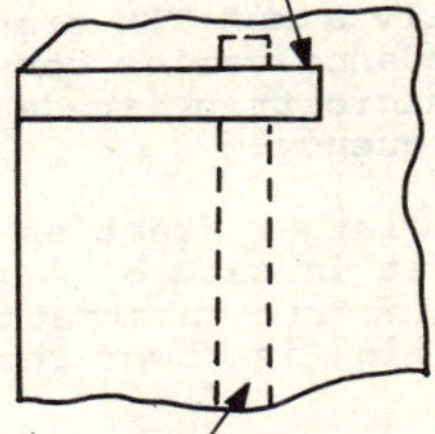

Fig. 4.6 A microstrip to slot line transition

Another type of transition suggested for use with a slot line is a microstrip to slot line transition as shown in Fig.4.6. A microstrip and slot line of equal characteristic impedances cross each other at right angles and extend approximately one quarter wavelength beyond the crossing point. A transition covering 30 percent bandwidth can be achieved in this way. The microstrip may also be terminated in an open circuited quarter wave radial stub. Using this technique, a transition with VSWR less than 1.1 over a frequency range of 8 to 10 GHz has been reported[8].

4.2.5 Comparison with the microstrip line

The distinguishing features of the slot line as compared to the microstrip line may be summarized as follows:

1. As pointed out in Sec.4.2.1, for a wave along the slot line both H_x and H_r components of the magnetic field are present. As a consequence, a region of elliptically polarised magnetic field exists near the slot. This is useful for the fabrication of non-reciprocal ferrite circuits discussed in Chapter Seven.

2. In a slot line, the electric field lines run across the slot. This enables discrete components (active as well as passive) to be shunt mounted across the line. On the other hand, in a microstrip line the electric field lines run from the strip to the ground plane and therefore shunt mounting of the components is not very convenient. The microstrip line configuration is suitable when the components are to be mounted in series with the line.

3. For a slot line, the characteristic impedance increases with the width of the slot whereas in the case of a microstrip the characteristic impedance decreases with the width of the strip. Thus

high impedance lines, which require a very thin line to be photoetched in microstrip configuration, are more easily achieved with the slot line.

4. The slot line is essentially a non-TEM transmission line and therefore characteristic impedance varies considerably with frequency. Comparatively, the microstrip impedance exhibits only a very small variation with frequency.

5. For a slot line there is a larger fraction of the power flowing in the air region than that in case of a microstrip line. In other words, for the same dielectric substrate the effective dielectric constant of the slot line is lower than that for the microstrip line.

4.3 COPLANAR WAVEGUIDE

As shown in Fig.4.2, a coplanar waveguide[3] (CPW) consists of a strip of thin metallic film deposited on the surface of a dielectric slab with two ground electrodes running adjacent and parallel to the strip on the same surface. Unlike conventional waveguides, CPW does not have a low frequency cut off because of the quasi-TEM mode of propagation. However, as in the case of slot line, the wave propagation along CPW is essentially non-TEM. The r.f.electric field between the centre conducting strip and the ground electrodes is tangential to the air-dielectric boundary and produces a discontinuity in displacement current density at the interface. As explained in Sec.3.4.1, this field configuration gives rise to an axial as well as transverse, component of r.f. magnetic field. These components provide the elliptical polarization needed for nonreciprocal ferrite circuits. Also, discrete components may be shunt mounted between the centre strip and either of the ground planes. Thus CPWs also exhibit the two major advantages of the slot line namely, application to the ferrite circuits and the shunt mounting of the components.

Characteristic impedance Z_o of CPW has been calculated[3] as a function of the ratio a_1/b_1, where $2a_1$ is the width of the centre strip and $2b_1$ is the distance between two ground planes. A zeroth-order quasi-static approximation is employed. Method of conformal transformation is used for the calculation of capacitance. The dielectric slab is assumed to be semi-infinite. The dielectric half-plane Z_1 in Fig.4.7a may be transformed to the interior of a rectangle in the Z- plane as shown in Fig.4.7b by conformal mapping techniques[9]. A suitable Schwarz-Christoffel transformation[3] yields

$$a/b = K(k)/K'(k)$$

where

$k = a_1/b_1$,

$K(k)$ = complete elliptic integral of first kind[10],

$K'(k) = K(k')$ and

$k' = (1 - k^2)^{1/2}$.

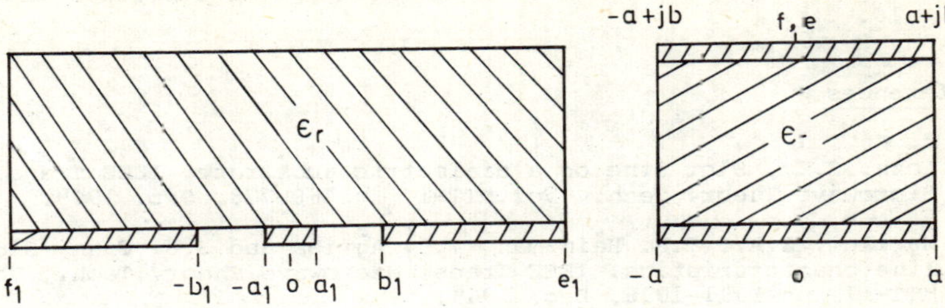

Fig. 4.7 Conformal transformation for coplanar waveguide

If the relative dielectic constant of the material filling the rectangle in the Z plane is ε_r, a uniform electric field E is set up in the capacitor with the top and bottom plates and the capacitance per unit length of the line including the empty half-space is

$$C = (\varepsilon_r + 1)\,\varepsilon_o\,(2a/b)$$

A zeroth-order quasi-static approximation has been employed[3] to estimate the phase velocity also. The approximation simply treats the CPW as a transmission line completely immersed in a dielectric with effective dielectric constant $(\varepsilon_r + 1)/2$. The resulting phase velocity is

$$v_{ph} = [2/(\varepsilon_r + 1)]^{1/2} c$$

where c is the velocity of light in free space. The characteristic impedance of the CPW may now be written as

$$Z_o = \frac{1}{C\,v_{ph}}$$

The theoretical calculation of impedance thus obtained is in fair agreement[3] with experimental measurement on transmission lines fabricated on substrates of relative dielectric constant $\varepsilon_r = 9.5$, 16 and 130.

The effect of change in thickness of the dielectric substrate becomes less critical with higher relative dielectric constants. In the CPW configuration the characteristic impedance changes by less than 10 percent when the thickness of the substrate is reduced from infinity to $b_1 - a_1$, the width of the slots, for infinitely large ε_r.

CPWs can be easily adapted to coaxial systems using miniature connectors. Broad-band (2-12 GHz) matching has been achieved[3] with a 50 ohm CPW deposited on a magnesium oxide substrate with $\varepsilon_r = 16$.

References

1. Cohn, S.B., Slot line on a dielectric substrate, IEEE Trans. Microwave Theory Tech., Vol.MTT-17, p 768-778, Oct. 1969.

2. Mariani, E.A., C.P. Heinzman, J.P. Agrios and S.B. Cohn, Slot line characteristics, IEEE Trans. Microwave Theory Tech., Vol. MTT-17, pp 1091-1096, Dec. 1969.

3. Wen, C.P., Coplanar Waveguide: a surface strip transmission line suitable for non-reciprocal gyromagnetic device applications, IEEE Trans. Microwave Theory Tech., Vol.MTT-17, No.12, p.1087, Dec. 1969.

4. Galejs, J., Excitation of slots in a conducting screen above a lossy dielectric half space, IRE Trans. Antennas and propagation, Vol. AP-10, pp 436-443, July 1962.

5. Ramo,S., and J.R. Whinnery, Fields and waves in modern radio 2nd Ed. New York: Wiley, 1953, pp 357-358.

6. Cohn, S.B., Slot line field components, IEEE Trans. Microwave Theory. Tech., Vol.MTT-20, pp 172-173, Feb. 1972.

7. Montegomery, C.G., R.H. Dicke and E.M. Purcell, Principles of Microwave Circuits, M.I.T. Rad. Lab. Ser., Vol.8, New York, McGraw-Hill 1948, p 230.

8. Robinson, G.H. and J.L. Allen, Slot line application to miniature ferrite devices, IEEE Trans. Microwave Theory Tech., Vol.MTT-17, No.12, pp.1079-1101, Dec. 1969.

9. Smythe, W.R., Static and Dynamic Electricity, New York: McGraw-Hill, 1950.

10. Jahnke, E. and F. Emde, Tables of Functions with Formulae and Curve, 4th Ed., New York, Dover Publication, 1945.

5

COUPLED MICROSTRIPS AND DIRECTIONAL COUPLERS

by

Nagendra Singh

Chapter Five

5.1 INTRODUCTION

There are two types of directional couplers which can be designed easily in microstrip configuration. These are the coupled-microstrip directional couplers and the branch-line couplers. In a coupled-microstrip line coupler, use is made of the natural coupling that exists between two microstrips placed side by side. A basic section of the coupled-microstrip coupler is one-quarter-wavelength long and the coupling is backward. The forward coupling should ideally be zero. But the two normal modes (even and odd modes) of a coupled pair of microstrips have different propagation velocities. This results in some forward coupling (resulting in poor directivity). The difference in the propagation velocities of the normal modes becomes greater as the coupling is made tighter. Therefore, the coupled-microstrip directional couplers may not give a wide bandwidth performance for tight coupling.

Branch line couplers are specially suitable for tight coupling. It is easy to design zero db couplers. Branch line couplers consist of two parallel microstrips coupled through a number of branch lines. The lengths of the branch lines and their spacings are all one quarter-wavelength long.

In this chapter the basic principles of the two types of couplers are presented. Wherever possible, design formulae and curves are given.

5.2 EVEN AND ODD MODE ANALYSIS

Any coupler can be analysed in terms of the even and the odd normal modes. To understand this, consider the directional coupler depicted in Fig.5.1. The symmetry of the junction can be helpful. Suppose the directional coupler in Fig.5.1 has a symm-

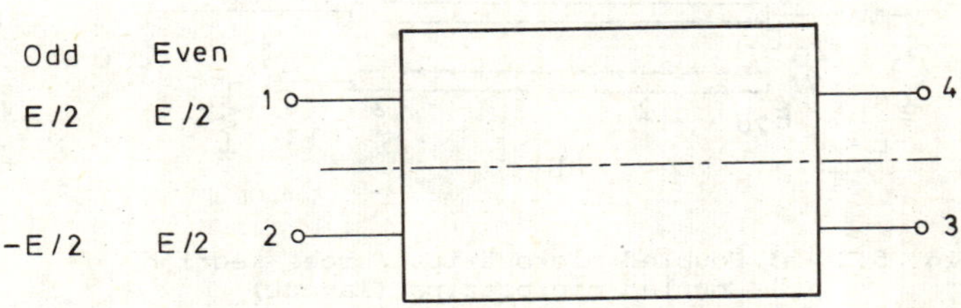

Fig. 5.1 A schematic of the directional coupler

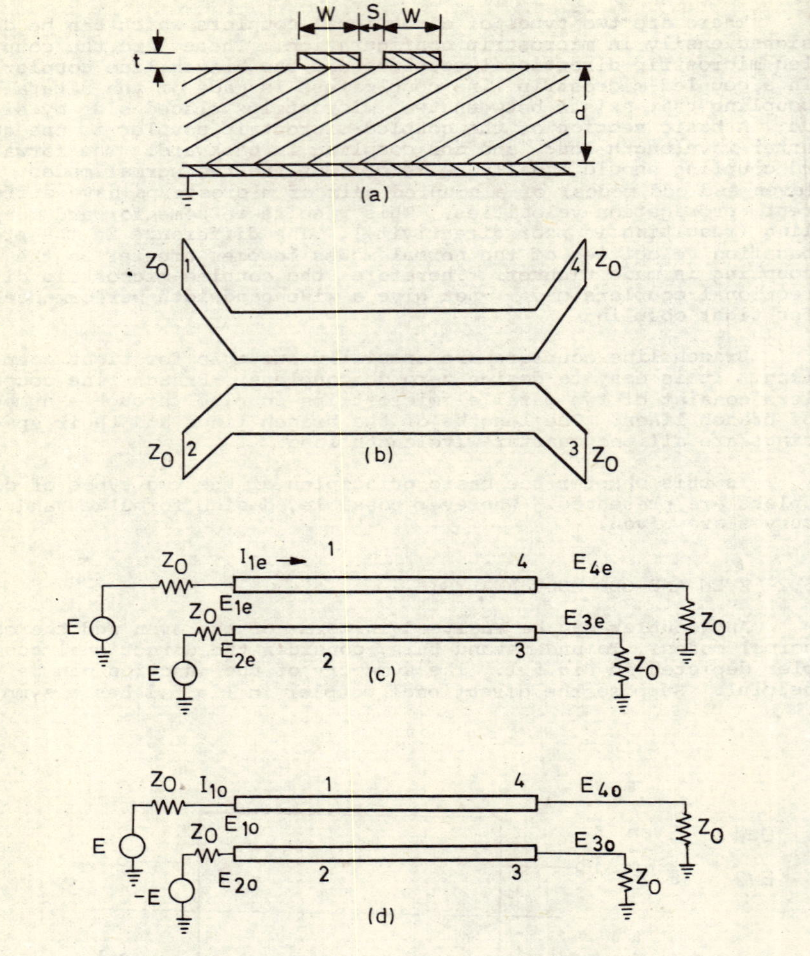

Fig. 5.2 (a) Coupled microstrips (cross section)
(b) Coupled microstrips (layout)
(c) Even mode of excitation
(d) Odd mode of excitation

etry plane indicated by the dotted line. If equal and inphase voltages are applied to the ports 1 and 2, a voltage maximum will occur along the line of symmetry. Thus, the junction may be open circuited at the symmetry plane without affecting the field distribution. A magnetic wall may be located at this plane. Similarly if two equal and opposite voltages are applied to ports 1 and 2 the voltage at the line of symmetry will be zero. Thus a short circuit (an electric wall) may be located in the symmetry plane without affecting the field distribution. These two modes of excitation of the junction are called the even and the odd modes. As an example the even and odd mode excitation of a coupled pair of microstrip lines is shown in Fig. 5.2. For each mode, the analysis reduces to that of a two port network which can easily be analysed. By superposition, the sum of two modes is equivalent to a single voltage applied to port 1.

5.3 THEORY OF COUPLED-MICROSTRIP DIRECTIONAL COUPLERS

5.3.1 Coupling coefficient

The basic coupled microstrip coupler is shown schematically in Fig. 5.2. The two coupled lines are terminated by input and output lines of characteristic impedance Z_o. The even and odd modes of excitation are shown in Fig. 5.2c and d. In order that the coupler be perfectly matched at all frequencies, it is necessary that the input impedance Z_{in} at the port 1 be always equal to Z_o. Superposing the two modes, one sees that the input impedance is given by

$$Z_{in} = (E_{1o}+E_{1e}) / (I_{1o}+I_{1e}) \tag{5.1}$$

It can be seen from Figs. 5.2c and d that

$$I_{1o} = E/(Z_o+Z_{1o}) \quad , \quad I_{1e} = E/(Z_o+Z_{1e})$$
$$E_{1o} = E\ Z_{1o}/(Z_{1o}+Z_o) \quad , \quad E_{1e} = E\ Z_{1e}/(Z_{1e}+Z_o) \tag{5.2}$$

where Z_{1e} and Z_{1o} are the input impedances for the coupled lines under the even and the odd mode excitation respectively. Using transmission line equations it can be easily shown that

$$Z_{1e} = Z_{oe}\ [(Z_o+jZ_{oe} \tan \theta_e)/(Z_{oe}+jZ_o \tan \theta_e)] \tag{5.3}$$

$$Z_{1o} = Z_{oo}\ [(Z_o+jZ_{oo} \tan \theta_o)/(Z_{oo}+jZ_o \tan \theta_o)] \tag{5.4}$$

where Z_{oe} and Z_{oo} are the even and odd mode characteristic impedances of the coupled lines, respectively, θ_e and θ_o are the electrical lengths of the microstrips for the even and odd modes i.e.

$$\theta_e = 2\pi fL/v_e \quad \text{and} \quad \theta_o = 2\pi fL/v_o.$$

Here L is the physical length of the coupled lines and v_e and v_o are the propagation velocities for the two modes. Calculations for Z_{oe} and Z_{oo} are given in the next section.

Assuming $\theta_e = \theta_o$, when eqns.(5.2) to (5.4) are substituted in eqn.(5.1), it is found that $Z_{in} = Z_o$ for all frequencies if,

$$Z_o = \sqrt{Z_{oe}\, Z_{oo}} \tag{5.5}$$

The resulting signals, at the four ports of the coupler can be obtained in terms of the even and odd mode reflection and transmission coefficients. Let Γ_e, Γ_o and T_e, T_o be the reflection and transmission coefficients respectively, for the even and odd modes. If E_1, E_2, E_3 and E_4 are the voltages appearing at the four ports,

$$E_1 = E(\Gamma_e+\Gamma_o)/2 \tag{5.6}$$

$$E_2 = E(\Gamma_e-\Gamma_o)/2 \tag{5.7}$$

$$E_3 = E(T_e - T_o)/2 \tag{5.8}$$

$$E_4 = E(T_e + T_o)/2 \tag{5.9}$$

The transmission and the reflection coefficients can be obtained in terms of a transfer matrix, as follows,

$$\Gamma = \frac{A + B/Z_o - CZ_o - D}{A + B/Z_o + CZ_o + D} \tag{5.10}$$

$$T = \frac{2}{A + B/Z_o + CZ_o + D} \tag{5.11}$$

where A,B,C and D are the elements of the transfer matrix. The transfer matrix relates the input voltage and current to the output voltage and current, and it is given by

$$\begin{vmatrix} A & B \\ C & D \end{vmatrix} = \begin{vmatrix} \cos\theta_i\,, & j\, Z_{oi}\, \sin\theta_i \\ (j/Z_{oi})\sin\theta_i\,, & \cos\theta_i \end{vmatrix} \tag{5.12}$$

where the subscript 'i' stands for 'e' or 'o' depending upon whether the even or the odd mode respectively are being considered.

Under the assumption that $\theta_o = \theta_e = \theta$ and using the matching condition in eqn.(5.5),one can easily show that

$$\Gamma_e = -\Gamma_o = \{\, j\, [(Z_{oe}/Z_{oo})^{1/2} - (Z_{oo}/Z_{oe})^{1/2}]\, \sin\theta\}/\Sigma \tag{5.13}$$

$$T_e = T_o = 2/\Sigma \tag{5.14}$$

where $\Sigma = 2\cos\theta + j[\sqrt{(Z_{oe}/Z_{oo})} + \sqrt{(Z_{oo}/Z_{oe})}]\sin\theta$ (5.15)

Then from eqns. (5.6) to (5.9) one obtains

$$E_1 = E_3 = 0$$

$$E_2/E = jC\sin\theta/[\sqrt{(1-C^2)}\cos\theta + j\sin\theta] \quad (5.16)$$

$$E_4/E = \sqrt{(1-C^2)}/[\sqrt{(1-C^2)}\cos\theta + j\sin\theta] \quad (5.17)$$

where

$$C = [Z_{oe}/Z_{oo}) - 1]/[(Z_{oe}/Z_{oo}) + 1] \quad (5.18)$$

It should be noted now that under the assumptions that $\theta_e = \theta_o$ and $Z_o = \sqrt{Z_{oo}Z_{oe}}$, port 3 is perfectly isolated (perfect directivity) from port 1. Port 1 and 2 are coupled. Thus it is a backward coupler. The forward coupling is zero. The backward coupling is given by

$$|E_2/E|^2 = C^2\sin^2\theta/(1-C^2\cos^2\theta) \quad (5.19)$$

This coupling is maximum when $\theta = \pi/2$, i.e., the sections of the microstrips are one quarter wavelength long. The midband voltage coupling is C and it is given in terms of the even and odd mode characteristic impedance by eqn.(5.18). Also, it is important to note that the outputs E_2 and E_4 are out of phase by 90° at all frequencies. This is a very useful property of the coupled-microstrip directional couplers.

In terms of the midband coupling, the even and odd mode impedances can be calculated from the relations,

$$Z_{oe} = Z_o\sqrt{(1+C)/(1-C)} \quad (5.20a)$$

$$Z_{oo} = Z_o\sqrt{(1-C)/(1+C)} \quad (5.20b),$$

Equations (5.20a,b) are useful for design calculations.

The assumption $\theta_e = \theta_o$ in the above analysis is not really true for microstrip lines. This is because of the fact that the propagation velocities for the even and the odd modes are not equal. An effect of this is poor directivity[1]. Therefore, the design formulas derived in this section will only serve as a first step in the process of designing coupled-microstrip directional couplers. Usually an optimum design may need several experimental iterations.

A large ratio of the even and odd mode propagation velocities can be used to advantage also. Daley[2] has demonstrated this for strip lines and has suggested a new type of directional coupler.

5.3.2 Bandwidth

The performance of the one-quarter-wavelength coupler is given by eqns.(5.17) and (5.19). This type of coupler can be used over a bandwidth of approximately one octave. Improved coupling bandwidth can be obtained by cascading several one-quarter wavelength sections. A bandwidth ratio upto 5:1 can be obtained by cascading three sections. A higher bandwidth can be obtained by cascading more sections. Cristal and Young[3] have published tables for couplers of 3,5,7 and 9 sections with the mean coupling 3.01,6,10 and 20 db.

Inspite of the large bandwidth of the symmetric multi-section couplers, they are not optimum in the sense of having maximum bandwidth for a given coupling tolerance. This optimum performance can be obtained by asymmetric coupled-microstrip couplers. Moreover, sometimes it is not possible in practice to realize tight coupling over a large bandwidth using symmetrical couplers[4]. This is because of the large value of the even mode characteristic impedance for the central section. This limitation is not so severe with asymmetric couplers. Levy[5] has described synthesis procedures for both the symmetric and asymmetric directional couplers.

5.4 CALCULATIONS FOR A COUPLED PAIR OF MICROSTRIPS

It is evident from the discussion in the last section that the coupling is determined by the even and the odd mode characteristic impedances. For a given coupling eqn.(5.20) determines the values of Z_{oe} and Z_{oo}. Now the task is to determine the physical parameters, (see Fig.5.2) the width (W/d) of the microstrips and the separation (S/d) between them. The thickness (t/d) is determined by considerations of skin-depth.

To determine W/d and S/d for known values of Z_{oe} and Z_{oo}, one needs to solve a pair of simultaneous eqns.[6]. This is a very laborious task. A convenient method of design is to use computer-generated curves [7,8]. Figure 5.3 gives curves for Z_{oe} and Z_{oo} as a function of W/d with S/d as a parameter. To find the width and the separation for a coupled pair of lines, one draws parallel lines for the two values of impedances horizontally through the plotted curves. A perpendicular line intersecting the horizontal lines at the same value of separation is the simultaneous solution for S and W. An example will illustrate this method. Suppose a 10 db coupler is desired, then 20 log C = -10 or C = 0.3126.

For 50 ohm input lines eqns.(5.20) give Z_{oo} = 36 ohm and Z_{oe}= 69 ohm. Then from Fig.5.3, S/d=0.1 and W/d=0.6. Figure 5.4 gives curves for S/d and W/d against coupling. These curves simplify procedure for determining the dimensions of the coupled microstrips.

It is important to note that the calculation of Z_{oe} and Z_{oo} are based on the assumption that the microstrip lines support

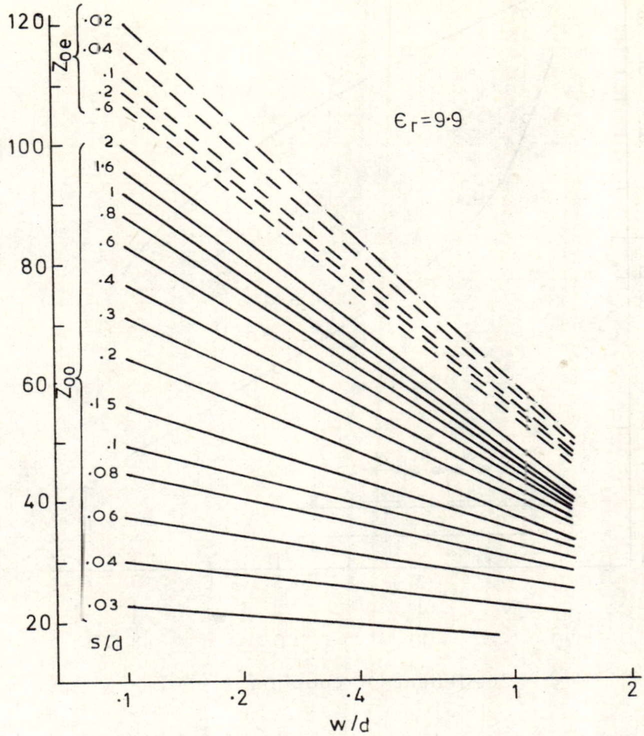

Fig. 5.3 Even and odd mode impedances as a function of W/d (Ref. [7])

only the TEM mode of propagation. Since microstrip is a device which contains two different dielectric media, the mode supported can not be a true TEM mode and is in general a hybrid mode [9]. But designs based on the TEM mode propagation have been found adequate when the operating frequencies are below 4 GHz.

5.5 BRANCH LINE COUPLERS

5.5.1 General

Branch line couplers consist of two parallel microstrip li-

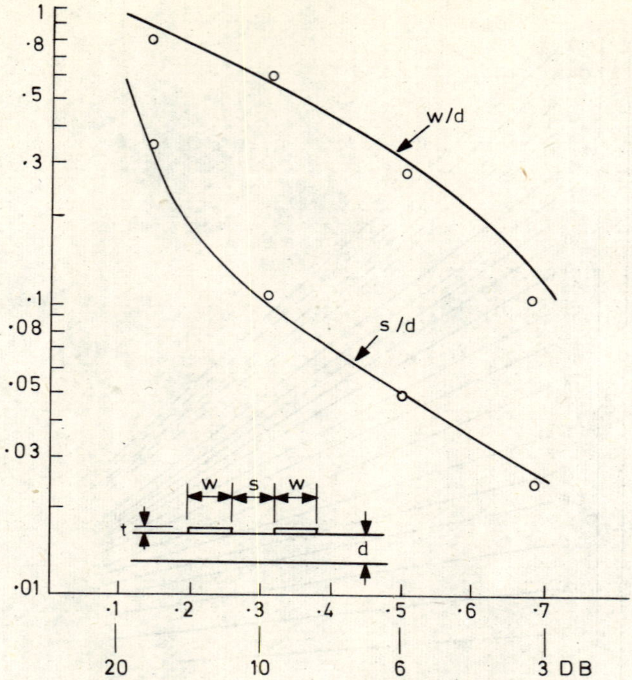

Fig. 5.4 s/d and w/d as a function of coefficient of coupling (Ref. [7])

nes coupled through a number of branch lines[10,11]. The lengths of the branch lines and their spacings are all one quarter guide wave length at the midband frequency. A schematic diagram of a branch line coupler is shown in Fig. 5.5a. The characteristic impedance of the main lines and the branch lines can be changed from section to section to improve the performance of the coupler. The significant properties of branch line couplers as summarized by Young[12] are :

1. The coupling between the two lines is through joining branch lines of finite lengths. This gives flexibility in designing

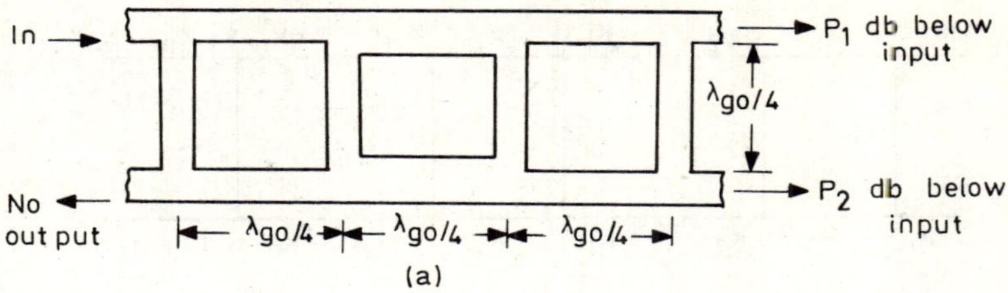

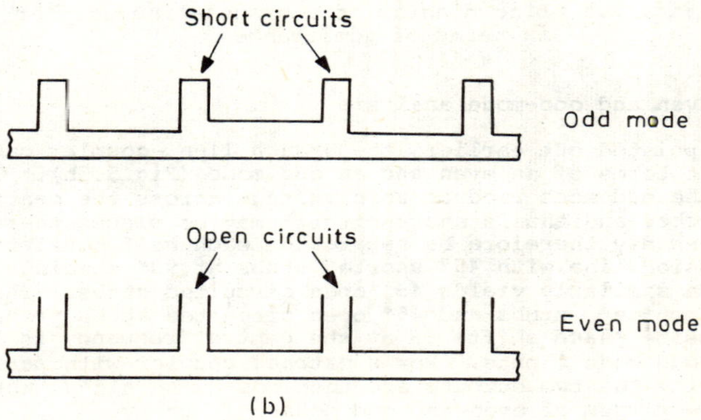

Fig. 5.5 (a) Branch line directional coupler
(b) Odd and even mode equivalent circuits

for example, special purpose chokes or filters can be placed in the branches.

2. The couplers are capable of handling high RF powers.

3. They are better suited for tight coupling (coupling coefficient less than 20 db) than for weak couplings. Zero db couplers are feasible over a large bandwidth. This order of tight coupling is not conveniently feasible with coupled line couplers.

A schematic diagram of the coupler in terms of the characteristic admittances of the main lines and the branches is shown in Fig.5.6. End to end symmetry of the coupler exists, i.e. $H_{n+1} = H_1$ etc., and $K_1 = K_n$ etc. All the admittances are assumed to be normalized with respect to the terminating admittances, i.e. $K_o = K_{n+1} = 1$.

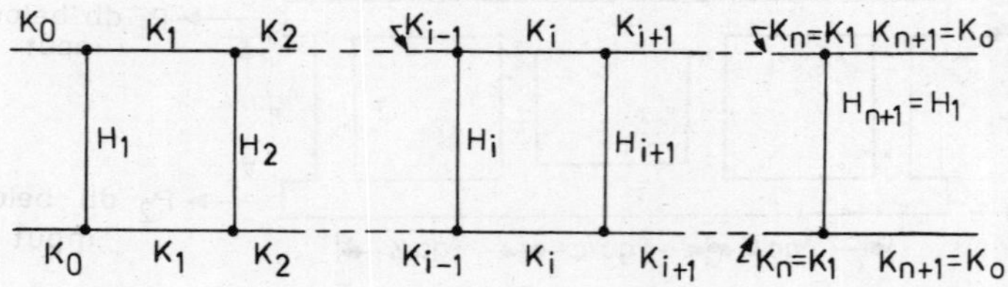

Fig. 5.6 Line diagram of a branch-line coupler in terms of admittances

5.5.2 Even and odd-mode analysis

As pointed out earlier, the branch line -coupler can be analyzed in terms of an even and an odd mode (Fig.5.5b). Two inputs of the odd mode produce zero voltage across the centre of all the branches and thus a short-circuit may be placed there. The two halves may therefore be separated, each half consisting of a transmission line with 45° shorted stubs at 90° spacings. The even mode similarly yields 45° open circuited stubs. The 45° short circuited stubs and 45° open circuited stubs produce equal and opposite phase shifts $\pm\theta$ at the centre frequency in the even and the odd mode inputs. For a matched coupler with perfect directivity the two outputs are then $\cos^2\theta$ and $\sin^2\theta$, when the input power (sum of even and odd modes) is unity.

When matched couplers are cascaded, they act as a single directional coupler with coupling of the combination as following:

$$P_{1comb} = 20 \log_{10} [\cos (\sum_{i=1}^{N} \theta_i)] \text{ db} \tag{5.21}$$

$$P_{2comb} = 20 \log_{10} [\sin (\sum_{i=1}^{N} \theta_i)] \text{ db} \tag{5.22}$$

where N is the number of cascaded stages and θ_i is the even or odd mode phase shift of the ith stage. As an example, a phase shift of $\theta = 30°$ yields a 6 db coupler, and a phase shift of 90° yields a 0-db. coupler. A cascade of the three matched 6-db couplers results in a 0-db coupler.

Matthaei et al.[13] have given a table showing relationships between H_i's and K_i's to yield perfect match and directivity at the midband.

5.5.3 Periodic branch line coupler

When the main lines of a branch-line coupler remain uniform ($K_i=K_o$, i=1 ,2 ,... ,n+1) and all the branches are identical ($H_i=H_2$ $2 < i < n$, $H_1=H_{n+1}$), the coupler is called a periodic branch-line coupler. The admittances H_1 and H_2 of a periodic coupler are given by the relation[14].

$$S_{n-1}(-H_2) = \begin{cases} \sqrt{(1-C_o^2)} & \text{for n odd} \\ C_o & \text{for n even} \end{cases} \quad (5.23)$$

and

$$H_1 = |\{\ |\sqrt{[1-S_{n-1}^2(-H_2)]}| - |S_{n-2}(-H_2)|\ \} / S_{n-1}(-H_2)\ | \quad (5.24)$$

where, C_o is the midband coupling, $S_n(X)$ is the Chebyshev polynomial[15] and n is the number of sections in the coupler. The above equations can be used for designing a periodic coupler.

Some idea about the frequency response of this type of coupler can be obtained from Table 5.1. In this table w_b is the fractional bandwidth given by

$$w_b = 2\left(\frac{\lambda g_1 - \lambda g_2}{\lambda g_1 + \lambda g_2}\right)$$

where λg_1 is the longest guide wavelength and λg_2 is the shortest guide wavelength over the band under consideration. Detailed characteristics of Branch-line couplers are given in[14].

The performance of couplers with an even number of branches is found to be superior to those with an odd number of branches.

Table 5.1 Maximum VSWR of 0-db couplers over stated bandwidths [11]

Fractional Bandwidth W_b	Couplers Six branch	Eight branch	Ten branch	Twelve branch	Fourteen branch
.16	1.1	1.06	1.04	1.03	1.02
.28	1.19	1.09	1.05	1.03	1.02
.48	1.21	1.1	1.12	1.09	1.05

5.5.4 Synchronous branch-line couplers

In a synchronous branch line coupler the main line admittances are not uniform. They can be changed from section to section. This gives an additional flexibility in designing an optimum coupler over a wide bandwidth. Young[12] has given curves which enable a designer to realise a broadband coupler with maximally flat or equal ripple VSWR or directivity characteristics. Levy and Lynd[16] have given a synthesis procedure which gives exact results for maximally flat characteristics and almost exact results for equal ripple characteristics.

In order to realize the desired characteristics, a synchronous branch-line coupler needs a smaller number of branches compared to a periodic branch-line coupler.

A synthesis procedure for an asymmetrical branch-line coupler is given by Lynd[17]. This allows one pair of terminating resistances to have a different value from the other pair. The resulting device is therefore both a branch-line directional coupler and an impedance transformer.

References

1. Levy, R., Directional Couplers, Advances in Microwaves, Vol. 1, Academic Press, 1966, p.186.
2. Daley, J.E., A strip-line directional coupler utilizing a non-Homogeneous Dielectric Medium, IEEE Trans. Microwave Theory Tech., Vol.MTT-17, 1969, pp 706-712.
3. Cristal, E.G. and L. Young, Tables of Optimum Symmetrical TEM-mode coupled-transmission-line directional couplers, IEEE Trans. Microwave Theory Tech., Vol.MTT-13, 1965, pp 544-558 .
4. Levy, R., op.cit., p.179.
5. Levy, R., op.cit., pp 172-185.
6. Schwarzmann, A., Microstrip equations, Electronics, October 2, 1967.
7. Schwarzmann, A., Approximate solutions for a coupled pair of microstrip lines in microwave integrated circuits, Microwave Journal, Vol.12, May 1969, pp.79-82.
8. Bryant, T.G. and J.A. Weiss, Parameters of microstrip transmission lines and of coupled pairs of microstrip lines, IEEE Trans. Microwave Theory Tech., Vol.MTT-16, 1968, pp 1021-1027.
9. Carr, D.G., and Davies, J.B., Computer-analysis of the fundamental and higher order modes in single and coupled microstrip, IEEE Trans. Microwave Theory Tech., Vol.MTT-20, 1972, pp 669-678.
10. Levy, R., op.cit., pp.155-161.
11. Matthaei, G.L., L. Young, and E.M.T. Jones, Microwave Filter, impedance-matching networks and coupling structures, McGraw-Hill, 1964, pp 809-841.
12. Young, L., Synchronous branch guide directional couplers for low and high power applications, IRE Trans., Microwave Theory and Tech., Vol.10, 1962, pp 459-475.
13. Matthaei, G.L., et al., op.cit., p.814.

14. Reed, T., The multiple branch waveguide coupler, IRE Trans. Microwave Theory Tech., Vol.MTT-6, 1958, pp 398-403.

15. Tables of Chebyshev polynomials $S_n(x)$ and $C_n(x)$, National Bureau of Standards Applied Mathematics Series No.9, United States Government Printing Office, 1952.

16. Levy, R., and L.F. Lynd, Synthesis of symmetrical branchguide directional coupler, IEEE Trans. Microwave Theory and Techniques, Vol.MTT-16, 1968, pp 80-89.

17. Lynd, L.F., Synthesis of asymmetrical branch-guide directional coupler - impedance transformers, IEEE Trans. Microwave Theory Tech., Vol.MTT-17, 1969, pp 45-48.

6

LUMPED ELEMENTS FOR MICs

by

K.C. Gupta

Chapter Six

6.1 INTRODUCTION

In conventional microwave circuits distributed elements (employing sections of transmission lines and waveguides) are used. This is because of the fact that the size of normally used lumped elements (resistors, inductors and capacitors) becomes comparable to the wavelength at microwave frequencies. However if the size of lumped elements can be reduced, they should be acceptable at these frequencies also. Lumped elements used in MICs are designed on this consideration. With the advent of photolithography and thin film techniques, the size of elements can be reduced such that these elements may be operated even upto X-band frequencies. In this chapter some features of design and applications of these lumped elements are discussed.

6.2 DESIGN OF LUMPED ELEMENTS

Design of lumped elements at microwave frequencies may be arrived at by considering them as a very short section (much smaller than a wavelength) of a TEM transmission line. Consider a transmission line with series inductance per unit length L, resistance per unit length r, shunt capacitance per unit length C and shunt conductance per unit length g. By transmission line formula, the input impedance of a length ℓ of the transmission line terminated in an impedance Z_L is given by

$$Z_{in} = Z_o \frac{Z_L \cosh \gamma\ell + Z_o \sinh \gamma\ell}{Z_o \cosh \gamma\ell + Z_L \sinh \gamma\ell} \tag{6.1}$$

where Z_o is the characteristic impedance of the transmission line given by

$$Z_o = \sqrt{[(r+j\omega L) / (g+j\omega C)]} \tag{6.2}$$

and γ is the propagation constant given by

$$\gamma = \sqrt{(r + j\omega L)(g + j\omega C)} \tag{6.3}$$

When $\gamma\ell \ll 1$, Z_{in} may be written as

$$Z_{in} = Z_o \frac{Z_L + Z_o \gamma \ell}{Z_o + Z_L \gamma \ell} \tag{6.4}$$

Z_{in} can be made inductive or capacitive by terminating a small section of the line in a short circuit (Z_L=0) or in an open circuit (Z_L=∞) respectively. Characteristics of 'lumped' inductors and capacitors thus obtained are discussed in the present section.

6.2.1 Inductors and resistors

When a small section ($\gamma\ell \ll 1$) of a transmission line is terminated in a short circuit, the input impedance as evaluated by eqn.(6.4) may be written as

$$Z_{in} = Z_o \gamma \ell = (r + j\omega L)\ell \tag{6.5}$$

This section of transmission line will behave like a resistor or an inductor depending on the value of r as compared with that of L. However it is not necessary to have both the conductors of a transmission line to obtain a resistor or an inductor. A single metallic strip on a dielectric substrate may be used for this purpose. In this case values of r and L correspond to that of this metallic strip. r is the a.c. resistance per unit length of a ribbon of rectangular cross section and L is the self inductance per unit length. Terman[8] has derived the formulas for r and L. The resistance r is given by

$$R_{ac} = r.\ell = \frac{K\ R_s\ \ell}{2(W+h)} \tag{6.6}$$

where R_s is the sheet resistance in ohms per square and W is the width and h is the thickness of metallic ribbon. K is a correction factor that takes into account crowding of current at the corners of the ribbon. Value of K is a function of W/h and varies from about 1.3 to 2 as W/h varies from 1 to 100. For a typical W/h =10, K is about 1.5.

Inductance of a rectangular strip is given by[8]

$$L.\ell = 5.08\text{x}10^{-2}\ \ell\ [\ln\{\ell/(W+h)\} + 1.193 + 0.2235\ (W+h)/\ell] \tag{6.7}$$

where $L.\ell$ is in nanohenries and all dimensions are in mils (10^{-3} inches).

The unloaded Q of the ribbon inductor may be written as

$$Q = \omega L\ell/R_{ac} \tag{6.8}$$

For a copper ribbon ($R_s = 2.61 \times 10^{-7}\sqrt{f}$) and for W/h = 10, one obtains

$$Q = 4.81 \times 10^7\ \sqrt{f}\ WL/K \tag{6.9}$$

From this expression it is found that values of Q greater than 100 can be obtained for wide ribbons ($\ell/W<15$) with an inductance of 1 nanohenry or greater at about 1 GHz. For these equations to hold good, the thickness h must be appreciably greater than thrice the skin depth. The ribbon will be practical for inductances of 2 nanohenries or less, and would have to be only 110 mils long (with $\ell/W=10$) in order to obtain a good Q. This length is less than one turn of 40-mil diameter spiral coil. For larger inductances many spiral turns are necessary and the above expression does not hold good.

An approximate expression for inductance of a spiral inductor with n turns may be written as

$$L_s(nH) = \frac{a^2n^2}{8a+11c} \text{ (all dimensions are in mils)} \quad (6.10)$$

$c = (d_o - d_i)/2$, and $a = (d_o + d_i)/4$ (i.e. average radius). d_i and d_o are minimum and maximum diameters respectively. Resistance of the spiral inductor may be written as

$$R = K' \pi na R_s/W \quad (6.11$$

where K' is a correction factor, similar to K for straight ribbon, and accounts for the current crowding. From the above two relations, Q for the spiral coil may be written as

$$Q = 2x10^{-9} f W n a / (K'R_s[8a + 11c]) \quad (6.12)$$

Some significant considerations involved in the design of inductance may be summarised as follows:

1. The spiral should have the widest ribbon possible (W large) while keeping the overall diameter (d_o) small. This implies that separation S between the turns should be as small as possible.

2. There should be some space at the centre of the spiral inductor to allow the flux lines to pass through (increasing the stored ed energy per unit length). It has been quoted[9] that d_i/d_o= 1/5 will optimize Q.

3. R_s increases as $f^{1/2}$, the Q of the inductors increases with the square root of frequency. This has been verified experimentally [1]. The measured values of Q are within 2/3 of the theoretical value with K'=1. An assigned value of K' (=1.5) gives reasonable agreement with the expression given above. Q values of about 100 have been achieved.

4. It has been shown [8,9,10] that for the same d_o, the Q of a circular spiral will be higher than that of a square spiral (10 percent), although the inductance is significantly less (20 percent).

5. Multi-turn coils have higher Q (because of higher inductance per unit area) but because of inter-turn capacitance they have lower resonance frequencies.

6.2.2 Capacitors

A lumped capacitor may be visualised as a small length of an open circuit line for which input impedance may be written as,

$$Z_{in} = Z_o \coth \gamma\ell \quad (6.13)$$

when $\gamma\ell<<1$ eqn.(6.13) may be simplified as

$$Z_{in} = Z_o/\gamma\ell = 1/(g\ell+j\omega C\ell) = g/\omega^2C^2\ell + 1/j\omega C\ell \quad (6.14)$$

Equivalent circuit will thus be a resistance $g/(\omega^2 C^2 \ell)$ in series with the capacitor $C\ell$. A second order approximation yields (ignoring L)

$$Z_{in} = Z_o/\gamma\ell + Z_o\gamma\ell/3 = g/\omega^2 C^2 \ell + 1/j\omega C\ell + r\ell/3 \qquad (6.15)$$

This implies a series circuit of two resistors and a capacitor. The third term is the series resistance representing the conductor loss, may be written as

$$r\ell/3 = 2R_s\ell/3W \qquad (6.16)$$

The factor 2 accounts for the loss on both plates of the capacitor. The other resistance component $g/(\omega^2 C^2 \ell)$ is dielectric loss and may be written as $1/(Q_d\, C\omega\ell)$ where Q_d is the dielectric Q given by $(g/\omega C)$. The total Q may thus be written as

$$1/Q = 1/Q_c + 1/Q_d \qquad (6.17)$$

where the conductor Q is

$$Q_c = 2W \,/\, 2\omega C\ell^2 R_s \qquad (6.18)$$

Q_c for a square capacitor ($W=\ell$) with copper plates and of total capacitance C_T (in pF) may be written as

$$Q_c = 2.9 \times 10^4 \,/[f^{3/2}(\mathrm{GHz})\; C_T(\mathrm{pF})] \qquad (6.19)$$

In order to have a large value of capacitance per unit area, it is desirable to decrease the distance between the two conductors (top and bottom) of the transmission line section. Since the normal thickness of substrates used for MIC work is of the order of 25 mils (or more), a metal-dielectric-metal structure is fabricated on the top surface of the MIC substrate. The dielectric used for this sandwich (normally silicon dioxide) is typically 0.02 to 0.04 mil (0.5 to 1.0 μm) thick. The technique of fabrication is discussed in the next section. This type of capacitor has been fabricated with dielectric Q of 600 at 1 GHz. Q at higher frequency is lower in accordance with eqn. (6.19). An alternative to this sandwich geometry is the interdigital configuration.

A typical geometry for the interdigital capacitor is shown in Fig.6.1. In this geometry, both the conductors of the capacitor are in the same plane which is the top surface of the dielectric substrate used for MICs. This type of capacitor can be fabricated by the techniques used in the fabrication of conventional microstrip circuits and does not require any additional processing steps.

An analysis for interdigital capacitor has been reported by Alley[6]. The capacitance between two sets of digits in interdigital structure is found by using the capacitance formula for coupled microstrip lines discussed in chapter 5. Since the overa-

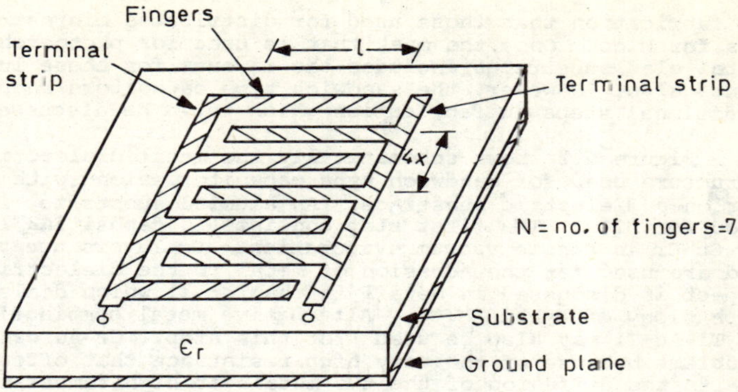

Fig. 6.1 Typical geometry for interdigital capacitors

ll size of the capacitor is small compared to the wavelength, formulas for the even-mode and the odd-mode capacitances may be used. As in the case of sandwich geometry both the dielectric loss and the conductor loss determine Q. The Q of a 2.83 pF capacitor of the type shown in Fig.6.1 (N=40, ℓ=44.8 mils, X=2 mils) and fabricated on a 24 mil thick alumina substrate (99.5 percent purity) has been measured[6] to be 677 at 1.92 GHz.

Because of the inductance contributed by the equivalent transmission line the interdigital capacitors show multiple resonances[6]. Change in the capacitance value with frequency depends on the ratio W/ℓ. The optimum value of W/ℓ may be found numerically such that the capacitance variation with frequency is minimum. As an example [6], W/ℓ=1.66 and N=24 makes the capacitance value constant upto a frequency of 3 GHz. Use of interdigital capacitor in low pass S-band filters has been reported[6].

Interdigital capacitors with Q values comparable with those for sandwich type have been fabricated directly on alumina and no additional dielectric is required. They appear to be lumped upto 3 GHz and values from 0.1 to 10 pF can be achieved. The interdigital approach for capacitors requires a relatively large area (500 milx70 mil for a 12 pF value as compared with 22 milx22 mil for 1μ m-thick SiO_2 capacitor). However, this type may be preferred for applications at frequencies below 3 GHz because only single layer of metallization is required.

6.3 FABRICATION OF LUMPED ELEMENTS

The technology for lumped element circuits has been described in references [3] and [4]. Fabrication of inductors and interdigital type capacitors does not require any additional steps in fabrication than those used for distributed microstrip circuits for MICs. Only the mask, that is used for photoetching the metal clad substrates, includes the layouts for these lumped elements also. However, the sandwich type capacitors require a few additional steps in fabrication which would be discussed now.

Figure 6.2 shows schematically the metal-dielectric-metal structure used for sandwich type capacitors along with an inductor on a dielectric substrate. An alumina substrate is normally used for MICs. The first step consists in depositing layers of Cr-Cu-Cr in single vacuum evaporation. Cr layers are very thin and are used for the adhesion of metal to the dielectric. This aspect is discussed in detail in chapter 11 which deals with the technology of Hybrid MICs. Alternative metal combinations such as Ti-Au-Ti may also be used for this step. Cr-Au can introduce problems because of the very high resistance that often develops due to the diffusion of the two metals at high temperatures[4]. The next step is a photoresist etch to define the bottom layer of sandwich type capacitors. Metal layer is etched clean over the remaining portion of the substrate.

The depositionof a SiO_2 film is the next step. SiO_2 film is grown by pyrolytic decomposition of silane*. SiO_2 layers so obtain have voltage breakdowns in excess of 10^7V/cm, can be etched precisely and have low dielectric loss when densified. The thickness of SiO_2 film is of the order of .5 to 1.0 µm. SiO_2 from unwanted portions of the substrate is removed by selective photoetching. The oxide is then densified by heating to a temperature over 400°C. The effect of densification on Q is as follows[7]: Q factor for deposited thin film SiO_2 is in the range 20 to 80. When densified to 450°C, Q increases to about 600. By densifying at 600°C, one can obtain Q greater than 5000. For comparison it may be noted that Q for fused quartz is about 10,000.

A thin layer of top metal is deposited (typically .04 to .08 mils) and the thick metal required for inductors and the rest of

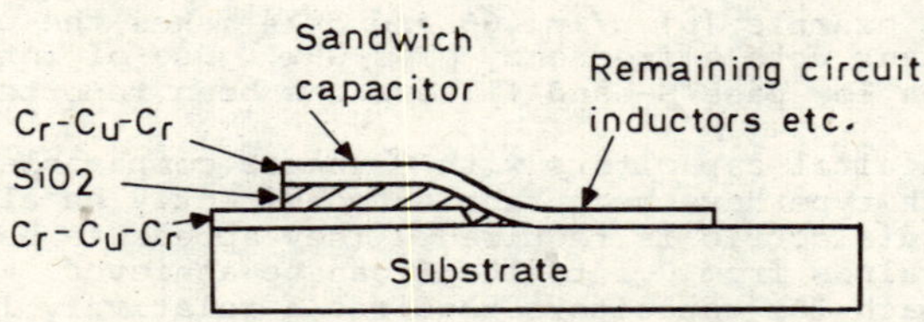

Fig. 6.2 Metal-dielectric-metal configuration for sandwich type capacitors

*This process is discussed in details in chapter 13 which deals with the technology of monolithic MICs.

the circuit is selectively electroplated. The thin unplated areas are etched clean. This technique is compatible with batch processing. It may be noted that the inductor has to be clear of the metal underneath. Resistive material can be incorporated as part of the adhesive layer instead of chromium or titanium.

6.4 CIRCUITS USING LUMPED ELEMENTS

The design procedures for lumped element circuits at microwave frequencies are similar to the corresponding methods for low frequencies. In fact it is a major advantage of using lumped elements that the circuit design and optimisation techniques developed for low frequency circuits may be extended to microwave frequencies also. Distributed element circuit design may be transformed to the lumped element circuits by using lumped equivalents for transmission line sections.

Several types of circuits using lumped elements have been reported in the literature. Some of these are described below:

S-band low pass filter[7]

Audio-frequency-design procedures for composite filters have been extended to microwave frequencies. The structure for a filter with the cut-off frequency at 2.8 GHz is shown in Fig.6.3. In this circuit all inductances are in nano-henries and capacitances in picofarads. Lumped element layout for this circuit is shown in Fig.6.4. Various elements are interconnected by 1 mil

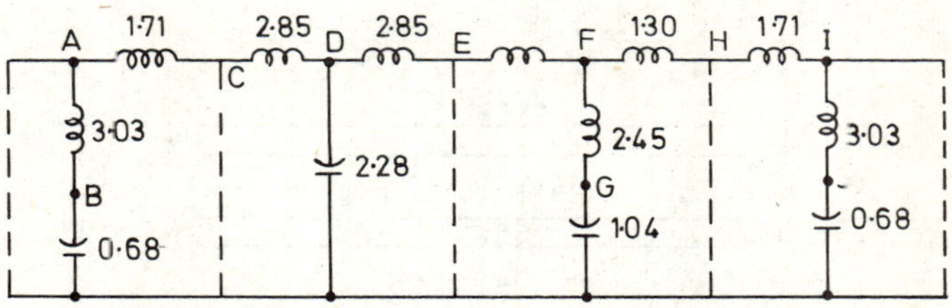

Fig. 6.3 Circuit for an S-band low pass filter (From Ref.[7]).

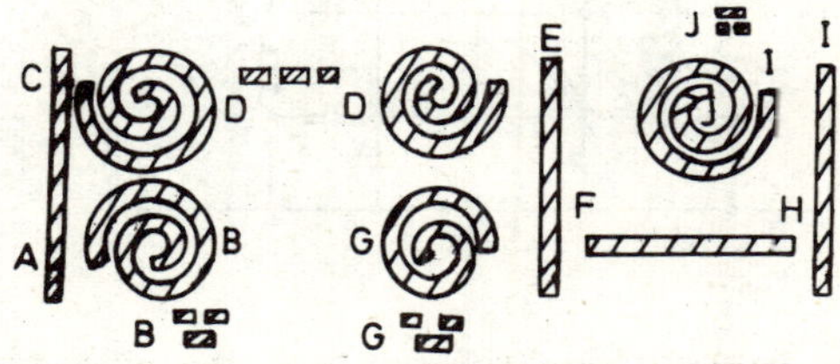

Fig. 6.4 Lumped element layout for low pass filter (From Ref. [7])

gold wires using usual bonding techniques. The number of such interconnections can be reduced by combining the adjacent elements together in the layout itself. In the work reported, one of the aims was to make measurements on separate filter sections and hence no effort was made towards reducing the number of interconnections.

Turns of the inductor coils are 6 mil wide and 0.5 mil apart. The filters were batch processed on a 0.75" x 1" x 0.006" alumina substrate. Size of a single filter is 0.135" x 0.160". Theoretical and experimental response curves for this filter agree reasonably well[7].

Branch line coupler[7]

The lumped equivalent of a branch line coupler may be arrived at by replacing the quarter wavelength sections of the main transmission lines and the branch transmission lines by equivalent pi-networks. A branch line coupler and the equivalent lumped network is shown in Fig.6.5. L_1 and C_1 may be calculated by the following relations,

$$j\omega L_1 = jZ_{o1} \sin \theta_1$$

and $$1/j\omega C_1 = -jZ_{o1} \tan (\theta_1/2) \qquad (6.20)$$

Similar relationships exist between L_2, C_2, Z_{o2} and θ_2.

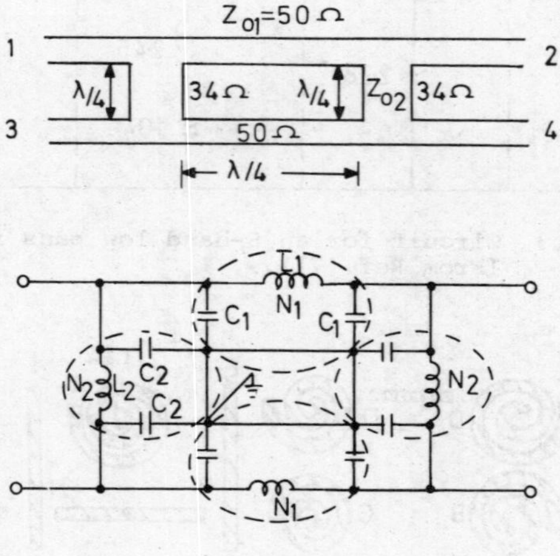

Fig. 6.5 Branch line coupler and the equivalent lumped network (From Ref.[7])

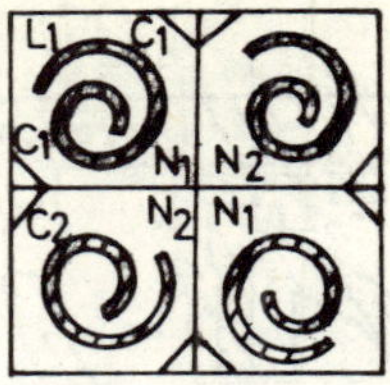

Fig. 6.6 Layout for lumped element coupler (From Ref.[7])

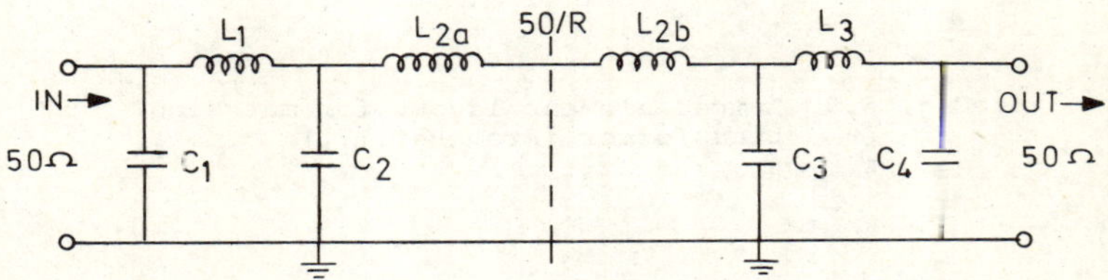

Fig. 6.7 Circuit for matching transformer (From Ref.[7])

Layout of lumped element branch line coupler is shown in Fig.6.6. This layout is such that when properly interconnected a multiple section low pass filter can also be obtained. For a centre frequency of 2.2 GHz the size of branch line coupler is 0.125" x 0. 125". In this case L_1 = 3.60 nH, L_2 = 2.57 nH, C_1 = 1.44 pF and C_2 = 2.04 pF . Performance of this 3-db quadrature hybrid is described in reference[7]. A minimum isolation of 10 dB over 1.8 to 2.5 GHz and a coupling of 3.5 dB (±0.5 dB) from port 1 to 3 and coupling of 5.5 DB (± 0.5 dB) from port 1 to 2 has been reported.

Impedance transformers at C band[7]

A matching transformer with Chebyshev response has been designed[7] using Mathaei's tables[11]. Circuit for two transformers (50 ohm to 50/R ohm and 50/R ohm to 50 ohm) combined 'back to back' is shown in Fig. 6.7.

In the circuit fabricated in MIC form, provision was made to adjust the values of various elements. The adjustment is accomplished by paralleling capacitors and paralleling or tapping strip inductors with bond wires. The circuit layout is shown in Fig.6.8. R can be varied from 15 to 20 by changing capacitors and inductors. The low-impedance plane is the dotted line. The combined (back to back) network would be matched with 50-ohms load and source impedances. This combined network forms a wide

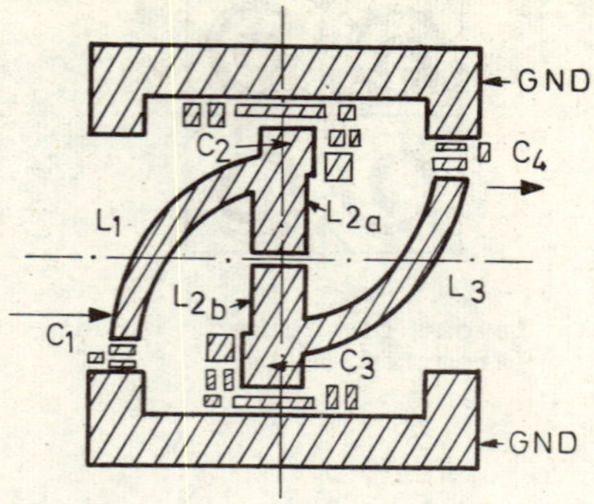

Fig. 6.8 Lumped element layout for matching transformer (From Ref.[7])

band Chebyshev filter centred at 6 GHz. Transmission and reflection characteristics of this filter have been reported[7]. The 3-db bandwidth of the measured transmission characteristic is 1.5 GHz (corresponding theoretical value is 3 GHz). The transmission loss at mid range was predicted to be around 0.6 dB. The measured lows is 0.7 dB.

This lumped element transformer can be used to transform an impedance of 50 ohm to the 3 ohm or less associated with a high power device. These results demonstrate the high performance capability of lumped elements for circuits through C band.

Transistor amplifier at S-band[7]

A lumped element S-band amplifier capable of delivering over 5 W of CW power and 30-dB gain at 2.55 GHz has been reported[7]. The amplifier consists of four cascaded single transistor amplifier stages. The low level stages are biased in a class-A operation, the third stage is a TA 7487 transistor in class C operation and the fourth stage incorporates a high power (5W) TA 7205 chip in a special package.

Circuit diagram for first three stages is given in Fig.6.9. The circuit layout was designed to accept a 1-W 15 mil x 15 mil transistor chip. Tuning for different impedance levels was accomplished by paralleling capacitors and paralleling or tapping strip inductors with bond wires. The circuit size is 115 mil x 115 mil (on 6 mil sapphire). Microstrips on teflon-glass laminate are bonded to input and output. Microstrip line on polyolefin or teflon glass provide reference planes for tuning the device for optimum performance. The amplifier delivered 7W of

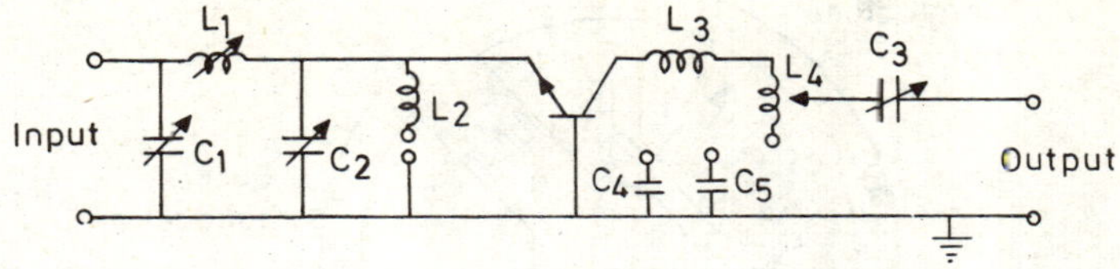

Fig. 6.9 Circuit for S-band amplifiers (From Ref.[7])

power at 2.25 GHz with a gain of 9 dB.

Tunnel diode amplifier at 4 GHz[12]

Figure 6.10 shows the circuit configuration used for a tunnel diode amplifier [12]. An inductor L resonates the tunnel diode chip. A stabilizing network L_3-C_3-R removes the negative conductance of the tunnel diode at all frequencies except that at which gain is required and a transforming net transforms the impedance of the transmission line (usually 50 ohms) to the value required to give the desired gain from the tunnel diode amplifier. The corresponding lumped element circuit[12] is shown in Fig.6.11. The circuit is deposited on an annular quartz disc (9 mm in dia and 0.5 mm thick) which can be placed across a 50 ohm coaxial line. The two capacitors in the circuit are of interdigital type and only one layer of metallisation is required.

The circuit yielded a 9 dB gain over the band 2.8 to 4.1 GHz. The peak gain was about 12 dB. The measured noise figure

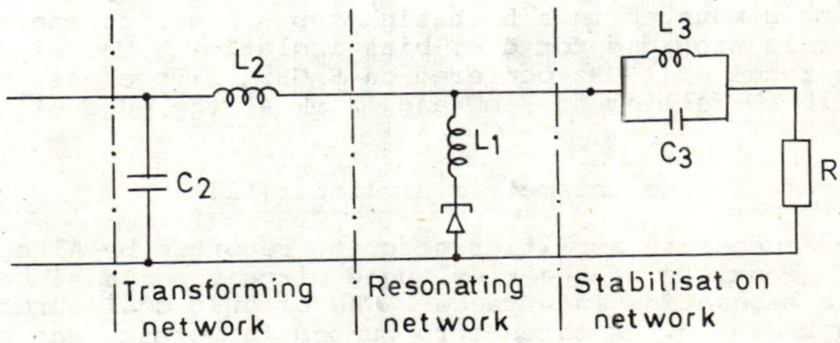

Fig. 6.10 Circuit for a Tunnel diode amplifier (From Ref.[12])

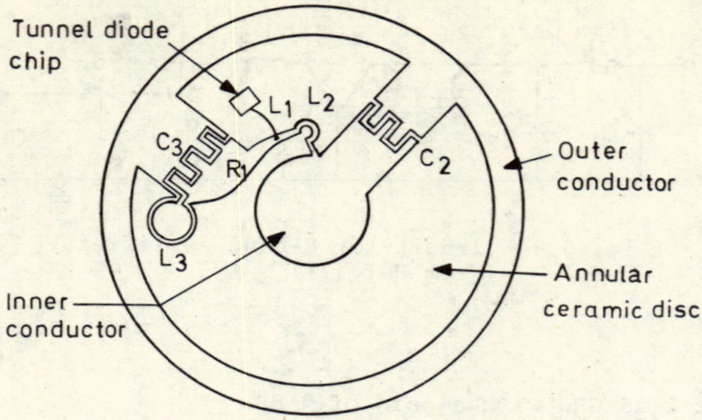

Fig. 6.11 Lumped element circuit for Tunnel diode amplifier (From Ref.[12])

is 6.2 dB whereas the theoretical value is 4.8 dB.

X-band varactor tuned Gunn oscillator

This circuit has also been achieved[12] by using the diodes (both the varactor and the Gunn) in chip form. Capacitors of interdigital type are used and therefore only a single layer of metallisation is required. A very simple circuit with the varactor chip in series with the Gunn chip is used. The layout is shown in Fig. 6.12. Two three-element lowpass filters are used for supplying dc bias to the Gunn and the varactor chip. The Gunn chip is mounted on a heat sink tap. A series sandwich type capacitor is provided for d.c. bias isolation. The circuit has a tuning range of 1 GHz centered on 8 GHz, with a peak output power of 16 mW falling to 7 mW and 12 mW at the edge of the tuning range.

A degenerate S-band parametric amplifier[12]

The degenerate amplifier circuit reported by Aitchison et al.[12] consists of a series tuned circuit containing a varactor and a resonating inductance. The circuit configuration is shown in Fig.6.13. A three-turn lumped inductance and a Schottky varactor are mounted on a 9-mm diameter ceramic disc. The circuit is fed directly from a 50 ohm impedance. The pump power at a frequency equal to twice that of the signal circuit resonant frequency, is fed into the parametric amplifier through a coaxial directional coupler. A gain of 8.0 dB and a bandwidth of 210 MHz was obtained at 3.1 GHz. The measured amplifier noise temperature was approximately 200°K.

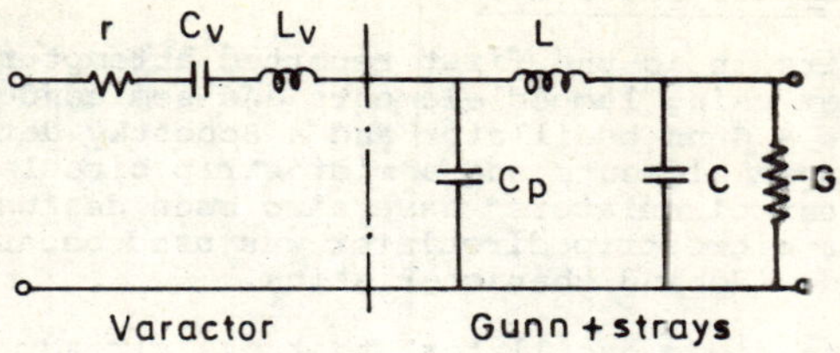

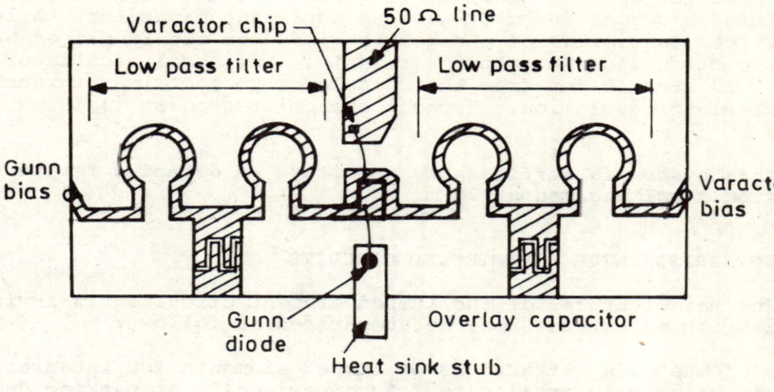

Fig. 6.12 Layout for a varactor tuned Gunn oscillator (From Ref. [12])

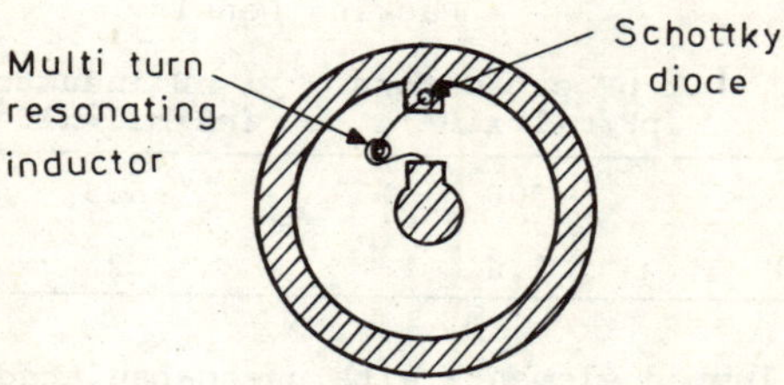

Fig. 6.13 Configuration for lumped element degenerate parametric amplifier (From Ref. [12])

An X-band doppler radar[12]

This circuit is the first reported attempt to make a complete subsystem using lumped elements and semiconductor chips. The circuit uses a Gunn oscillator and a Schottky detector constructed from lumped elements and a microstrip circulator. Although lumped element circulators* have also been designed by the same group[12], a microstrip circulator was used because of its availability and wideband characteristics.

The Gunn diode oscillator, that has already been described, is incorporated after removing varactor tuning facility. The schottky detector circuit consists of a reactive matching network containing a series inductor with an isolating capacitor. A low pass filter consisting of a capacitor-inductor tee is placed between the diode and the output terminals. The local oscillator power (10 dBm) is fed from the circulator to the mixer through the circulator isolation. Typical transmitted power is 10 mW at 10 GHz.

This system is sufficiently sensitive to detect a return signal from a walking man at 30 m.

6.5 COMPARISON WITH DISTRIBUTED CIRCUITS

The main features of the lumped element circuits vis-a-vis the distributed circuits may be summarised as follows:

1. One of the main attractions of lumped elements for integration is their relatively small size. A comparison[7] of packing densities is made in Table 6.1. Because of high packing density, lumped elements are strong contenders for monolithic microwave integrated circuits.

Table 6.1 Comparison of packing densities

	Packing Density	
Elements	1 pF capacitors ($pF/mil^2 x10^4$)	1 nH inductors ($nH/mil^2 x10^4$)
Lumped	300	5.5
Distributed	1.6	2

2. Combination of lumped elements with unencapsulated semiconductor devices leads to a significant reduction in price and also improves the bandwidth[12]. This improvement arises not only from the use of unencapsulated diodes but also from the use of lumped resonant structures instead of distributed resonant structure [12].

* A lumped element circulator is described in Chapter 7 on Non-reciprocal components in MICs .

3. There are more processing steps required for the fabrication of sandwich type capacitors than for microstrip circuits. Also the lumped element circuits require higher mask accuracies than those for microstrip circuits. This makes up for some of the cost reduction which results because of the saving in substrate area.

4. Difficulties in grounding the elements may be alleviated with lumped element circuits, as a ground can be placed near. This provides an advantage over microstrip circuits. However, new types of transmission lines (like slot line) do permit convenient shunt mounting of elements.

5. Lumped element circuit design employs procedures commonly used at lower frequencies. Thus circuit optimisation methods used at lower frequencies can be extended to microwave frequencies.

References

1. Daly, D.A., S.P. Knight, M. Caulton and R. Ekholdt, Lumped elements in microwave integrated circuits, IEEE Trans. Microwave Theory Tech., Vol. MTT-15, Dec. 1967, pp.713-721.

2. Caulton, M., S.P. Knight and D.A. Daly, Hybrid integrated lumped element microwave amplifiers, IEEE Trans. Electron Devices, Vol. ED-15, July 1968, pp. 459-466.

3. Caulton, M., The lumped element approach to microwave integrated circuits, Microwave Jour. Vol.13, May 1970, pp.51-58.

4. Caulton, M. and H. Sobol, Microwave integrated circuit technology - A survey, IEEE J. Solid-State Circuits, Vol.SC-5, Dec. 1970, pp.292-303.

5. Sobol, H., Technology and design of hybrid microwave integrated circuits, Solid-state Technol., Vol. 13, Feb. 1970, pp.49-57.

6. Alley, G.D., Interdigital capacitors and their applications to lumped element microwave integrated circuits, IEEE Trans. Microwave Theory Tech., Vol. MTT-18, December 1970, pp.1023-1033.

7. Caulton, M., B. Hershenov, S.P. Knight and R.E. Debrecht, Status of lumped elements in microwave integrated circuits - Present and Future, IEEE Trans. Microwave Theory Tech., Vol. MTT-9, July 1971, p. 588-599.

8. Terman, F.E., Radio Engineering Handbook, McGraw-Hill, New York, 1943.

9. Dill, H.G., Designing inductors for thin film applications, Electronic Design, pp. 52-59, Feb. 17, 1964.

10. Grover, F.W., Inductance Calculations, Princeton, N.J. Van Nostrand, 1946.

11. Mathaei, G.L., Tables of Chebyshev impedance-transforming networks of low pass filter form, Proc. IEEE, Vol. 52, Aug. 1964, pp. 939-963.

12. Aitchison, C.S. et al., Lumped-circuit elements at microwave frequencies, IEEE Trans. Microwave Theory Tech., Vol. MTT-19, Dec. 1971, pp.928-937.

7

NON-RECIPROCAL COMPONENTS FOR MICs

by

K.C. Gupta

Chapter Seven

7.1 TWO APPROACHES: FERRIMAGNETIC INSERTS AND FERRIMAGNETIC SUBSTRATES

Microwave systems and sub-systems use a variety of non-reciprocal components like circulators, isolators, non-reciprocal phase shifters and switches. Design of these components is invariably based on non-reciprocal properties of magnetised ferrimagnetic materials.

There are two distinct methods of incorporating ferrimagnetic components in MICs[1,3]. The first consists of a ferrimagnetic disk (or 'puck') cemented into the usual nonmagnetic ceramic substrate. Components using this method are referred to as 'plug-in' type. This scheme is a logical extension of stripline circuits and had been widely used in the initial stages of MIC work. The second method uses an all ferrimagnetic substrate. However, only those portions of the substrate where the ferrimagnetic action is required are magnetised. Thus, the active ferrimagnetic area is equal to that of the magnet. In the first method the active area is equal to the area of ferrimagnetic 'puck' itself.

7.2 FERRIMAGNETIC SUBSTRATES

Properties of ferrimagnetic materials required for microwave circuits are well documented in the literature[4,5]. For MICs a thin slab of metallised ferrimagnetic material is required. The microstrip circuitry is photoetched on the substrate. For microwave circuit work, the r.f. losses should be minimum. In choosing a ferrimagnetic substrate consideration must be given to both the dielectric and the magnetic losses (in addition to the conductor loss and the radiation loss). The dielectric loss of most ferrimagnetic materials is comparable to the dielectric loss of good microwave dielectric materials. The magnetic losses in ferrimagnetic materials are frequency dependent. When using a ferrimagnetic substrate in an unmagnetised state in an MIC, the ferrite must be chosen such that the following inequality is satisfied to achieve low magnetic loss

$$\omega > (4\pi M_s + H_{anis})\gamma \qquad (7.1)$$

where ω is the frequency of operation, γ is the effective gyromagnetic ratio (=2.8 MHz/Oe), $4\pi M_s$ is the saturation magnetization and H_{anis} is the anisotropy field of the material.

The design criterion for choosing a ferrimagnetic material for a circulator (for operation below reasonance) is generally given by

$$\gamma 4\pi M_s/\omega \simeq 0.6 \qquad (7.2)$$

The substrate materials thus selected meet the restriction for low magnetic loss given by eqn.(7.1). This allows the circuits, which do not use the ferrimagnetic interactions, to be incorporated on the ferrimagnetic substrate along with the nonreciprocal components.

Conductor loss is also important in the case of ferrimagnetic (as much as for pure dielectric) substrates. The metals commonly used for conductors are gold, silver and copper. A thin layer of chromium (200-400°A) is generally vacuum deposited to provide good adherence to the substrate surface. On the top of this chromium layer the conductor metal is evaporated and plated to the desired thickness. Some factors that minimise the conductor loss are:

(i) The substrate should be thick and the microstrip impedance as low as possible for a given substrate thickness (this results in uniform distribution of r.f. current on the conductor).

(ii) The thickness of the conductor should be at least 2 to 4 skin depths at the lowest operating frequency.

(iii) Surface roughness of the substrate should be better than 10 μin and the substrate should be flat within 1 mil/in.

Various properties of typical ferrimagnetic substrates[3] are shown in Table 7.1. Alumina and magnesium titanate are also included for comparison of dielectric loss.

Table 7.1 Comparison of substrates (From[3])

Material	Dielectric constant	Dielectric loss tangent	$4\pi M_s$ (Gauss)	Linewidth ΔH (Oe.)
Mg-Mn Ferrite	13.00	0.0005	2100	500
Ni-Zn Ferrite	9.4	0.0010	3800	110
YIG	16.00	0.0005	1780	40
YIG (Al)	15.5	0.0005	1325	40
	14.5	0.0005	900	40
	14.0	0.0005	200	40
Alumina (99.5% Al_2O_3)	9.4	0.0002	-	-
Magnesium titanate (D-16 from Trans. Tech Inc.)	16.00	0.0002	-	-

7.3 MICROSTRIPS ON FERRIMAGNETIC SUBSTRATE

The concepts of effective dielectric constant and 'filling factor' for a microstrip on dielectric substrate have been discussed earlier in chapter 3. In the case of ferrimagnetic substrate, the relative permeability of the substrate is also different from unity and the concepts of 'effective permeability' and 'magnetic filling factor' may be introduced in a similar manner. Design formulas for microstrips on magnetic substrate are based on these concepts[2,7,8].

For the calculation of 'effective relative permeability' and 'magnetic filling factor', it is not necessary to go through the mathematical derivation similar to that in the case of dielectric substrates. The required information can be obtained by using a duality relationship for dielectric and magnetic substrates. The duality relationship is derived from the duality of ε and $1/\mu$ in Maxwells equations[7]. This duality amounts to replacing ε by $1/\mu$ and ε_{eff} by $1/\mu_{eff}$ in the formulas for the dielectric case. On the basis of this duality, a magnetic filling factor can be defined as

$$q_m = (\mu_{eff}^{-1}-1) / (\mu^{-1}-1) \qquad (7.3)$$

Magnetic losses in ferrimagnetic substrates can be taken into account in a similar manner. For dielectric losses[9] one defines an effective dielectric loss tangent δ_{deff} in the following manner:

$$\tan \delta_{deff} = q_{d\ell} \tan \delta_d \qquad (7.4)$$

Here δ_d is the loss tangent for the substrate and $q_{d\ell}$ is a filling factor for loss given by

$$q_{d\ell} = q_d \, \varepsilon_r/\varepsilon_{eff} = (1-\varepsilon_{eff}^{-1})/(1-\varepsilon_r^{-1}) \qquad (7.5)$$

where ε_{eff} is the effective dielectric constant for the substrate. For magnetic substrates, the filling factor of the magnetic loss tangent $q_{m\ell}$ and the effective value of this loss tangent take the following form:

$$q_{m\ell} = q_m \, \mu_{eff}/\mu = (1-\mu_{eff})/(1-\mu) \qquad (7.6)$$

$$\tan \delta_{meff} = q_{m\ell} \tan \delta_m \qquad (7.7)$$

Assuming a pure TEM mode, one can write the following formulas for characteristic impedance Z_o, guide wavelength λ_g and total substrate loss per wavelength $\alpha\lambda_g$ for microstrip on a substrate exhibiting both dielectric and magnetic properties;

$$Z_o = Z_o' \sqrt{(\mu_{eff}/\varepsilon_{eff})} \tag{7.8}$$

$$\lambda_g = \lambda_o / \sqrt{(\varepsilon_{eff}\ \mu_{eff})} \tag{7.9}$$

$$\alpha\lambda_g = 27.3\ (\tan\delta_{deff} + \tan\delta_{meff})\ \text{dB} \tag{7.10}$$

where λ_o corresponds to wavelength in free space, and Z_o is the characteristic impedance when $\mu_r=\varepsilon_r=1$, the value of Z_o can be calculated from the capacitance per unit length for an air dielectric or from Wheeler's expressions by taking $\varepsilon_r=1$.

The formulas described above apply (strictly speaking) only to substrates exhibiting a scalar permeability. For ferrites which are gyromagnetic, these relations must be modified. For gyromagnetic substrates the effective permeability will depend on the frequency of operation, as well as the magnitude and direction of the biasing field or magnetisation.

For TEM mode along a microstrip on a gyromagnetic substrate biased along the direction of propagation, the relative permeability can be characterised by a two-dimensional tensor of the form[8]

$$\mu = \begin{bmatrix} \mu & -j\kappa \\ j\kappa & \mu \end{bmatrix} \tag{7.11}$$

For a demagnetised substrate, $\kappa=0$ and $\mu=\mu_{dem}$ can be approximated by the following expression[8,10];

$$\mu_{dem} = [1 + 2\sqrt{\{1-(\omega_m/\omega)^2\}}\ ]/3 \tag{7.12}$$

where $\omega_m = (4\pi M_s)$ and ω is the angular frequency.

For partially magnetised ferrites (e.g. for latched substrates), the following expressions for elements of tensor permeability have been proposed[8]. These expressions are based on extensive experimental data.

$$\mu = \mu_{dem} + (1-\mu_{dem})\ (M/M_s)^{3/2} \tag{7.13}$$

and

$$\kappa = \gamma 4\pi M/\omega \tag{7.14}$$

where $4\pi M$ is the magnetisation. If the substrate is latched into remanence, the magnetisation takes on its remanent value $4\pi M_r$.

Finite difference technique has been extended to solve for field distributions in microstrips on gyromagnetic substrates and to obtain values for the inductance per unit length as a function of W/d ,μ and κ . By curve fitting the graphs of this inductance plotted against these parameters, the following analytic approximation for the relative permeability $\mu = \mu_{mag}$ has been suggested (without claiming any physical basis for the same) [8].

$$\mu_{mag} = (\mu^2 - \kappa^2)/[\mu\{1-(\kappa/\mu)^2\sqrt{(d/W)}\ \ln[1+\mu/(\mu^2-\kappa^2)]/7\}] \quad (7.15)$$

whre μ and κ are as defined in eqn. (7.11)

The expressions for μ_{dem} and for μ_{mag} for the two states of operation can now be used to evaluate the effective permeability and the magnetic filling factors.

Experimental results for guide wavelength and attenuation along the microstrip on ferrite and garnet substrates have been presented by Masse and Pucel[8]. Good agreement has been obtained between experiment and theory. In particular the observed increase in wave attenuation at frequencies near ω_m is fully explained when the frequency dependence of the characteristic impedance is taken into account. At frequencies higher than ω_m typical value of attenuation is 0.1 db per guide wavelength.

7.4 MICROSTRIP CIRCULATORS

Ferrimagnetic circulator is one of the most versatile and widely used non-reciprocal component in microwave systems. Circulators can be used as duplexers, isolators (when one of the ports is terminated in a thin film load), switches, modulators, phase shifters and input/output circuits for single port circuits such as negative resistance amplifiers and locked oscillators.

There are two types of circulators. First type, designed for fixed bias operations, normally uses a permanent magnet. The second type is designed for switched bias (latching) operation. These employ an electromagnet. Both of these types have been fabricated in MIC form.

The design procedure for the microstrip circulators is the same as that for the balanced stripline circulators[6]. The most significant step in the design is the calculation of the radius of the ferrite disc (in case of plug-in-type circulator) or that of magnet (in case of circulator on the ferrite substrate). For a three port circulator, this radius R may be determined from the following equation[11]

$$R = 1.84/[\omega_o\sqrt{(\varepsilon_o \varepsilon_r \mu_o \mu_e)}] \quad (7.16)$$

where ω_o is centre frequency, ε_o is free space permittivity, ε_r is relative dielectric constant of ferrite, μ_o is free space permeability and μ_e is scalar effective permeability. μ_e is given by

$$\mu_e = (\mu^2 - \kappa^2)/\mu \quad (7.17)$$

$$\mu = 1 - p\sigma/(1-\sigma^2), \quad \kappa = p/(1-\sigma^2) \qquad (7.18)$$

where $p = |\gamma|M_s/\omega$ is normalised saturation magnetisation and $\sigma = |\gamma|H_{dc}/\omega$ is normalised biasing magnetic field, which implies that $\sigma = H_{dc}/H_o$ and $p = M_s/H_o$, where H_o $(=\omega/|\gamma|)$ is the field required for gyromagnetic resonance in the infinite ferrite medium.

For an all ferrimagnetic circulator, the circulator junction radius calculated from eqn.(7.16) is interpretted as the radius of the magnet providing the static magnetic field. A metallic disc having slightly smaller radius is used above the junction as part of the tuning structure. Typical configuration for such a circulator is shown in Fig.7.1. As pointed out earlier the material is generally chosen to satisfy eqn.(7.2).

Minimum transmission loss between the two ports (1 to 2) is determined by the copper, the dielectric and the magnetic losses of the input strips and the junction. The maximum isolation between the two ports (2 to 1) is primarily dependent on the impedance match at the isolated port[3]. The impedance matching structures are derived experimentally using standard smith chart calculations. The circuit geometry thus includes the metallic disc above the junction plus impedance matching artifices such as quarter wave transformer, a tapered strip or tuning stubs. All these form parts of the mask used in photoetching process. The magnet can be located above or below the substrate.

The performance of a typical microstrip circulator on an all ferrimagnetic substrate is given in reference[3]. Using a YIG substrate 0.055 in thick, an isolation greater than 20 dB, a VSWR less than 1.2 and an insertion loss less than 0.8 dB has been reported for a frequency range from 8.5 to 9.9 GHz.

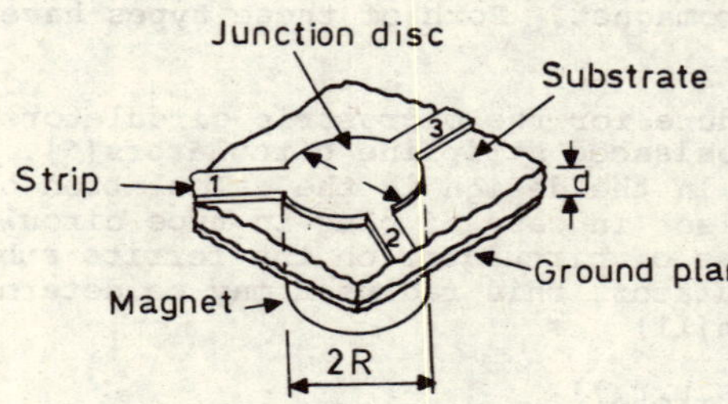

Fig. 7.1 A microstrip circulator

Power handling capacity of this type of circulator depends on the circuit design as well as on the basic properties of the YIG material. In the circuit design, any technique which reduces the magnitude of the r.f. field in the material will increase the power handling capacity. This can be accomplished by lowering the impedance or by increasing the substrate thickness. Among the material properties, intrinsic line width and value of $4\pi M_s$ are significant. Increasing the intrinsic linewidth or decreasing the $4\pi M_s$ of the material will increase the peak power threshold. Reduction in $4\pi M_s$ can be accomplished by substituting aluminium ions in the material. Increase in the intrinsic linewidth can be achieved by selected ion substitution. It has been pointed out in [3] that the peak power threshold can be increased by using a substrate containing small dysprosium (Dy) substituted in YIG. These results are summarised in Table 7.2.

TABLE 7.2. Power handling capacity vs. Dy content[3].

Frequency 8.5 GHz.

Substrate thickness 0.055 in.

Percentage of Dy	Insertion loss (dB)	Limiting Threshold (Peak Power)
0	0.3	60 watt
2	0.4	1.5 KW
5	0.7	8.5 KW
10	0.9	Not measurable

Performance of circulators with plug-in design is comparable to those with all ferrimagnetic substrates. Performance of an X-band plug-in circulator is described in[1].

Latching Circulators

Microstrip circulators can be designed to operate without an external d.c. magnetic field. To accomplish this, the ferrimagnetic material geometry is arranged such that closed-loop magnetic flux paths exist. If the material is then subjected to a magnetic field pulse, the magnetisation of the sample will attain some maximum value and fall back to a remanent value (determined by the hyteresis loop of the material) when the bias field pulse is removed. A bias field pulse of reversed polarity will reverse the direction of remanent magnetisation. Thus, by reversing the polarity of the switching pulse the direction of circulation in the circulator can be changed. This permits the device to be used as a single pole double throw switch.

One of the techniques of fabricating latched microstrip circulator uses platinum wire loops embedded in a sheet of ferrimagnetic material. A circulator using this technique[4] has been reported at a centre frequency of 9 GHz and with a bandwidth of about 400 MHz. Insertion loss over this band has been reported to be less than 1.5 dB with an isolation greater than 17 dB.

7.5 ISOLATORS

Microstrip isolators may be obtained by terminating one port of a three port circulator in a matched load. The termination generally consists of a chromium layer which is deposited initially in all circuits for adherence. More than one type of satisfactory terminations have been reported[4]. A tapered chromium film about 0.6 inches long with a surface impedance of about 270 ohms exhibited a VSWR of less than 1.2 over the 5.5 to 11.0 GHz band. A more compact design is shown in Fig.7.2. In this design the chromium film acts as a lumped 50-ohm resistor terminated in a quarter-wave open stub. The VSWR of this termination is less than 1.2 over the 4.5 to 6.0 GHz band. Another design uses a 50 ohm lumped resistor shunt mounted across the strip line as shown in Fig.7.3. A transmission line equivalent circuit is also shown in the figure.

Resonance type of isolators have also been proposed in the microstrip geometry[4]. In this type of isolators the operation is based on the presence of a circularly polarised r.f. magnetic field at air dielectric interface near the microstrip. If the microstrip is loaded with a ferrite strip as shown in Fig.7.4 and magnetised in transverse direction, nonreciprocal attenuation occurs at a frequency equal to ferrimagnetic resonance frequency. A C-band isolator based on this principle is described in reference[3]. However, this type of isolator has not been popular because of the complicated geometry and the requirement of a high

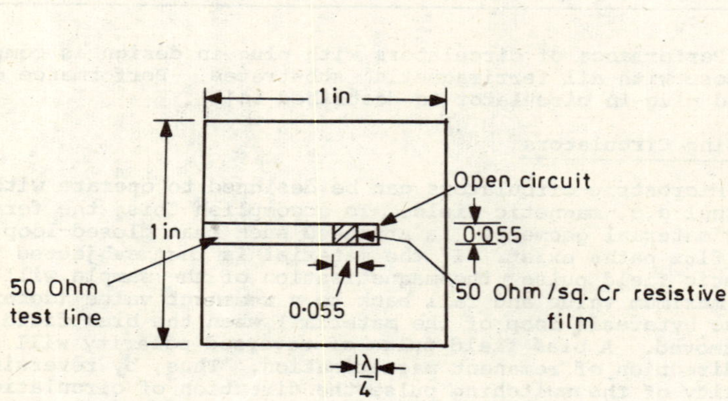

Fig. 7.2 Isolator termination with series mounted resistor

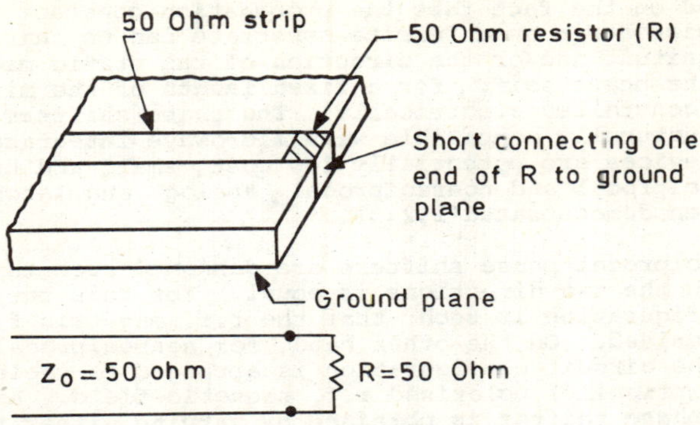

Fig. 7.3 Isolator termination with shunt mounted resistor

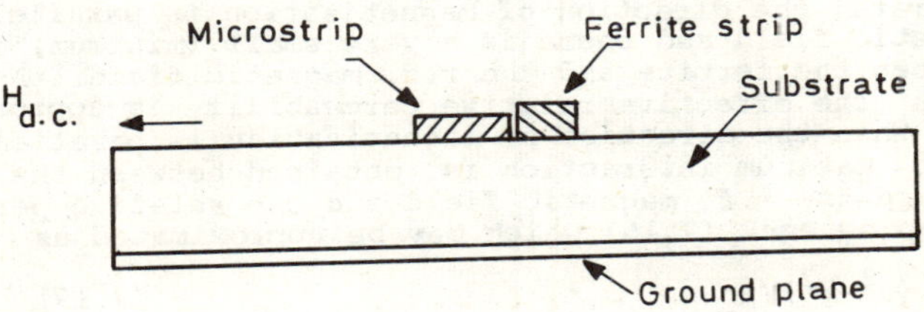

Fig. 7.4 Resonance type isolator

magnetic field in the plane of the circuit. The resonance type of isolator can be designed more conveniently by using a slot line on a ferrimagnetic substrate. As discussed in Chapter 6 a slot line provides a natural region of elliptical polarisation of the r.f. magnetic field in the dominant mode of propagation. However no practical isolators using this geometry have been reported so far.

7.6 PHASE-SHIFTERS

7.6.1 Use of ferrites for phase shifters:

The design of phase shifters using ferrimagnetic substrates is based on the fact that the propagation constant of a wave along a microstrip on a ferrite substrate can be changed by varying the magnitude and/or the direction of the static magnetic field. Thus, the phase shift, for a fixed length of the microstrip line, can be controlled electrically. The phase shifters thus obtained are structurally compatible with microwave integrated circuits. These devices are potentially low cost, small and highly reliable. Both reciprocal and nonreciprocal, analog and latching versions have been demonstrated[1,2,3].

Reciprocal phase shifters are designed such that the phase shift in the two directions is equal. For this purpose microstrip configuration is such that the r.f. magnetic field is linearly polarised. On the other hand, for nonreciprocal phase shifters the circuit configuration is arranged to yield a circularly (or elliptically) polarised r.f. magnetic field. Analog control of the phase shifter is obtained by varying either the magnitude or the direction of the magnetisation vector. Digital latching control is achieved by using high remanence materials and pulsing the substrate by a current pulse. The basic operation of reciprocal and nonreciprocal phase shifters is discussed below:

7.6.2 Reciprocal phase-shifters

When a ferrite substrate (with microstrip) is magnetised in a direction perpendicular to the plane containing the microstrip and the substrate, the direction of magnetisation is parallel to the r.f. magnetic field and there is a very small (minimum) interaction between the ferrite and the r.f. magnetic field. As a result of this, the effective relative permeability is approximately unity. When the direction of magnetisation is parallel to the microstrip, maximum interaction is obtained between the ferrite medium and the r.f. magnetic field and the relative permeability is given by eqn. (7.15) which may be approximated as

$$\mu_{mag} = (\mu^2-\kappa^2)/\mu \qquad (7.19)$$

In the frequency range of interest ($\omega>\omega_m$), and with small applied magnetic field, it can be shown that

$$\mu_{mag} \simeq 1- (\omega_m / \omega)^2 \qquad (7.20)$$

The different values of relative permeability for the two mutually perpendicular directions of magnetisation yield two different values of the phase constant for wave propagation along the microstrip line. If the magnetic filling factor, eqn.(7.3) is taken as 0.5, we have, for magnetisation parallel to the microstrip[2],

$$\beta \simeq \omega\sqrt{\mu_o\varepsilon_o\varepsilon_{reff}}\ [\{2-2(\omega_m/\omega)^2\}/\{2-(\omega_m/\omega)^2\}]^{1/2} \qquad (7.21)$$

and for magnetisation perpendicular to the microstrip

$$\beta = \omega\sqrt{(\mu_o \varepsilon_o \varepsilon_{reff})} \tag{7.22}$$

Thus, the fractional differential change in propagation constant obtained when the ferrite magnetisation direction is changed between the two states described above is given by

$$\Delta\beta/\beta = 1 - \{[2-2(\omega_m/\omega)^2]/[2-(\omega_m/\omega)^2]\}^{1/2} \tag{7.23}$$

A good agreement between the above expression and the experimental results has been reported[2] when the fractional change in phase constant is less than 0.1.

A microstrip circuit layout for the reciprocal phase shifter based on the above principle is shown in Fig.7.5. The microstrip conductor is loosely meandered in order to minimize the physical size of the device. The adjacent strips should be, however, separated by at least five times the ground plane spacing to avoid the interaction between r.f. magnetic field of adjacent conductors. As explained in the next subsection, this interaction leads to non-reciprocal effects. For the phase shifter shown in Fig.7.5, two states of magnetisation (corresponding to the two values of phase shifts) are obtained by sending current through wires passing through holes 1,1' and 2,2' respectively.

7.6.3 Non-reciprocal phase shifters

For non-reciprocal phase shifters, a microwave circuit which produces a circularly polarised magnetic field (or at least an elliptically polarised field) is required. A circularly polarised magnetic field region can be obtained by using a meander configuration when the adjacent lines of the meander are tightly coupled and when the line lengths are a quarter wavelength at the centre frequency as shown in Fig.7.6. Because of the coupling between the two adjacent lines of the meander, the two components of the magnetic field at point A (in Fig.7.6) are spatially orthogonal. Since the lines are a quarter wavelength long, the current flowing at the centre of line 2 is delayed by 90° in time relative to the current flowing at the centre of line 1. Thus, the resulting magnetic field at A is circularly polarised. However, at points away from the centre of the lines, the sense of polarisation varies from circular through elliptical to linear at the ends. If the ferrite medium is magnetised along the directions of meander line sections, a non-reciprocal differential phase shift is obtained.

Although satisfactory theoretical model has not yet been reported for the impedance and propagation constant of a meander line on a magnetised ferrite substrate, this type of phase shifter is being developed and improved. Performance considerably better than that of reciprocal phase shifters (described in sec.7.6.2) has been reported[3]. In a single bit form the substrate is less than 1 in. by 1 in. by 50 mil thick and provides phase shift of 360° + 10° with less than 2.40 dB of loss for 10 percent band centred at 5.75 GHz[3].

A latching device can be obtained if the ferrite substrate

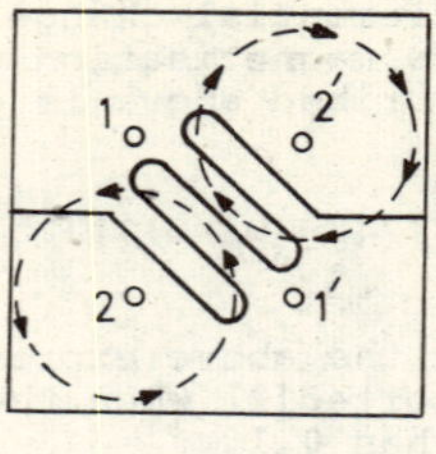

Fig. 7.5 Reciprocal phase shifter on ferrite substrate

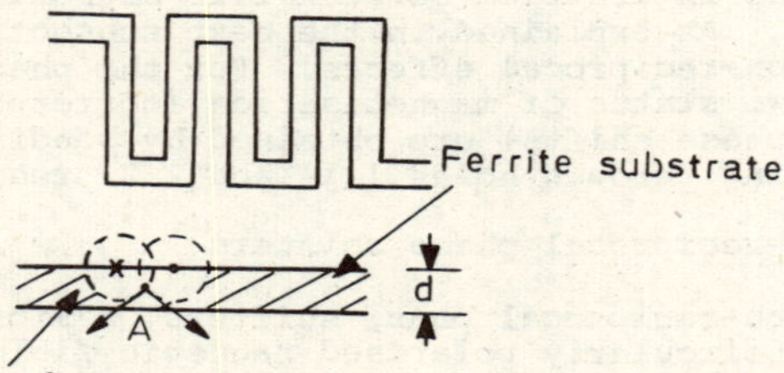

Fig. 7.6 Magnetic field configuration for non-reciprocal phase shifters

is sufficiently thin and small wires are inserted through holes in the substrate. In this case the substrate becomes a planar toroid and the current pulses in the wires can be used to alter the direction and the magnitude of ferrite magnetisation.

References

1. Allen, J.L., and D.R. Taft, Ferrite elements for Hybrid Microwave Integrated Circuits, IEEE Trans. Electron Devices, Vol.ED-15, No.7, July 1968, pp.467-472.

2. Roome, G.T. and H.A. Hair, Thin Ferrite Devices for microwave integrated circuits, IEEE Trans. Electron Devices, Vol.ED-15, July 1968, pp.473-481.

3. Harrison, R.R. et al., Ferrimagnetic Parts for Microwave Integrated Circuits (invited paper), IEEE Trans. Microwave Theory Tech., Vol.MTT-19, No.7., July 1971, pp.577-587.

4. Von Aulock, W.H., Ed., Handbook of Microwave Ferrite Materials, New York, Academic Press, 1965.

5. Lax, B., and K.J. Button, Microwave Ferrites and Ferrimagnetics, New York, McGraw Hill, 1962.

6. Fay, C.E., and R.L. Comstock, Operation of the ferrite junction circulator, IEEE Trans. Microwave Theory and Tech., Vol. MTT-13, pp. 15-27, Jan 1965.

7. Pucel, R.A. and D.J. Masse, Microstrip propagation on magnetic substrates Part I-Design theory, IEEE Trans. Microwave Theory Tech., Vol.MTT-20, 1972, pp.304-308.

8. Masse, D.J., and R.A. Pucel, Microstrip propagation on magnetic substrates, Part II-Experiment, IEEE Trans. Microwave Theory Tech., Vol.MTT-20, No.5, May 1972, pp. 309-313.

9. Pucel, R.A., D. Masse and C.P. Hartwig, Losses in microstrip, IEEE Trans. Vol MTT-16, pp 342-350, June 1968, also correction to Losses in microstrip IEEE Trans. (corresp.) Vol.MTT-16, p 1064, Dec. 1968.

10. Denlinger, E.J., A frequency dependent solution for microstrip transmission lines, IEEE Trans. Microwave Theory Tech., Vol.MTT-19, p 30 , Jan. 1971.

11. Von Aulock, W.H., and C.E. Fay, Linear ferrite devices for microwave applications, 1968, Academic Press, N.York, p.116.

8

ACTIVE DEVICES FOR MICs - I

by

M.M. Hasan* and D.J. Roulston**

* Sections 8.1 and 8.2
** Sections 8.3 to 8.6

Chapter Eight

The present chapter and the next chapter describe various types of semiconductor devices commonly used in microwave integrated circuits. Schottky barrier diodes, microwave transistors (including FETs), varactors and PIN diodes are discussed in the present chapter. The next chapter will deal with TEDs, Impatt diodes, Barritts and surface wave amplifiers.

8.1 SCHOTTKY-BARRIER DIODES

When a metal-semiconductor contact gives rectifying characteristics, the junction is called a Schottky-barrier. The metal-semiconductor contact giving a Schottky-barrier is a very simple structure. It has various useful applications such as in detectors and mixers at microwave frequencies and at lower frequencies as a clamp in bipolar and field-effect transistor circuits including ICs[1,2]. Point-contact backward and tunnel diodes are also used as mixers and detectors from UHF to millimeter wave frequencies, though before the arrival of Schottky diodes, point-contact diodes were mostly used for these functions.

Schottky-barrier diodes are made using the well-established planar technology[3]. Generally, n-type silicon or n-type GaAs are used. For cut off frequencies in the range of X-band and above, GaAs Schottky diodes are preferred. Silicon Schottky diodes are fabricated on n-type thin epitaxial layer (0.5 - 1.0 micron thick) grown on n^+ substrate. A suitable SiO_2 layer is thermally grown or deposited on the semiconductor wafer. Windows for Schottky junctions are made using photolithography techniques and desired metal is evaporated, sputtered or plated. Another photolithography process is used to remove the metal from the oxide surface. Junction diameter may be as small as 5 microns. Figure 8.1 shows two possible structures of Schottky diode . The

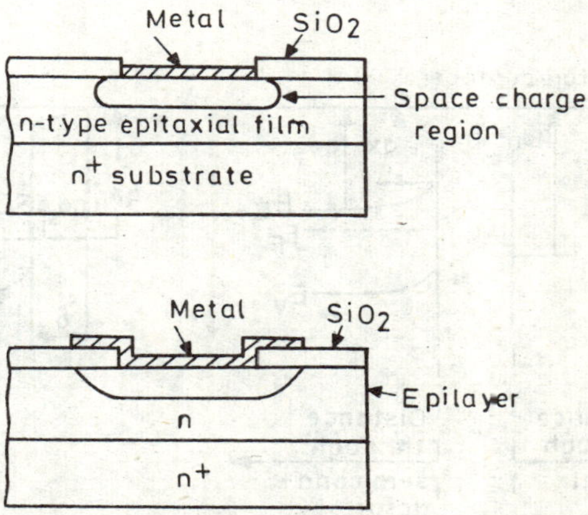

Fig. 8.1 Two possible structures of Schottky diode

metal extending over the SiO_2 immediately around the junction is preferred because it ensures a clean interface between the metal edge and the semiconductor, and results in low 1/f noise. Metals used for Schottky junctions are selected so as to yield a low metal-semiconductor barrier.

8.1.1 Band diagram of metal-semiconductor

Figure 8.2 shows band diagram of metal-semiconductor (n-type). Figure 8.3 shows band diagrams of Schottky barrier under conditions of (a) thermal equilibrium (i.e. no bias), (b) forward bias, and (c) reverse bias. ϕ_{ms} is the barrier height existing between the metal and the semiconductor. qV_D is the total band bending in the semiconductor and W is the space charge layer width. The band bending occurs in such a manner that it retards the flow of majority carriers into the metal; thus a "barrier" is present at the contact. This barrier is formed because of the difference in the work functions of metal and semiconductor and the surface states in the gap of the semiconductor. In Fig.8.2

ϕ_m = work function of metal
X = electron affinity of semiconductor
ϕ_s = work function of semiconductor
δ = gap distance between metal and semiconductor
Δ = contribution due to surface states
ϕ_n = gap between the bottom of the conduction band and the Fermi level.

When a metal-semiconductor contact is made, δ is estimated of the order of 0.5 to 5A° and this interfacial layer is transparent to carriers crossing between the metal and the semiconductor. The maximum barrier height seen by the carriers is known as the metal-semiconductor barrier height[4,5].

$$\phi_{ms} = \phi_m - X - \Delta \qquad (8.1)$$

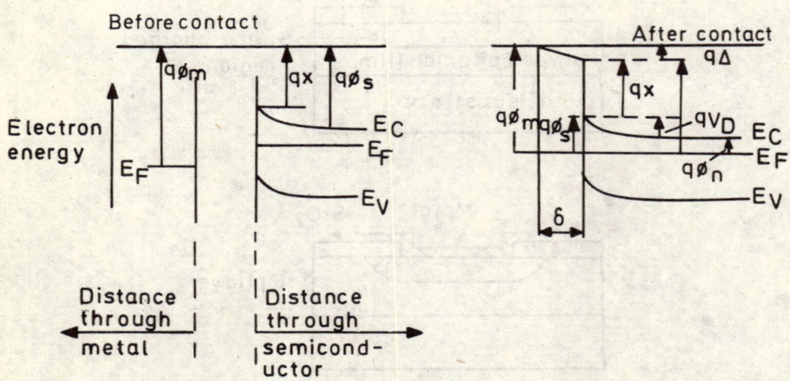

Fig. 8.2 Band diagram of metal-semiconductor(n-type) junction

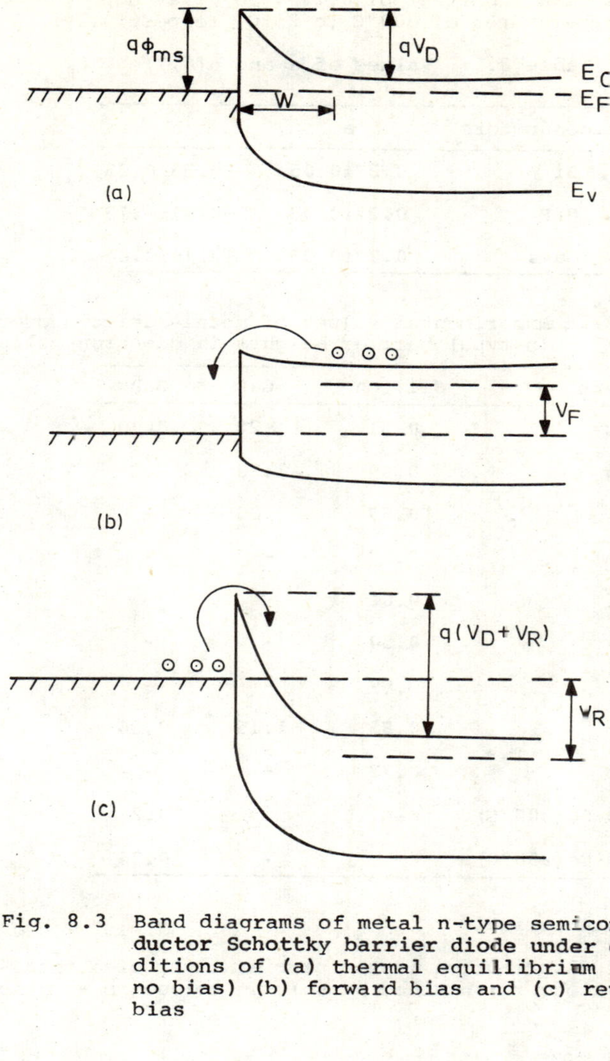

Fig. 8.3 Band diagrams of metal n-type semiconductor Schottky barrier diode under conditions of (a) thermal equillibrium (i.e. no bias) (b) forward bias and (c) reverse bias

Cowley and Sze[6] have derived an expression for ϕ_{ms} as $\phi_{ms} = a\,\phi_m + b$, where a and b reflect the effects of surface states and X. Table 8.1 gives values of a and b. Table 8.2 gives values of ϕ_{ms} for various metals on n-type semiconductor. Titanium on n-type Si gives the smaller barrier height ($\approx$0.5eV). Recently Moroney and Anand[2] have obtained Au-Ge Schottky diodes on n-type GaAs with a barrier of 0.27eV to 0.9eV depending on fabrication temperatures of 300°C to 25°C, respectively.

Table 8.1: Values of a and b[6,7]

Semiconductors	a	b
Si	0.27±0.05	-0.55±0.22
GaP	0.27±0.03	-0.01±0.13
GaAs	0.07±0.05	+0.49±0.24

Table 8.2 : Experimental values of metal-semiconductor (n-type) barrier heights in electron volts

Metal	Silicon	GaP	GaAs
Au	0.81	1.28	0.90
Ag	0.69	1.20	0.88
Al	0.67	1.05	-
Mo	0.60	-	-
Ni	0.66	-	-
Ti	0.50	-	-
W	0.69	-	-
Pt	0.85	1.45	0.86
Pd	0.72	-	-
Au-Ge(300°C)	-	-	0.27
Au-Ge(200°C)	-	-	0.35

8.1.2 C-V Characteristic

Potential distribution in the barrier layer of a metal-n-type semiconductor diode is obtained by solving Poisson's equation

$$d^2V/dx^2 = -\rho/\varepsilon \tag{8.2}$$

where $\varepsilon = \varepsilon_r\varepsilon_o$ and ρ is the charge distribution. The barrier layer is similar to the one-sided step p^+-n junction.

$$W = \sqrt{[2\varepsilon(V_D+V_R)/qN_d]} \tag{8.3}$$

Capacitance per unit area is

$$C = \varepsilon/W = \sqrt{[q\varepsilon N_d/2(V_D+V_R)]} \quad (8.4)$$

This can be written as:

$$1/C^2 = 2(V_D+V_R)/q\varepsilon N_d \quad (8.5)$$

Here V_D is the semiconductor-metal barrier height(eV), V_R is applied reverse bias, N_d is donor concentration of the semiconductor and q, ε are constants. The plot of $1/C^2$ vs. V_R will be a straight line. The intercept will give V_D and the slope will give N_d. Capacitance-voltage measurements have been used to determine the semiconductor doping and the barrier height.

Current-voltage characteristics

The current transport in metal-semiconductor barriers is mainly due to the majority carriers in contrast to p-n junctions where it is due to the minority carriers. The V-I characteristic of a Schottky-barrier diode are explained by the following approaches[4,5,7,8,9]:

(1) Thermionic Emission Theory

(2) Diffusion Theory

(3) Thermionic Emission-Diffusion Theory

When the space charge layer width is relatively thick (for example, $N_d < 10^{17}$ cm^{-3} for n-Si), the only means of electron transport at the interface is by thermionic emission over the barrier. With forward bias (metal positive on n-Si)(Fig.8.3b), the barrier from the semiconductor to the metal is reduced and many electrons with adequate thermal energies can climb the barrier resulting in a rapid increase in the forward current. This current is expressed as:

$$J_{s-m} = A^*T^2 \exp(-q\phi_{ms}/kT) \exp(qv_a/kT) \quad (8.6)$$

where A^* = Richardson Constant = $4\pi m^*k^2/h^3$ (8.7)
T = Temperature in °K, and V_a = applied voltage

When the diode is reverse biased as in Fig.8.3(c) (metal negative), the barrier for the electrons moving from the metal into the semiconductor remains the same, the current flowing into the semiconductor is thus unaffected by the applied voltage. It must therefore be equal to the current flowing from the semiconductor into the metal when $v_a = 0$. This is obtained from the above expression of J_{s-m} as

$$J_{m-s} = -A^* T^2 \exp(-q\phi_{ms}/kT) \quad (8.8)$$

The total current density is given by the sum of J_{s-m} and J_{m-s}:

$$J = A^*T^2 \exp(-q\,\phi_{ms}/kT) \exp(qv_a/kT) - 1$$

$$= J_{oT} \exp(qv_a/kT) - 1 \qquad (8.9)$$

where $J_{oT} = A^*T^2 \exp(-q\phi_{ms}/kT)$ (8.10)

In general, the Schottky diode current-voltage equation is written as

$$I = I_s \exp(q\,V_a/nkT) - 1$$

where n = ideality factor.

For Schottky barriers on a lightly doped semiconductor, thermionic emission approach is adequate for V-I characteristic. Equation (8.9) is similar to V-I characteristics for p-n junctions. However, the expressions for the saturation current density are quite different. We would make the following observations:

(1) In Schottky barriers, the current is carried by the majority carriers and there is no storage delay time, thus Schottky barriers are ideal for high frequency and switching applications.

(2) For diodes of similar area, Schottky barriers have much larger currents at the same forward bias than a p-n junction. For a particular Schottky barrier diode, forward drop is 0.2 volts for a current of 1 µA whereas it is more than 0.5 volt for the p-n diode. By increasing the doping in the semiconductor, the turn-on voltage can be further reduced. This low turn-on-voltage feature makes Schottky barriers useful in various clamping operations. When the doping in the semiconductor is high, the space-charge layer width is thin and "tunneling" through the barrier is possible in addition to thermionic emission over the barrier. **This added tunneling current is called thermionic field emission[9]** If the semiconductor doping is very heavy, the space-charge layer width is so thin that carriers at the Fermi level can tunnel through the barrier. This tunneling is called field emission. When this situation is achieved, metal-semiconductor contact becomes ohmic (example - Al on n^+-silicon).

8.1.3 Performance of Schottky-barrier diodes[4]

Figure 8.4 shows the equivalent circuit diagram of a Schottky barrier diode. C_j is the depletion layer capacitance of the semiconductor, and R_j and R_s are resistances. L_s and C_c are the package parameters. Figure 8.5 shows the variation of $R_j + R_s$ with biasing for Schottky barrier diodes made with Si and GaAs and for silicon point-contact diodes. Schottky devices showing such characteristic are classified as Schottky-barrier varistors and they are designed for performing detection and frequency mixing functions. Though point contact diodes have been used for microwave detection and mixing Schottky-barrier varistors are rapidly replacing them. Schottky-barrier diodes have lower series resistance, lower power loss, lower noise figures, greater dynamic range and greater resistance to pulse burnout. Furthermore, unlike point-contact diodes, Schottky diodes can be readily

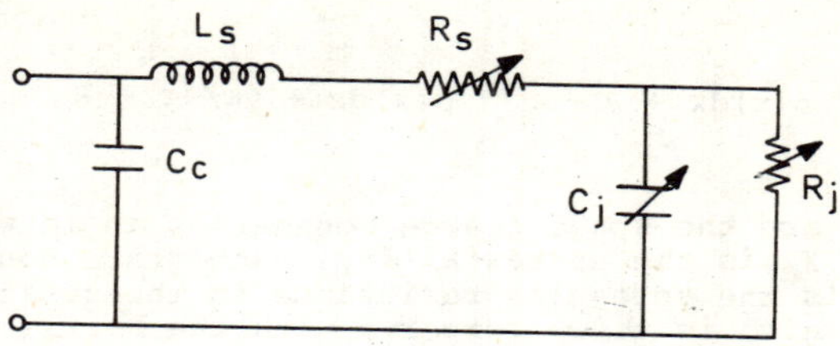

Fig. 8.4 Equivalent circuit diagram of a Schottky barrier diode

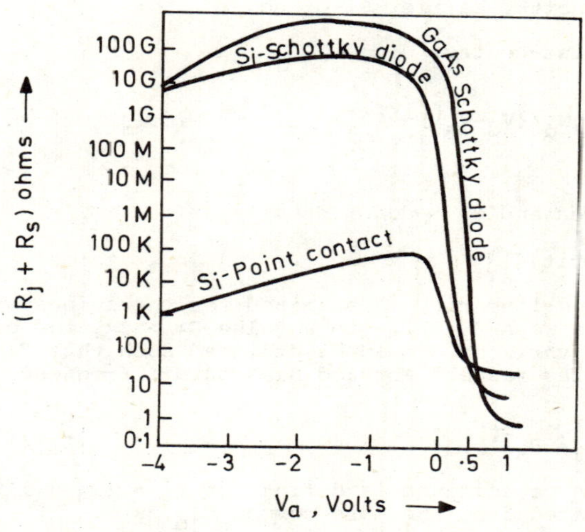

Fig. 8.5 Variation of $(R_j + R_s)$ with bias (From Ref. [4])

(From 'Microwave Semiconductor Devices and their Circuit Applications' by H.A. Watson. Copyright 1969, McGraw-Hill Book Company. Used with permission of McGraw-Hill Book Company)

used in MICs. Schottky diodes are free of injected minority carriers (when $\phi_{ms} < 2/3\ E_g/q$) and are therefore free of charge-storage capacitance also.

8.1.4 Components of the equivalent circuit[4]

When the diode is forward biased with sufficient voltage, R_j may become comparable to R_s or even smaller than R_s. This value of $R_s = R_{sf}$. The thin epitaxial layer on heavily doped substrate may not have uniform doping. Therefore, R_{sf} is expressed as:

$$R_{sf} = A^{-1} \int_{W_f}^{W_o} \rho(x)dx + A^{-1} \int_{W_o}^{X_e} \rho(x)dx + \rho_b/4r + R_c \qquad (8.11)$$

where W_f and W_o are the space charge boundaries in forward bias and at $V_a = 0$. X_e is the epitaxial layer-substrate boundary; the third term is the spreading resistance in the substrate of resistivity ρ_b which is thick compared with the barrier radius r, and R_c is the resistance of the wafer and barrier contacts. The first term comprises a large portion of the total R_{sf}.

$$R_j = \frac{nkT}{qI} \quad ; \text{ (when } V_a > \frac{kT}{q}) \qquad (8.12)$$

n = 1.1 for Schottky-barrier Si diode

= 1.4 for Point-contact diode

$$C_j = A[0.5qN_d/(V_d-V_a)]^{1/2} \qquad (8.13)$$

where A = barrier area and N_d = donor density.

8.1.5 Figures of merit[4]

The resistive non-linearity is examined for the swings of the driving signal between $R_j = R_{sf}$ to a value $R_j >> Z_o$, the circuit impedance. The varistor should be designed such that C_j is minimized over the bias range. Forward bias cutoff frequency is defined as:

$$f_{cf} = 1/(2\pi C_{jf} R_{sf}) \qquad (8.14)$$

Forward-bias junction capacitance is defined at $V_a = (V_d-0.1)V$, and it can be calculated from the more common C_{jo} as

$$C_{jf} = C_{jo} \sqrt{10\ V_B} \qquad (8.15)$$

The figure of merit given by eqn.(8.14) correlates well with conversion loss and noise figure in simple down-converter circuits.

Another figure of merit for down-converter diodes is defined as N_R. N_R is the ratio of the available noise power from the diode to that available from a pure resistance of value $(R_j + R_{sf})$, i.e.

$$N_R = (P_{noise})_{diode} / (P_{noise})_{R_j + R_{sf}} \qquad (8.16)$$

Shot noise of a temperature-limited diode is represented by R_j in parallel with a current generator with a mean square current given by:

$$\overline{i^2} = 2qIB \tag{8.17}$$

where I is the diode current, and B is the receiver bandwidth. The power available to a matched load using Eq. (8.17) is:

$$P_{shot} = \overline{i^2}R_j/4 = nkT_e/2 \tag{8.18}$$

Combining the shot noise with the thermal noise from R_{sf} and dividing by kTB, where kTB is the noise power available from a pure resistance, we get

$$N_R = [(n/2)R_j+R_{sf}] / (R_j+R_{sf}) \tag{8.19}$$

For small local oscillator drive and for an applied bias less than $(V_D - 0.1)$ volts, $R_j >> R_{sf}$ and $N_R = 0.5$ can be obtained if n=1. However, the high value of R_j introduce high loss in the signal path.

Low N_R can be achieved by using suitable circuit design and drive level. f_{cf} is a function of doping and epitaxial layer thickness. Use of thinner epitaxial layers and improved planar technology would increase f_{cf}. The performance of GaAs Schottky-barrier diodes is showing considerable improvement over the silicon diodes.

8.2 MICROWAVE BIPOLAR TRANSISTORS

8.2.1 Introduction

Most modern microwave transistors are silicon planar epitaxial diffused npn structures with emitter geometries that combine long emitter edges Z (for high current capabilities) with narrow emitter strip widths L (to reduce emitter current crowding and base resistance). The commonly used emitter geometries [10] are (Figures 8.6 and 8.7).

1) Interdigitated for small signal and power transistors, and

2) Overlay for power transistors only.

Interdigitated transistor is the best structure for S-band and C-band small signal applications. The power outputs obtainable from a single microwave transistor are at present approximately as follows:

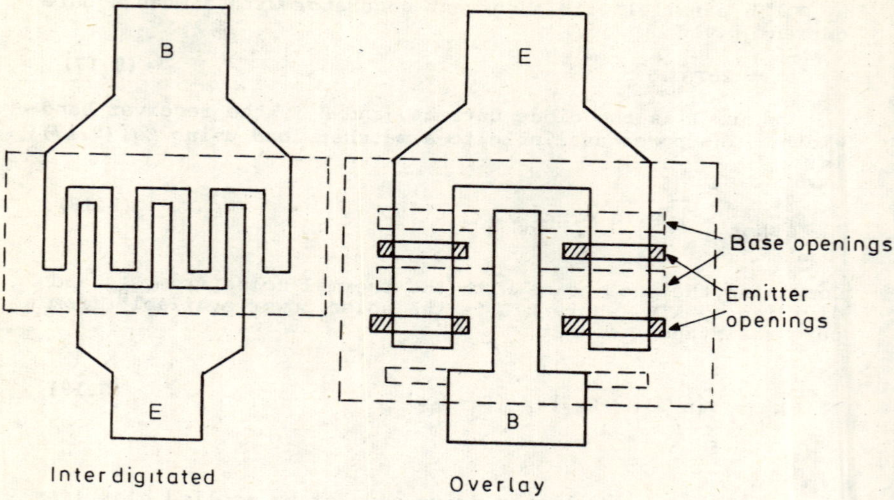

Fig. 8.6 Interdigital and overlay structures

50 watts at 500 MHz, 35 watts at 1.3 GHz, 10 watts at 2.3 GHz and 5 watts at 4 GHz. The best amplifier noise figures range from about 2 dB at 1 GHz to 4.0 dB at 2 GHz and 5 dB at 4 GHz.

For microwave transistors, the dimensions of the active regions must be shrunk to exceedingly small values and this must be accompanied by extremely low wafer and package parasitics. The fabrication processes, like solid-state diffusion, photolithography, thin-film metallization etching, and packaging operations are pushed to the limits of their capability. Small signal microwave transistors recently reported by Archer[11] have the following special features; (i) one-micron emitter stripes, (ii) arsenic diffused emitters and (iii) ohmic contacts using silicon doped aluminium. Reduction in emitter width reduces base resistance and improves the maximum frequency of oscillation f_{max}. Arsenic diffusion for emitter eliminates the emitter dip effect and enables better control of base width resulting in reduced base width and base transit time. A further reduction in base width is reported by using ion-implantation for p-n junction formation.

8.2.2 Gain Band-Width Product, f_T

f_T is an important figure-of-merit for high frequency transistors and is defined as the extrapolated frequency at which the common-emitter short-circuit current gain is unity. The frequency f_T is related to the physical structure of the transistor through

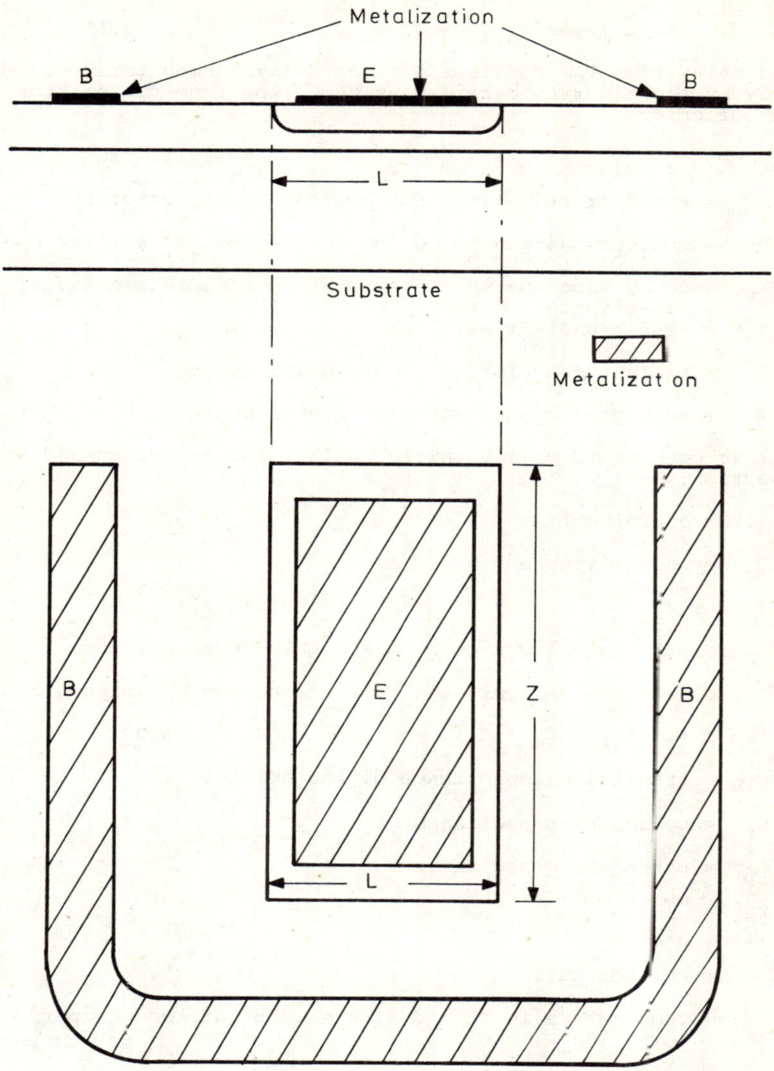

Fig. 8.7 Details of single strip structure

the emitter to collector delay time, T_{ec}, by [12,13]

$$f_T = 1/(2\pi\ T_{ec}) \tag{8.19}$$

Carrier delay time T_{ec} consists of five delays encountered sequentially by the minority carriers as they flow from the emitter to the collector.

$$T_{ec} = T_e + T_{eb} + T_b + T_d + T_c \tag{8.20}$$

where T_{ec} = total transit time from emitter to collector

T_e = emitter charging time due to the neutral emitter charge

T_{eb} = delay time due to the emitter base depletion layer capacitance

T_b = base transit time

T_d = collector-depletion layer charging time

T_c = collector-capacitance-resistance charging time

These five components of the carrier delay time are discussed in this section.

(a) Emitter charging time

T_e may be written as

$$T_e = Q_e/I_c \tag{8.21}$$

where Q_e is the neutral charge in the highly doped emitter.

(b) Delay due to the emitter-base capacitance may be written as

$$T_{eb} = r_e\ (C_{TE} + C_c + C_p) \tag{8.22}$$

C_{TE} = transition capacitance of the emitter

C_c = collector capacitance

C_p = parasitic capacitance

$$r_e = \frac{kT}{qI_E} \tag{8.23}$$

I_E = emitter current

T = temperature in °K and kT/q = .025V at room temperature (300°K)

(c) Base transit time

Base transit time, T_b, is the time required for the injected minority carriers to cross the base region. Planar transistors have a non-uniform impurity distribution in the base. For the computation of base transit time, a knowledge of exact impurity profile in the base is essential.

By neglecting recombination in the neutral base region (a generally valid assumption), by considering that the current gain is high so that the hole current in the base is small compared to the electron current (in an NPN transistor) and by assuming low level injection conditions, it can be shown that the base transit time T_b is given by:

$$T_b = Q_s/I_n \tag{8.24}$$

$$= \int_0^W [1/N_B(x)] \left[\int_x^W \{N_B(z) / D_N(z)\} dz \right] dx \tag{8.25}$$

For a graded-base transistor, calculation of T_b using Eqn.(8.25) requires the use of a digital computer[14].

Due to the dependence of D_n on N_B and the fact that the impurity profile reaches a peak (Figure 8.8) before decaying, no simple and accurate analytic expression can be obtained for T_b. However, exact computations[20] indicate that to a good approximation

$$T_b = W^2 / \bar{D}_n F \tag{8.26}$$

where F = 2 and $\bar{D}_n$ is the diffusion constant corresponding to the average conductivity of the base region. Thus diffused base transistors have in general a transit time which is close to that obtainable with a uniformly doped base transistor with the same base conductivity.

(d) Collector depletion-layer transit time

For microwave transistors, epitaxial layer in the collector would be of lower doping than the base material. Therefore, the entire depletion layer is assumed in the collector region. For a step junction approximation, the depletion layer x_d is given as:

$$x_d = [2\varepsilon(V+V_b)/qN_c]^{1/2} \tag{8.27}$$

For silicon transistor:

$$x_d \simeq 3.6 \times 10^3 \, [(V+V_b)/N_c]^{1/2} \text{ cm.} \tag{8.28}$$

where

N_c = carrier concentration in the epitaxial layer and

V_b = built-in voltage.

Transit time across the depletion layer may be written as

$$T_d = x_d/v_s \tag{8.29}$$

where v_s = scatter-limited saturation velocity of electrons in silicon.

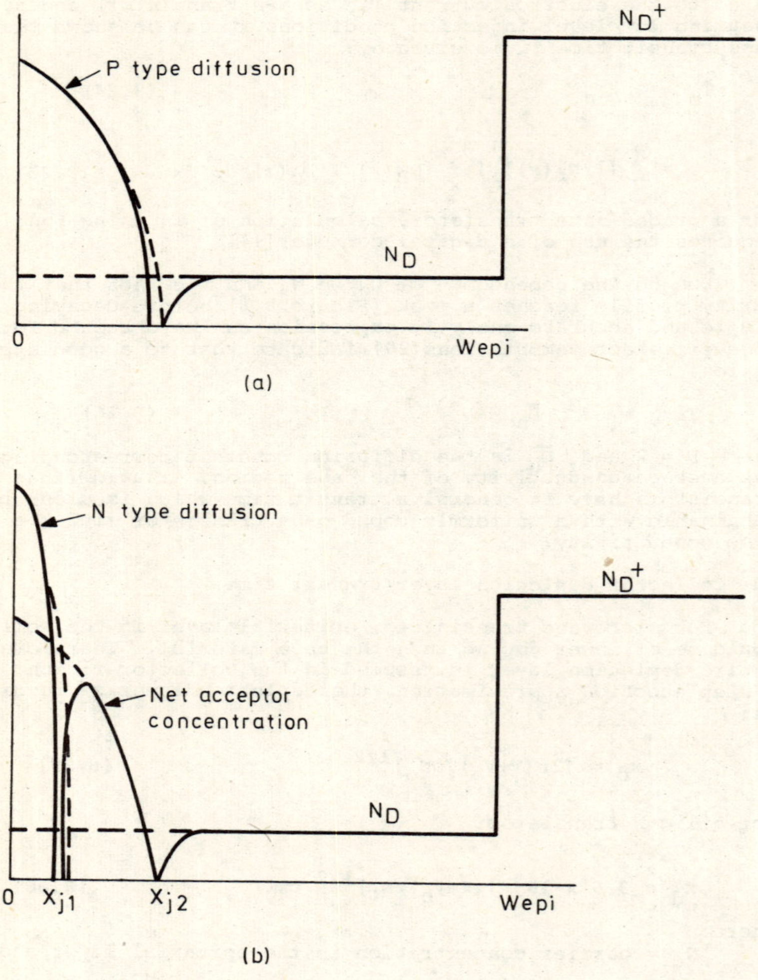

Fig. 8.8 Typical impurity profile
(a) after p-type diffusion
(b) after both p and n-type diffusion

For the computation of cut-off frequency T_d is taken equal to $x_d/2v_s$ [13].

(e) Collector charging time

T_c may be written as

$$T_c = r_c C_c \quad (8.30)$$

where r_c is the collector series resistance and C_c the collector capacitance. For epitaxial transistors, r_c is very small and T_c can be neglected.

Capacitances

The two junction capacitances involved in computation of f_T are discussed below:

(a) E-B Junction Capacitance

Using step junction approximation, the transition capacitance may be written as

$$C_{TE} \approx A_E \left(\frac{\varepsilon q N_{BE}}{2(V+V_b)} \right)^{1/2} \quad (8.31)$$

for silicon

$$C_{TE} \approx A_E x 2.92x10^{-4} \, [N_{BE}/(V+V_b)]^{1/2} \text{ pF}; \; (A_E \text{ in cm}^2) \quad (8.32)$$

(b) C-B Junction Capacitance

The C-B junction capacitance can be classified in two parts, i.e.

(i) under the emitter

(ii) other regions of the base.

The part under the emitter cannot be reduced and it is denoted by C_i. The other part is denoted by C_o where $C_c = C_i + C_o$

$$C_c = A_c x 2.92x10^{-4} \, [N_c/(V+V_b)]^{1/2} \quad (8.33)$$

There are three important geometrical considerations in establishing a low T_{ec} [16]. These are the emitter area A_E, (in eqn.(8.31)) the base width W(in eqn.(8.26)) and the collector depletion layer width x_d(in eqn.(8.29)). If one operates the transistor at an emitter current density commensurate with the emitter area, the emitter capacitance loses its limiting significance in T_{ec}. One may achieve high f_T independent of the emitter junction area.

8.2.3 Maximum Frequency of Oscillation

The power gain of a transistor is expressed as

$$G = f_T/(8\pi f^2 r'_{bb}\ C_c) \tag{8.34}$$

where r'_{bb} is the base resistance.

The maximum frequency of oscillation, f_{max} is defined from eqn. (8.34) when G = 1.0 as:

$$f_{max} = [f_T / 8\pi r'_{bb}\ C_c]^{1/2}$$

$$= 1 / [4\pi(r'_{bb}\ C_c\ T_{ec})^{1/2}] \tag{8.35}$$

For a microwave double-diffused transistor (Fig. 8.7)

$$f_{max} = 1/[4\pi(r''_b\ C_o + r'_b\ C_i)T_{ec}]^{1/2} \tag{8.36}$$

where r''_b and r'_b are components of r'_{bb} as explained below:

Base Resistance:

In a planar transistor, base resistance is usually part of an RC transmission line.

r'_{bb} may be broken down into the following four regions:

(i) base resistance under the emitter r'_b

(ii) base resistance between the emitter and the p^+ region

(iii) base resistance in the p^+ region itself.

(iv) contact resistance between the metal and the p^+ region.

The calculation of base resistance will depend upon the structure of the transistor but at low and moderate currents the value of the intrinsic component (i) r'_b is given by:

$$r'_b = L/(12\sigma_b\ W_B) \tag{8.37}$$

In order to derive useful power gain at microwave frequencies, eqn.(8.36) requires minimum r'_b , r''_b and capacitances. Thus, though a large-area device may have an inherently high f_T, the base resistance must be minimized. The design of microwave transistors is dominated by these geometrical considerations: the base thickness and depletion layer width, which controls the f_T (and hence T_{ec}), and the electrode spacing and areas, which determine the r'_b , r''_b , C_i and C_o [10, 17, 18].

It should also be noted that at high collector currents two important effects occur which cause f_T (and hence f_{max}) to fall-off rapidly:

(i) base widening when the current density exceeds a value J_K, where $J_K = q\, v_{th}\, N_C$ (8.38)

v_{th} is the scatter limited drift velocity of carriers and N_C the collector(epitaxial layer) doping level.

(ii) Emitter current crowding when the base current exceeds a value I_{BC}:

$$I_{BC} = kT/q\, r'_{bb'} \qquad (8.39)$$

It should be noted that when I_B exceeds I_{BC}, the current density at the edge of the emitter exceeds a value $I_C/(LZ)$ where I_C is the total collector current. The base widening effect is thus enhanced by current crowding. The implications of eqns.(8.38) and (8.39) are that both the base and collector doping levels should be kept as high as is compatible with the other design requirements. (current gain, collector-base breakdown voltage etc.). In any case, the f_T versus I_C curve will fall off at high currents due to the above effects and at low currents due to the $r_e\, C_{TE}$ term in eqn. (8.22).

8.2.4 Modeling of Microwave Transistor [14, 17, 18, 19, 20]

Recent attempts at modeling of microwave transistors have used one of three basic approaches:

(1) Linear Two-Port Model

(2) Device Physics Model using numerical methods [2]

(3) Equivalent Circuit Model based on physical operation of the device.

Approach (1) uses black-box method and physical processes are not considered explicitly. Small signal ac measurements (Y or Z) are made at the ports. In the microwave frequency range, S parameters are used.

Approache(2) requires informations about doping, geometry and applied bias. Poisson and diffusion equations have to be solved to obtain the model.

Approach (3) occupies a position intermediate between approaches (1) and (2). Resistors, capacitors, diodes and dependent current or voltage sources are commonly used as elements of the model.

Detailed discussion on these approaches for transistor modeling are available in several references [14, 17, 18, 19, 20].

8.3 JUNCTION FIELD EFFECT TRANSISTORS

8.3.1 Introduction

The field effect transistor is a much simpler device than the bipolar transistor, at least in as much as determination of its characteristics using first order theory is concerned. Indeed, it is also a much simpler device to study using numerical techniques. The main reason for this simplicity lies in the fact that the junction field effect transistor (we shall not be concerned with the insulated gate or MOS structure here) is essentially a majority carrier device. The semiconductor equations are thus simplified due to both the fact that only one type of carrier need be considered and, since we are concerned only with majority carriers, the lifetime of the carriers does not enter into the equations.

Let us consider the structure shown in Fig. 8.9. We have a semiconductor bar with an abrupt junction on the top and bottom

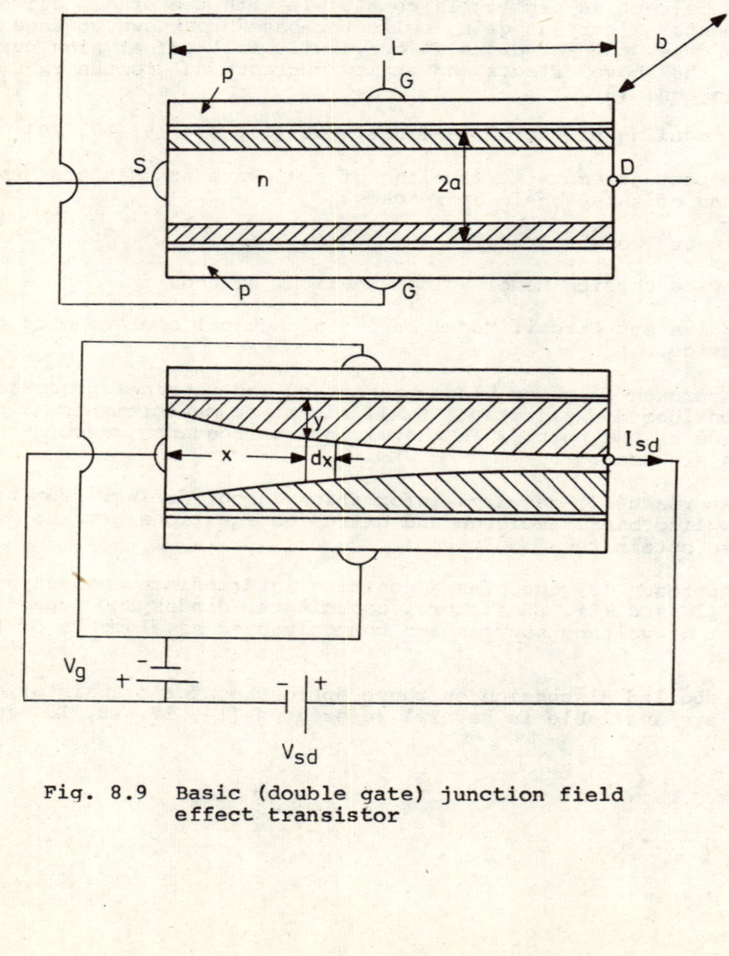

Fig. 8.9 Basic (double gate) junction field effect transistor

and a pair of contacts at either end. If the impurity concentration is sufficiently high (for a given channel width 2a) the bar will appear almost as a pure resistor for low currents (since the space charge regions will occupy only a very small fraction of the total width 2a).

However if we now apply bias to the structure as shown in the figure, and start increasing the voltage between the source and the drain contacts, the voltage across the two space charge layers will increase as we proceed from the source to the drain side of the bar. It is quite obvious that if the channel remains purely resistive, at a certain voltage the two space charge layers will meet at the drain side of the bar. In other words a high resistivity region will be encountered. This will cause the voltage current characteristic to be non-linear as shown in Figure 8.10. At low voltages the characteristic will be almost linear and will in fact behave as a resistor of value:

$$R_C = L/(2ab\sigma) \quad (8.40)$$

where b is the width (perpendicular to the plane of paper) and σ is the conductivity. It is also possible to write down by inspection the voltage V_p which will make the two space charge layers meet. This is analogous to the punch through voltage in diodes and transistors. We see that:

$$V_p = (q\ N_D\ a^2)/2\varepsilon \quad (8.41)$$

where N_D is the donor density in the semiconductor bar. This is referred to as the pinch off voltage.

In order to obtain some simple analytic expressions relating source-drain current and voltage to gate voltage we shall make some basic assumptions. These are the assumptions originally made by Shockley[21].

First, we shall assume that the boundary between the space charge layers and the neutral channel is abrupt. This is the same assumption that was made in treating the bipolar transistor.

Secondly, we shall assume that the channel width is changing slowly with respect to distance, x. This is the gradual channel approximation and leads to the results being valid for (relatively) large ratios of L/a.

In fact, this second assumption is effectively the same as saying that we are considering the field to be changing slowly in the x direction, i.e.

$$dE_x/dx \simeq 0 \quad (8.42)$$

This, stated in other words, means that we are assuming space charge neutrality in the "neutral" channel. It is analogous (although the analogy should not be pushed too far) to the quasi neutrality assumption we have been making up to now in dealing with other devices.

The third approximation concerns the mobility. In the region where the channel becomes highly resistive, the voltage drop per

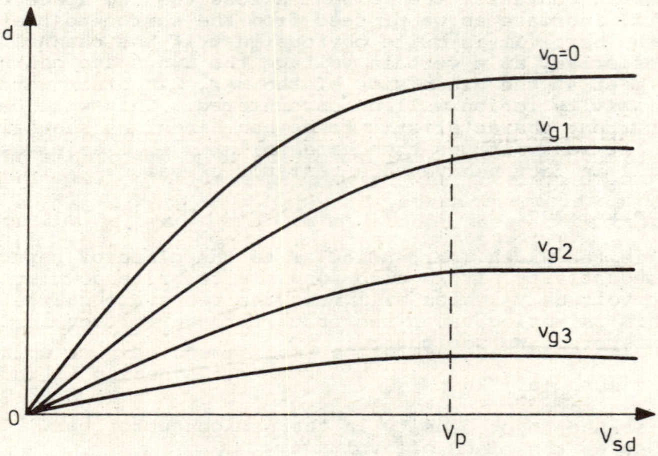

Fig. 8.10 Typical I-V characteristics

unit length, x, can become very large, i.e. the electric field E(x) can become large, (note that this is not the same as saying that the derivative dE/dx is large). In the simple analysis to follow, we shall assume that over the greater part of the channel, the mobility remains constant.

8.3.2 Theoretical expressions for the d.c. characteristics

The space charge width δ is given by the solution of Poisson's equation as:

$$\delta = \sqrt{(2\varepsilon\ V_j/qN_D)} \tag{8.43}$$

where V_j is the potential drop across the space charce layer. The resistance per unit length in the x direction is given by:

$$dR = dx/[2(a-y)\sigma b] \quad (8.44)$$

Neglecting the internal barrier potential we can relate the space charge layer voltage V_j at any point x to the gate voltage and the resistive voltage drop in the conducting channel:

$$V_j = V_g + I_{sd} R_x \quad (8.45)$$

$$= V_g + I_{sd} \int_o^x dR \quad (8.46)$$

$$= V_g + (I_{sd}/2b) \int_o^x [1/(a-y)\sigma]\, dx \quad (8.47)$$

where V_g is the gate voltage and I_{sd} is the source drain current.

Combining eqns.(8.47) and (8.43) and differentiating we obtain an expression for the source-drain current I_{sd}:

$$I_{sd} = V_{sd}\ (2bqaN_D\mu_n/L)[1-(2V_p/3V_{sd})\{(V_g+V_{sd})/V_p\}^{3/2} + (2V_p/3V_{sd})(V_2/V_1)^{3/2}] \quad (8.48)$$

Let us consider the case where the applied gate voltage V_2 is equal to zero. This gives a maximum drain current:

$$I_{sd} = V_{sd}\ 2bqaN_D\mu_n[1-2(V_{sd}/V_p)^{1/2}/3]/L \quad (8.49)$$

It is interesting to note that the derivative of I with respect to V_{sd} is exactly zero when V_{sd} is equal to the pinch off value given by eqn. (8.41), V_p.

This maximum, or saturated, value of I_{sd} is given by setting $V_p = V_g + V_{sd}$:

$$I_{sd}\ \text{(at pinch off)} = (V_p/3R_c)[1-3(V_g/V_p)+2(V_g+V_p)^{3/2}] \quad (8.50)$$

By differentiating this expression with respect to V_g we obtain the transconductance g_m of the device:

$$dI_{sd}/dV_g|_{\text{at pinch off}} = [\surd(V_g/V_p)-1]/R_c \quad (8.51)$$

The maximum value of the transconductance is seen to occur for an applied gate voltage of zero. This value is simply:

$$g_{m(max)} = 1/R_c = 2q\mu_n\ N_D b\ (\tfrac{a}{L}) \quad (8.52)$$

In other words we obtain the result that the maximum value of the transconductance is the reciprocal of the open circuit (zero bias) channel resistance [21]. We also see from eqn.(8.52) that the trans-conductance (which determines the gain of the device at low frequencies) is proportional to the ratio a/L. This result implies that if we alter the dimensions of a device which satifies the gradual channel approximation in order to improve its performance, we will eventually reach a condition where the original approximation is no longer valid[22]. In fact the same is also true of the constant mobility assumption and it is especially this latter assumption which limits the usefulness of the results obtained above.

Figure 8.11 shows a simplified equivalent circuit representation of the transistor. The capacitances may be determined by integrating the depletion layer width over the whole channel length. Note that the capacitance C_{gd} is comparable to (e.g. one third) C_{gs}.

8.3.3 Charge analysis and high field conditions

Charge method of analysis

We have seen that although it was possible to derive simple analytic expressions for the JFET, these were based on some very restrictive assumptions, the most important of which concerned the mobility. It is in fact quite easy to derive a very general expression relating transconductance, g_m, gate capacitance, c_g, and channel transit time τ_{sd}.

Let us consider the total charge in the channel associated with the flow of current. From the definition of drift current in a base of length L and section A, with a voltage V applied, we can always write:

$$I = V.A\sigma/L \tag{8.53}$$

where

$$\sigma = q\,\mu_n\,N_D \tag{8.54}$$

$$= q\,N_D\,v/E \tag{8.55}$$

where v is the velocity of the carriers and E the electric field. Writing the conductivity in this form will enable us later to consider the case where the mobility is not constant. The current I, is thus given by:

$$I = (q\,N_D\,AL)\ (v/L) \tag{8.56}$$

$$= Q/t \tag{8.57}$$

Where Q is the total charge associated with the transport of carriers and t the total time they take to travel the length L.

Applying this formula to the field effect transistor and calling the total transit time in the channel τ_{sd}, we can write the change in charge ΔQ corresponding to a given change in source drain current I_{sd}:

$$\Delta Q = \tau_{sd}\,\Delta I_{sd} \tag{8.58}$$

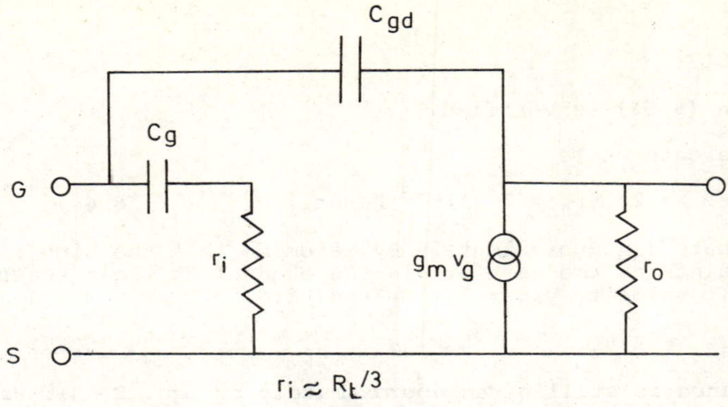

Fig. 8.11 Small signal equivalent circuit

The change in Q is brought about by varying the dimensions of the conducting channel due to application of a change ΔV_g in the gate voltage. Now, based on the fact that a change in the majority carrier charge Q in the conducting channel is compensated on a one to one basis by a change in the number of ionized impurity atoms in the space charge layer, we can write the gate capacitance C_g as:

$$C_g = \Delta Q / \Delta V_g \tag{8.59}$$

Combining eqns. (8.58) and (8.59) we obtain a very simple and basic relationship for the transconductance g_m

$$\begin{aligned} g_m &= \Delta I_{sd}/\Delta V_g \\ &= C_g/\tau_{sd} \end{aligned} \tag{8.60}$$

Note that eqn. (8.60) is valid for any field condition and hence for any mobility. It assumes only that the JFET is divided into space charge and quasi neutral regions.

8.3.4 Application of the general formula

The gradual channel approximation is valid for

$$\sigma a^2/L \ll 2\varepsilon v_{th} \simeq 2 \times 10^{-5} \text{ F/sec.} \tag{8.61}$$

Under this condition the transit time is given by:

$$\tau_{sd} \simeq \int_0^L dx/v \simeq L^2/\mu V_p \tag{8.62}$$

where V_p is the pinch off voltage. The gate-source capacitance is given by

$$C_g \simeq 4\varepsilon bL/a \tag{8.63}$$

and equation (8.52) is verified.

For the case where

$$\sigma\, a^2/L >> 2\varepsilon\, v_{th} \simeq 2 \times 10^{-5} \text{ F/sec.} \tag{8.64}$$

we can evaluate τ_{sd} approximately by assuming that the high field condition maintains the carriers in the channel at their scatter limited drift velocity $v_{th} \simeq 10^7$ cm/sec. i.e.

$$\tau_{sd} \simeq L/v_{th} \tag{8.65}$$

The capacitance is still given approximately by eqn.(8.63). We see therefore that the transconductance g_m given by eqn.(8.60) increases with increasing 'a' for small values of a^2/L but decreases with 'a' for large values of a^2/L. The maximum value of g_m will occur for[24]:

$$a^2/L \simeq 2\varepsilon\, v_{th} = 2 \times 10^{-5} \text{ F/sec.} \tag{8.66}$$

This condition would therefore appear to represent an optimum criterion for JFET design.

The cut-off frequency of the JFET is determined by the gate capacitance-channel resistance time constant.

$$f_c = (2\Pi\ R\ C_g)^{-1} \tag{8.67}$$

$$\simeq [4\Pi\ (\varepsilon/\sigma)\ (L/a)^2]^{-1} \tag{8.68}$$

$$\simeq 1/2\Pi\ \tau_{sd} \tag{8.69}$$

The maximum frequency of oscillation is related directly to f_c but involves parasitic elements[23], since the device described by the equivalent circuit of Fig. 8.11 is unstable at all frequencies, unlike the bipolar transistor. In practice, f_{max} for the JFET may be 2 to 4 times f_c.

8.3.5 Practical JFET structures

Figure 8.12 shows a simple structure using epitaxial silicon. p type silicon is grown on an n substrate and this substrate forms one gate. The second gate (which will normally form the input) is formed by effecting a diffusion of donor atoms as shown. The source and drain contacts are formed in the same manner as the base contacts of a bipolar transistor.

The second structure shown in Fig. 8.12 is widely used for GHz JFETs. It consists of a moderately doped p (or n)epitaxial layer on a highly resistive substrate. The gate is formed by either a Schottky junction or a shallow diffused n (or p) layer. Using standard mask techniques, channel lengths less than 5 microns are quite feasible. A 5 micron value of 'L' with an optimum geometry as indicated in eqn.(8.66), gives for 1 ohm cm material, a=1 micron, f_c = 3 GHz with a transconductance of about 0.1 Amps/ **Volt per cm of gate width. It is important to realise, however,**

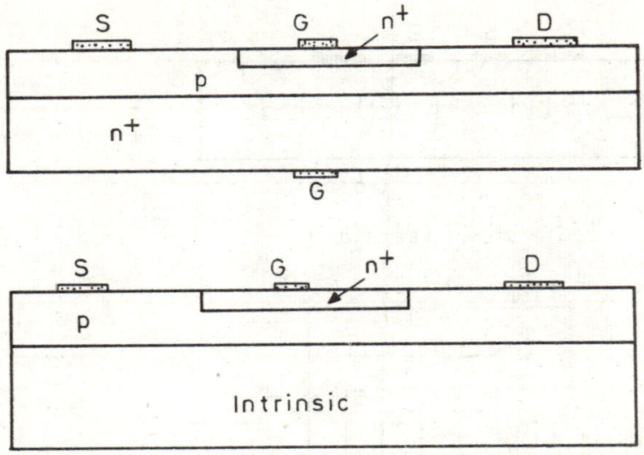

Fig. 8.12 Two microwave JFETs

that the gate cannot be infinitely long (wide) in the 'b' direction (perpendicular to the page). The finite resistance limits the value of b. A value of about 30 microns for a 1 micron value of L is typical for a diffused gate. This places a serious restriction on the power handling capabilities of this type of structure, since it is not possible to connect several devices in parallel using standard planar technology; one of the three contacts would have to cross the other.

The structure shown in Figure 8.13 the 'buried gate' JFET overcomes this disadvantage. The gates are first diffused into an epitaxially grown layer of n type silicon, to a distance of less than one micron. A layer of n type silicon is then grown to a depth of about 2 microns. As the n layer is grown, the p atoms in the gates relocate in the new epitaxial layer, giving a final gate geometry roughly as shown. Connection is made to the buried gates by diffusing a 'wall' of p^+ as shown in Fig.8.13 The source and drain contacts are formed in the normal way. Fig. 8.13 also shows the final metalization pattern, for a typical device.

Above a few GHz, GaAs is used instead of Si due to its higher mobility (eqn.8.62) giving a higher cut-off frequency.

8.4 VARACTOR DIODES AND FREQUENCY MULTIPLIERS

8.4.1 Introduction

The nonlinear capacitance characteristics of reverse bias

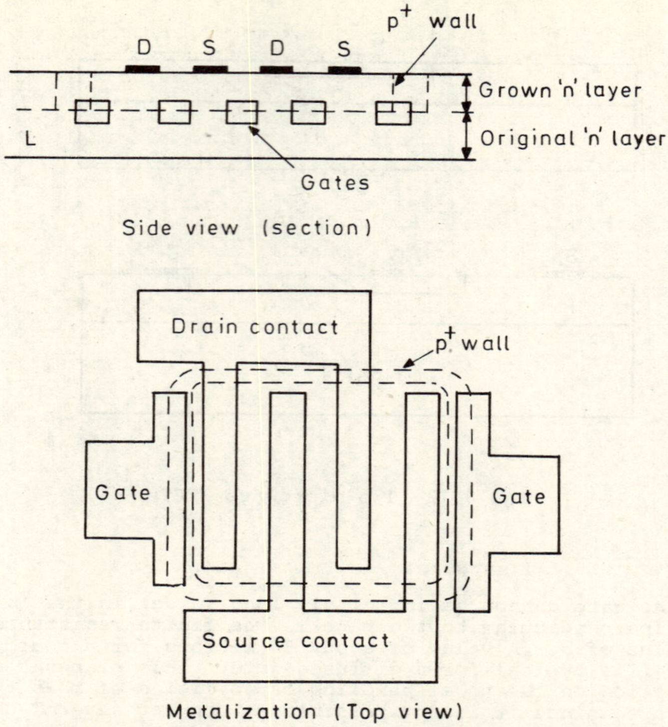

Fig. 8.13 Buried gate microwave power JFET

p-n junctions have been used since about 1960 as a means of obtaining r.f. power at frequencies considerably higher than the oscillation frequency of solid state sources such as transistor oscillators and, more recently, avalanche and Gunn devices. The depletion layer capacitance-voltage nonlinearity was used exclusively at the beginning. It was subsequently realised that driving the diode into forward conduction could, under certain conditions, increase the power handling capacity of the diode with little change in the power conversion efficiency, the effect being used in this case is the charge storage/step recovery effect. Present day frequency multipliers can be divided into two main categories:

- varactor frequency multiplier for x2 or x3 mainly

- step-recovery diode frequency multiplier for x4 and higher orders.

In the first category, the varactor can be either nominally driven[25] (pure varactor effect) or overdriven[29] (some forward conduction during each cycle, but not more than 180° in general). In the second category the name 'varactor multiplier' is sometimes erroneously applied, but in fact, the multiplication mechanism is almost entirely due to the fast recovery of minority carriers and subsequent sudden drop in diode current, after nearly 360° of conduction at the input frequency.

Both devices are, of course, $p^+ - n' - n^+$ structures. The only difference lies in the centre layer width and doping level. It is important to realise however that the step-recovery mechanism (to be dealt with in section 8.6) requires a carefully controlled impurity profile and that the input frequency must be greater than a minimum value determined by the minority carrier lifetime. The high order step-recovery diode multiplier is also more difficult to align and is more susceptible to instabilities than the pure varactor multiplier.

The operation of the pure varactor multiplier is limited only by the RC cut-off frequency of the diode, where R is the series resistance determined by the doping level and centre layer width, and C is the depletion layer capacitance. It is thus easier to fabricate, in general, a 'good' varactor than it is to fabricate a 'good' step-recovery diode. Since high frequencies **imply small dimensions, it is probable that technological limitations** will force the varactor to continue to be used in many cases, and in all cases where X2 multiplication only, is required (remember that higher order multiplication can always be obtained **by cascading two or more X2 circuits; 'idler' techniques can also** be used but tend to be tricky to set up).

8.4.2 Properties of the varactor frequency multiplier

The basic circuit, shown in Figure 8.14, consists of an input filter and matching transformer, an output circuit of the **same type, and the varactor coupling the input to the output. Either the series or shunt form may be used for the doubler; in** the pure varactor mode the results are essentially the same.

Consider the nonlinear capacitance-voltage characteristic of the diode:

$$c(v) = C_o \, (1 + v/V_B)^{-\gamma} \tag{8.70}$$

where C_o is the capacitance at a reverse bias V_B (which includes the built in barrier potential of about 0.6V for silicon) and γ is 1/2 for an abrupt junction and 1/3 for a linearly graded junction. If the total h.f. voltage across the diode consists of the fundamental and second harmonics as determined by the filters **at ω_1 and $2\omega_1$ eqn. (8.70) may be expanded in a Taylor series about V_B. We have also the relation:**

$$i = dq/dt = (dq/dv)/(dv/dt) = c(v)\, dv/dt \tag{8.71}$$

where $v = V_1 \sin \omega t + V_2 \sin (2\omega t + \phi_2)$ (8.72)

It is possible to combine Eqns. (8.70), (8.71) and (8.72) and

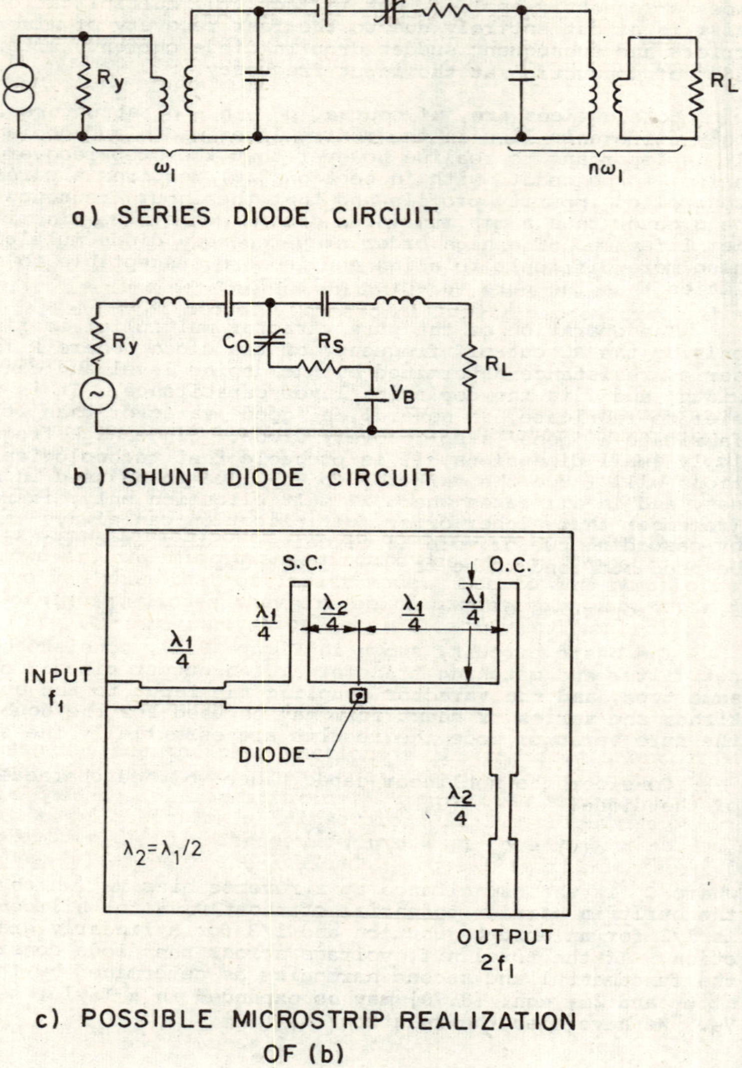

Fig. 8.14 Basic varactor diode frequency multiplier circuits (a and b) and a typical microstrip version (c) of the shunt diode circuit

select the terms giving rise to currents at ω_1 and $2\omega_1$. Applying Kirchoff's laws at the output node leads to the result that maximum output voltage is obtained for $\phi 2 = 0$ and the voltage 'gain' of the multiplier V_2/V_1 is given by:

$$V_2/V_1 = (\gamma/2)\ (V_1/V_B)\ \omega_1\ C_o/G_{2t} \tag{8.73}$$

where G_{2t} is the total shunt conductance at $2\omega_1$ (including the transformed diode loss due to R_5). The total power dissipated in G_{2t} is

$$W_{out} = V_2^{\ 2}\ G_{2t}/2 = (\gamma^2/8)(V_1/V_B)^2(\omega_1 C_o)^2 V_1^{\ 2}/G_{2t} \tag{8.74}$$

The maximum output power is limited by the diode breakdown voltage V_{br} and is obtained when $V_1 \approx V_{br}/2$ and $V_1/V_B \approx 1$. This gives:

$$W_{out} \approx (\gamma^2/32)\ (\omega_1 C_o)^2\ V_{br}^2/G_{2t} \tag{8.75}$$

The input conductance G_{eq} at ω_1 may be obtained in a similar way to eqn. (8.73) by applying Kirchoff's law at the input frequency:

$$G_{eq} = (\gamma/2)\ (V_1/V_B)\ \omega_1 C_o\ (V_2/V_1) \tag{8.76}$$

The power conversion efficiency η, may be defined as:

$$\eta = (W_{out} \text{ at } 2\omega_1\ /\ W_{available} \text{ at } \omega_1) \tag{8.77}$$

$$= (W_{out} - V_2^{\ 2}\ G_{2d}/2)\ /\ (W_{out} + V_1^{\ 2}\ G_{1d}/2) \tag{8.78}$$

where G_{1d} and G_{2d} are the equivalent shunt loss conductance of the diode at ω_1 and $2\omega_1$ respectively.

This may be shown to be optimum when the following condition is satisfied:

$$G_{2t} = \gamma(V_1/V_B)\ (\omega_1 C_o) \tag{8.79}$$

For this value of load the efficiency is simply:

$$\eta \approx \exp\{-8/[\gamma(\omega_c/\omega_1)(V_1/V_B)]\} \tag{8.80}$$

where $\omega_c = 1/R_s C_o$ (8.81)

Note that substituting eqn. (8.78) into eqn. (8.75) gives:

$$W_{out\ max} \approx (\gamma/32)\ V_{br}^{\ 2}\ \omega_1\ C_o \tag{8.82}$$

It is clear that although the power converted can be increased by increasing the value of the capacitance (i.e. the surface area of the diode), impedance considerations prevent indefinite increase of C_o, the operating point capacitance. The power output is thus limited ultimately by the breakdown voltage of the diode. In the overdriven mode[27, 28] using the charge storage effect, the power given by eqn.(8.74) can be increased by roughly an order of magnitude with little change in the efficiency.

8.4.3 Properties of the varactor diode

Assuming a uniformly doped centre layer, integration of Poisson's law gives; for an abrupt junction:

$$V = qNW^2/2\varepsilon \tag{8.83}$$

$$E = 2V/W \tag{8.84}$$

where W is the depletion layer width

V is the total depletion layer voltage

E is the maximum value of electric field (at the junction)

N is the doping level

Considering breakdown voltage conditions, we can write:

$$V_{br} = E_{br}^2 \varepsilon/2qn \simeq 10^{17}/N \text{ for Si.} \tag{8.85}$$

Also, assuming that the diode is designed so that punch-through occurs at V_{br} we have:

$$W = 2V_{br}/E_{br} = 10^{-5}\ V_{br} \tag{8.86}$$

As an example, for 1 ohm cm. material, V_{br} is approximately 10 volts and $W \simeq 1$ micron.

The RC cut-off frequency is given at $V \simeq V_{br}/2$, by:

$$\omega_c^{-1} = RC \simeq \rho\varepsilon \simeq V_{br}/10^{13} \tag{8.87}$$

Note that as a rough guide, at an output frequency equal to one tenth of the cut-off frequency, the efficiency is -3 dB or 50%, from eqn. (8.80).

A technique quite often used by manufacturers is to connect two varactors in series. For identical diodes, this will double the power handling capacity. This 'stacked pair' technique has the disadvantage of raising the thermal resistance since both diodes can not be grounded to a good heat sink.

8.4.4 Step-recovery diode multipliers

Figure 8.15 shows the step-recovery diode in a purely resistive circuit. It is easy to see that if the input frequency is higher than the reciprocal of the carrier lifetime, then nearly all the injected charge will be recovered. If, furthermore the centre layer width of the diode is narrow compared to a diffusion length (cf. section on PIN diodes), the recovery will be quite abrupt, as shown in the figure. Frequency multipliers using this type of circuit (with the addition of input and output filters/transformers) have been extensively studied in the literature (e.g. see[27]). Unfortunately, no simple analytical results exist. Instead of dwelling on this circuit, let us pass on to the more interesting (from both an academic and a practical viewpoint) inductive circuit shown in Figure 8.16[26, 30]. The operation of this circuit is explained by considering the waveforms shown in Fig. 8.17.

The diode switches on when

$$L\, di/dt = V_B \tag{8.88}$$

or, since

$$i_g = I_g \cos \omega t \tag{8.89}$$

we have:

$$\sin \omega t_1 = V_B/(\omega L I_g) \tag{8.90}$$

as the condition for switching on. The diode will remain forward biased until a time t_3 given by:

$$Q_d = \int_{t_1}^{t_3} i_d\, dt = 0 \tag{8.91}$$

where Q_d is the charge stored in the diode (notice that Q_d increases until time t_2, then decreases). If t_3 is slightly less than $t_1 + 2\pi$ as shown on Fig. 8.17, a sharp pulse will be generated across R (Fig. 8.16). The pulse risetime is determined by the step-recovery time of the diode in the circuit and the pulse decay time is determined by the LR time constant. Referring to Figure 8.17, it is obvious that if the power contained in the narrow pulse is filtered, a simple multiplier is obtained. This circuit is particularly attractive for high order (e.g. X 10) multiplication and efficiencies better than -10 dB per decade are obtainable with good diodes. Analytic results are not available for the general case (although[26] gives results for a specific case) but computer solutions indicate that even in the ideal case of an abrupt recovery spike, the efficiency is limited to a finite value (less than 100%). This type of circuit is very attractive from a stripline point of view when high power output is not a main requirement.

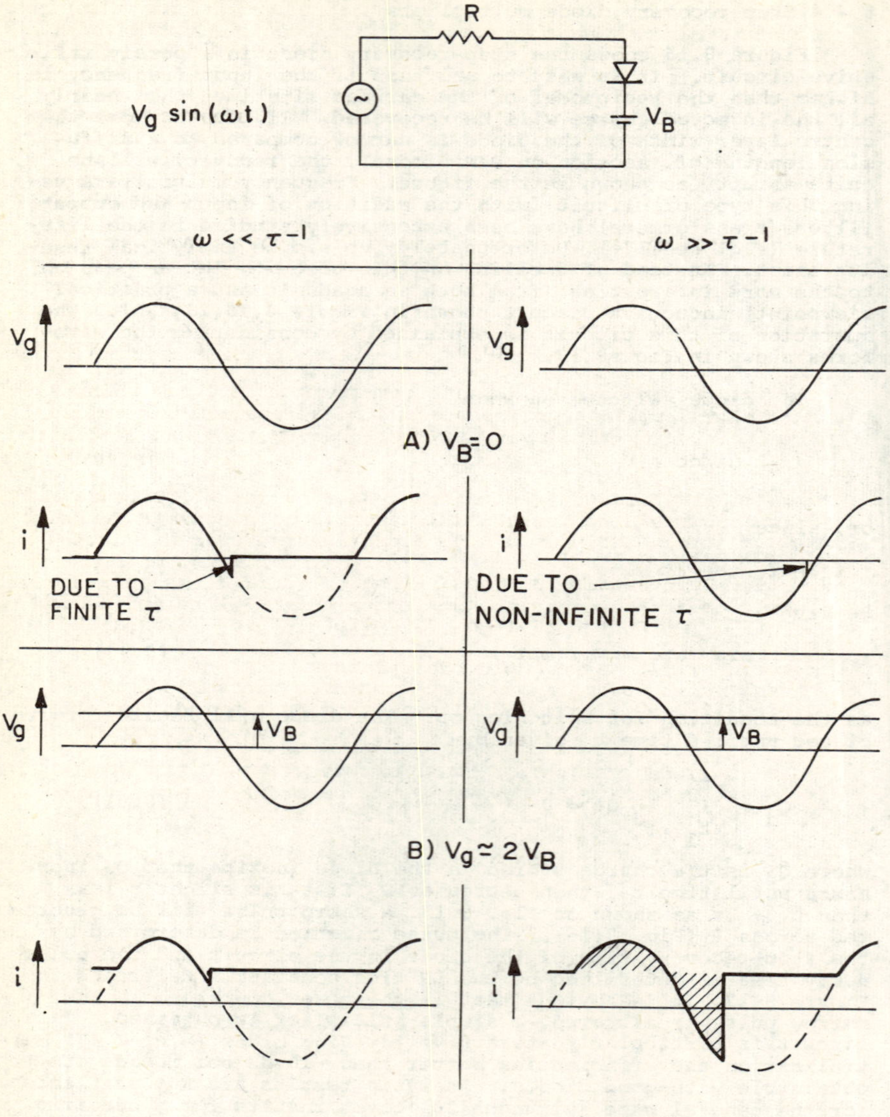

Fig. 8.15 Waveforms obtained with the charge storage step recovery diode in a resistive circuit at low frequencies and at high frequencies

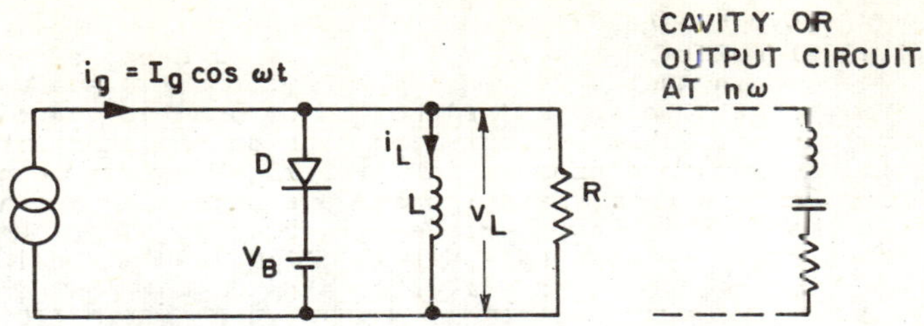

Fig. 8.16 The basic "pulse mode" or "inductive" step recovery diode frequency multiplier

A brief discussion of the properties of the step-recovery diode is given in Section 8.6.

8.5 PARAMETRIC DIODES AND AMPLIFIERS

8.5.1 Principle of operation

Like varactor diodes, parametric diodes use the non-linear depletion layer characteristic of a reverse (or zero) biased p-n junction. For parametric amplification, however, one is invariably dealing with small signal powers and the diode can therefore be highly doped on both sides of the junction, thus giving a very high cut-off frequency. Manley and Rowe (1958) showed in a very general way that a non-linear reactance, when driven from a sinusoidally varying voltage (or current) at a frequency f_p, with a signal applied at a frequency f_s less than f_p is capable of giving power gain if a current and voltage are allowed to exist at a frequency f_p-f_s or f_p+f_s. In the first case the difference frequency (referred to as the 'idler') voltage and current 'beat' with the 'pump' (f_p) to produce a negative resistance at the signal frequency (f_s); a negative resistance is also produced at the idler frequency. In the second case, the power available at the frequency $f_p + f_s$ is greater than the available signal power at f_s and the input resistance at f_s is positive. In the first case, the power gain can be infinite (i.e. the circuit can be made to oscillate); in the second case the power gain is limited to a value $(f_p+f_s)/f_s$.

In both the situations described, for an ideal non-linear reactance, the circuit behaves as an amplifier with no noise. The parametric diode used well below its cut-off frequency approaches an ideal non-linear reactance and is capable of giving stable gains of the order of 20 dB with noise temperatures as low as 20°K at room temperature (noise temperatures as low as 2°K have been reported in special applications[32]).

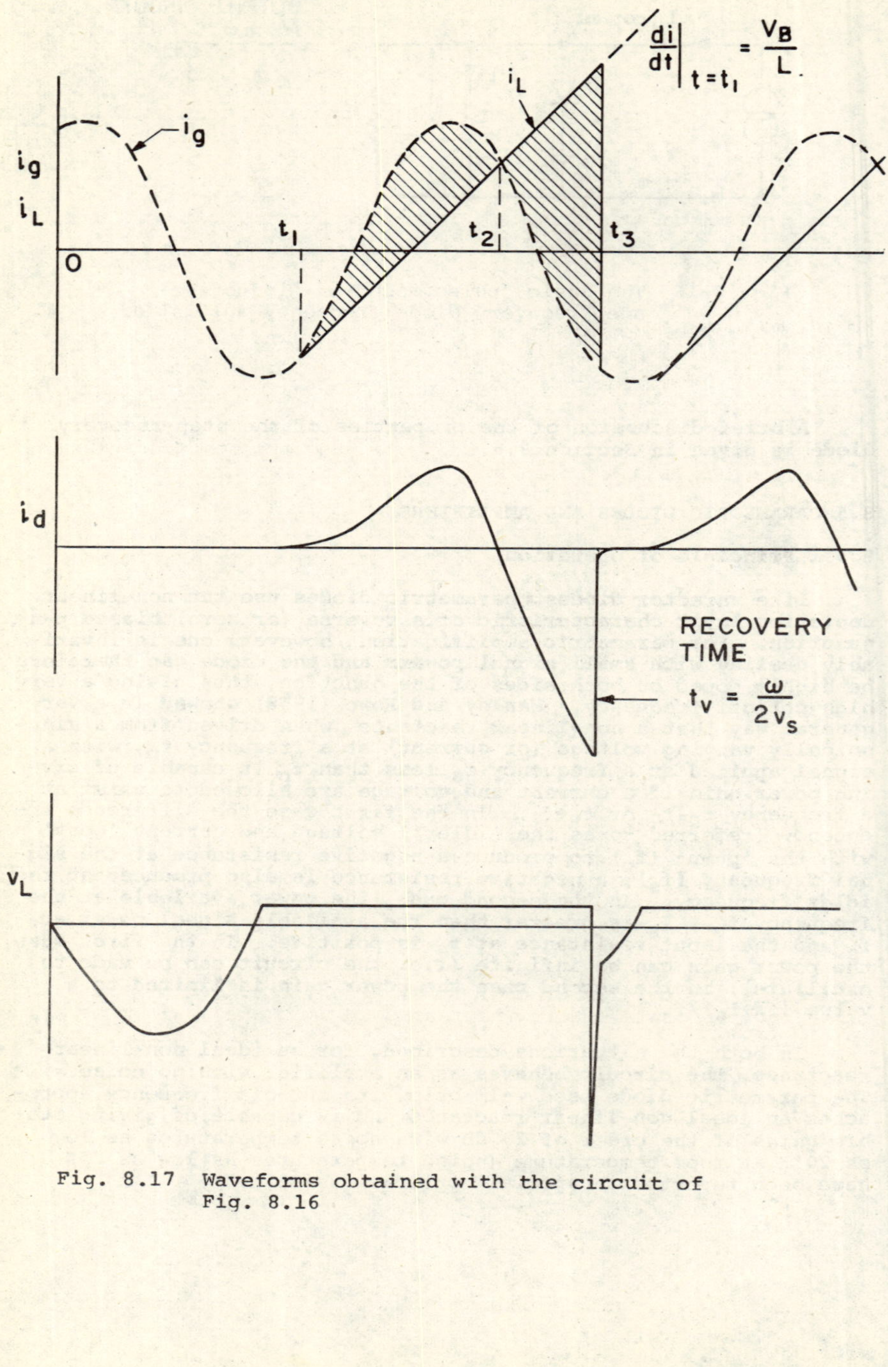

Fig. 8.17 Waveforms obtained with the circuit of Fig. 8.16

8.5.2 Circuit performance

Starting with eqns. (8.70) and (8.71), the basic parametric amplifier can be analysed in a manner very similar to the frequency multiplier, by separating the current into its three components, at the signal frequency f_s, the idler frequency f_3 and the pump frequency $f_p = f_s + f_3$ (see [31]). The following equations apply to the series diode case shown in Fig.8.18 but using standard series to parallel transformations, the results are equally valid for the shunt diode case (as for the varactor doubler). For designing low noise amplifiers, we require the following relations [33]:

Input negative conductance

$$G_{eq} = (\gamma V_1/2V_B)\ (\omega_s C_o)^2 (\omega_3/\omega_s)/G_3 \quad (8.92)$$

$$= (\gamma V_1/2V_B)\ \omega_s C_o Q_3 \quad (8.93)$$

where Q_3 = Q factor at idler frequency = $\omega_c/\omega_3 = f_c/(f_p-f_s)$ and

$\omega_c = 1/(R_s C_o)$, neglecting circuit losses.

G_3 is the conductance at ω_3 due to diode loss. V_1, V_B, γ and C_o are as defined in Sec.8.4.

Noise figure

$$F = 1 + \omega_s\omega_3/[\omega_c^2\ (\gamma V_1/2V_B)^2] + \omega_s/\omega_3 \quad (8.94)$$

Power gain

$$G = 4/(1-G_{eq}/G_g)^2 \quad (8.95)$$

For many applications, the bandwidth is also of importance. It can be shown that the 3 dB bandwidth at the signal frequency is given by:

$$\Delta f \simeq (Q_1Q_3/\sqrt{G})/(Q_1Q_3/B_1 + n^2 C_a Q_1 Q_3/C_o B_3) \quad (8.96)$$

$$\simeq B_1B_3/[\sqrt{G}(B_1+B_3)] \quad (8.97)$$

$$n = 1-\omega_s^2 L_s C_o$$

where Q_1 and Q_3 are the diode Q factors at the signal and idler frequencies respectively, B_1 and B_3 are the circuit bandwidths at the two frequencies with G = 1, L_s is the series inductance and C_a the case capacitance.

The effect of C_a and L_s on the effective input conductance are discussed in Sec.10.3.2.

For most applications the diode will be self biased with

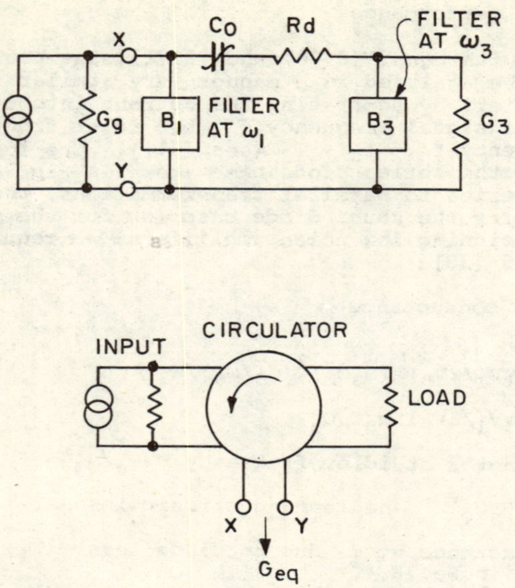

Fig. 8.18 (a) Basic parametric amplifier circuit
(b) Position of circulator

an appropriate value of bias resistor, and the peak pump voltage V_1 will be approximately equal to the bias voltage V_B. The value of γ is 1/2 for abrupt junctions and 1/3 for linearly graded junctions. Its value can be readily obtained by plotting log C versus log V, where V is the total voltage (sum of terminal voltage plus built-in barrier potential).

It should be noted that for a given diode and signal frequency, there is an optimum pump frequency giving the minimum noise figure. At very high GHz frequencies, however, it must be realised that a convenient power source may not be available for this frequency. In fact, it can be shown quite simply that the pump power required at f_p is :

$$W_{pump} \simeq (V_B{}^2/2)\ (\omega_p C_o)^2 R_s \simeq (V_B{}^2/2)\ \ \omega_p C_o (\omega_p/\omega_c) \qquad (8.98)$$

Note that in practice, the negative conductance G_{eq} will be nearly equal to the generator conductance referred to the diode terminals at the signal frequency (this being the condition for high gain). It may also be noted that the optimum pump frequency is given by:

$$f_3 = (\gamma/2) f_c$$

$$f_p = f_s + f_3$$

The corresponding minimum value of noise figure is:

$$F_{opt} = 1 + 4f_s/(\gamma f_c)$$

Fortunately, the curve of F versus f_p is fairly flat in the region of the minimum, and frequencies less than half the optimum can be used with an increase only of the order of 0.1 dB in the noise figure. As an example, consider an abrupt junction parametric diode with $C_o = 1$ pF, $R_d = 0.6$ ohm, i.e. $f_c = 100$ GHz. For a signal frequency of 1 GHz, the optimum pump frequency is 26 GHz and the corresponding noise figure is 1.08 or a noise temperature of 24°K (0.33 dB). For a pump frequency of 10 GHz and the same signal frequency of 1 GHz, the noise figure rises to 1.12 or a noise temperature of 35°K (0.5 dB).

Parametric diode construction

In order to have a low series resistance, a high doping level is obviously necessary. It is difficult to obtain abrupt junctions in this case using standard diffusion techniques. High frequency parametric diodes will therefore normally have an approximately graded junction characteristic with $\gamma \simeq 1/3$. Mesa structures are normally used to reduce capacitance due to edge effects to a minimum. Details are given in Section 8.6 (on PIN diodes). The parametric diode is characterised by its very narrow centre layer and breakdown voltage.

8.6 PIN DIODES

8.6.1 Introduction

In this section we shall consider the theoretical behaviour of the generalised PIN structure which finds many microwave applications [34, 40].

It should be noted at this point that the PIN diode is identical to a p^+-n-n^+ diode operating under high level injection conditions, i.e. with a free carrier concentration in the centre region which is much greater than the background impurity concentration. In fact, most so-called PIN diodes have a residual impurity concentration (p or n type) of the order of $10^{14}/cm^3$. At practical operating current densities this concentration is completely masked by the injected electron and holes as shown in Fig. 8.19.

Furthermore, it should be noted that in the following theory, we shall assume abrupt junctions at both ends of the centre layer as shown in Fig. 8.19. In fact, this assumption turns out to be perfectly valid in all cases where the centre layer is wide (compared to the diffused layer).

Finally, it should be noted that the following treatment for

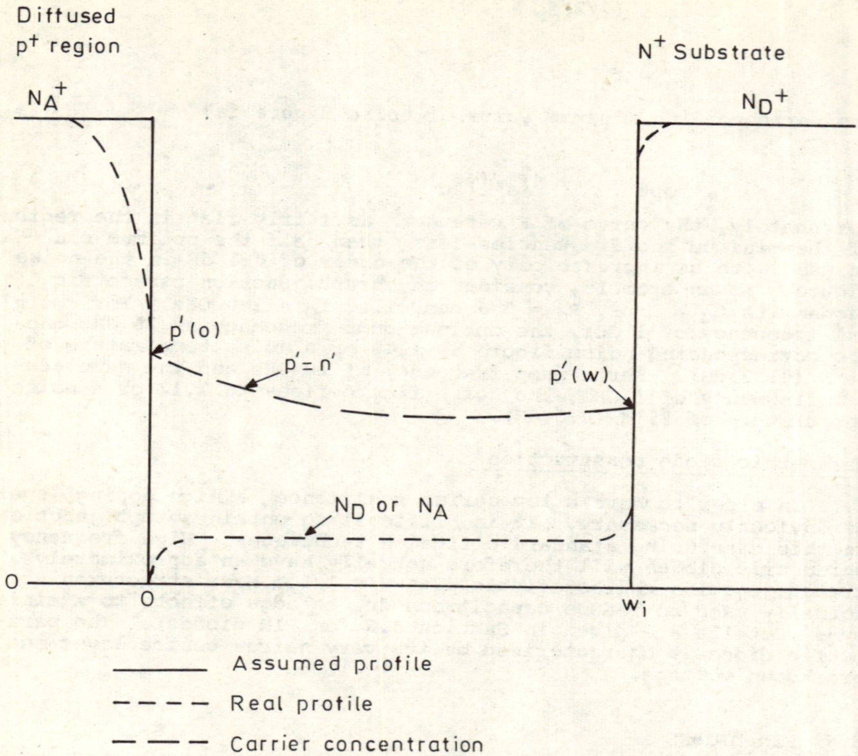

Fig. 8.19 General PIN structure

the generalised PIN structure covers many different applications. By altering the dimensions and residual centre layer doping level, the structure can be optimised for various microwave applications: PIN diode switch, PIN diode modulator, PIN diode limiter, step-recovery (snap-off) diode (for harmonic generation and frequency division), varactor diode. It is worth noting that the theory of the microwave PIN structure is identical to that for low frequency power rectifiers[35, 38].

We shall deal firstly with the forward bias steady state behaviour, secondly with the 'slow' transient turn-on and turn-off behaviour. The very fast turn-off transient behaviour will be treated at the beginning of the next section.

8.6.2 Steady state behaviour

In the centre layer of the PIN structure we have, at every point x, p = n. The continuity equation for electrons:

$$dJ_n/dx = qn/\tau_n \tag{8.99}$$

may be combined with the continuity equation for holes:

$$dJ_p/dx = -qp/\tau_p \tag{8.100}$$

with $\tau_p = \tau_n = \tau_{no} + \tau_{po} = \tau$ (8.101)

Equation (8.101) is obtained from the Shockley Read recombination equation.

After substituting the expressions for hole drift and diffusion current into eqns. (8.99) and (8.100) and setting n = p, we obtain :

$$p/\tau = D\, d^2p/dx^2 \tag{8.102}$$

where D is the ambipolar diffusion constant given by:

$$D = 2D_pD_n/(D_p+D_n) \tag{8.103}$$

It may be shown that the current crossing the p-i and n-i junctions are entirely hole and electron currents respectively provided that the p and n regions are highly doped.

We may therefore equate the electron current to zero at the p-i junction and obtain:

$$dp/dx|_{x=0} = -J_F/2qD_p \tag{8.104}$$

and we may equate the hole current to zero at the n-i junction to obtain in a similar manner:

$$dp/dx|_{x=w_i} = J_F/2qD_n \tag{8.105}$$

With these boundary conditions, the solution of eqn. (8.102) can be easily shown to be:

$$p = (LJ_F/2q)\ \{ \cosh(x/L)/[D_n \sinh(w_i/L)] + \cosh[(w_i-x)/L]/[D_p \sinh(w_i/L)] \} \quad (8.106)$$

for $w_i \ll L$, this simplifies to:

$$p = L^2 J_F/Dqw_i \quad (8.107)$$

This result is equivalent to saying that the carrier concentration diagram is 'rectangular', with a height p and a width w_i. The current density is:

$$J_F = Q/TA \quad (8.108)$$

where $Q = q\ A\ p\ w_i$ (8.109)

The relationship between p and the applied voltage V_a is given by applying the Boltzmann relations across each junction. Defining V_{aj} as the fraction of the applied (terminal) voltage occuring across each junction gives:

$$V_{aj(P-I)} = V_{aj(N-I)} = V_a/2 \quad (8.110)$$

The I-V characteristic is thus given by:

$$I = I_o[\exp(V_a/mV_t)-1] \quad (8.111)$$

where

$$I_o = q\ A\ n_i\ w_i/T \quad (8.112)$$

$$m = 2$$

$$V_t = .025\text{V at room temperature.} \quad (8.113)$$

Note that eqns. (8.111) and (8.112) are valid provided that the current I is greater than a value I_{HLI} corresponding to the onset of high level injection. I_{HLI} is given approximately by:

$$I_{HLI} = q\ A\ N_D\ w_i/\tau \quad (8.114)$$

where N_D is the residual background doping level (acceptor or donor) in the centre layer. Below this current, the device behaves as a p^+-n-n^+ structure. It can be shown that to a good approximation eqn. (8.111) is still valid if we redefine:

$$m = 1 \quad (8.115)$$

$$I_o' = q\ A\ n_i^2\ w_i/(\tau_{po} N_D) \quad (8.116)$$

The I-V characteristic is thus as shown on Fig. 8.20 (we have omitted discussion of space charge recombination at low forward bias, since this does not affect the high frequency high current operation of any of the microwave devices under discussion). It is useful to note the approximate relation:

$$I_o' \, I_{HLI} \simeq I_o^2 \tag{8.117}$$

Finally let us note that for steady state reverse bias, the solution to Poisson's law gives a breakdown voltage V_{br}:

$$V_{br} = (\varepsilon/2q)\ [E_{br}^2/N_D] \tag{8.118}$$

$$E_{br} \simeq 2.10^5 \text{ V/cm}$$

This holds only for the case where the value of V_{br} given by eqn. (8.118) is less than the punch through voltage V_{pt} where:

$$V_{pt} = q\ N_D\ w_i^2/2\varepsilon \tag{8.119}$$

For the case of light doping, the breakdown voltage will be given, to a good approximation, by:

$$V_{br}' \simeq E_{br}\ w_i \tag{8.120}$$

8.6.3 'Slow' transient turn-on and turn-off behaviour of PIN structures

We will restrict the following discussion to the common case of 'constant current' switching. Referring to the circuit of Fig. 8.21 we will consider firstly the application of a forward step of current density J_f, followed, after steady state conditions have been attained, by a reverse current step $-J_r$. The current density J_r will exist for the duration of the constant current storage phase t_s; after this time, an exponential decay of time constant t_r will be assumed. A detailed discussion of a fast decay which can occur under certain conditions will be discussed in Section 8.6.4.

Figure 8.22(a) shows the way in which the carrier concentration in the centre layer builds up as a function of time and distance[41]. The dotted curves represent the exact case. The solid diagrams represent a simple (but useful) approximation which will enable us to deduce some of the important transient characteristics of the diode. Due to the space charge neutrality condition, the hole and eletron concentrations are equal at every point, i.e. p = n. We can thus write, during the turn-on transient:

$$J_f = J_p = -2q\ D_p(dp/dx)|_{x=0} = 2q\ D_n(dn/dx)|_{x=w_i} \tag{8.121}$$

Considering, for simplicity, the case where $D_p = D_n$, it is apparent that for very short times $t \ll t_c$ after application of the current step J_f, the concentrations are approximately triangular in shape, as shown in Fig.8.22(a). The area under the curves represents the instantaneous stored charge, Q, which is also, for short times, given by:

$$Q \approx J_f t \tag{8.122}$$

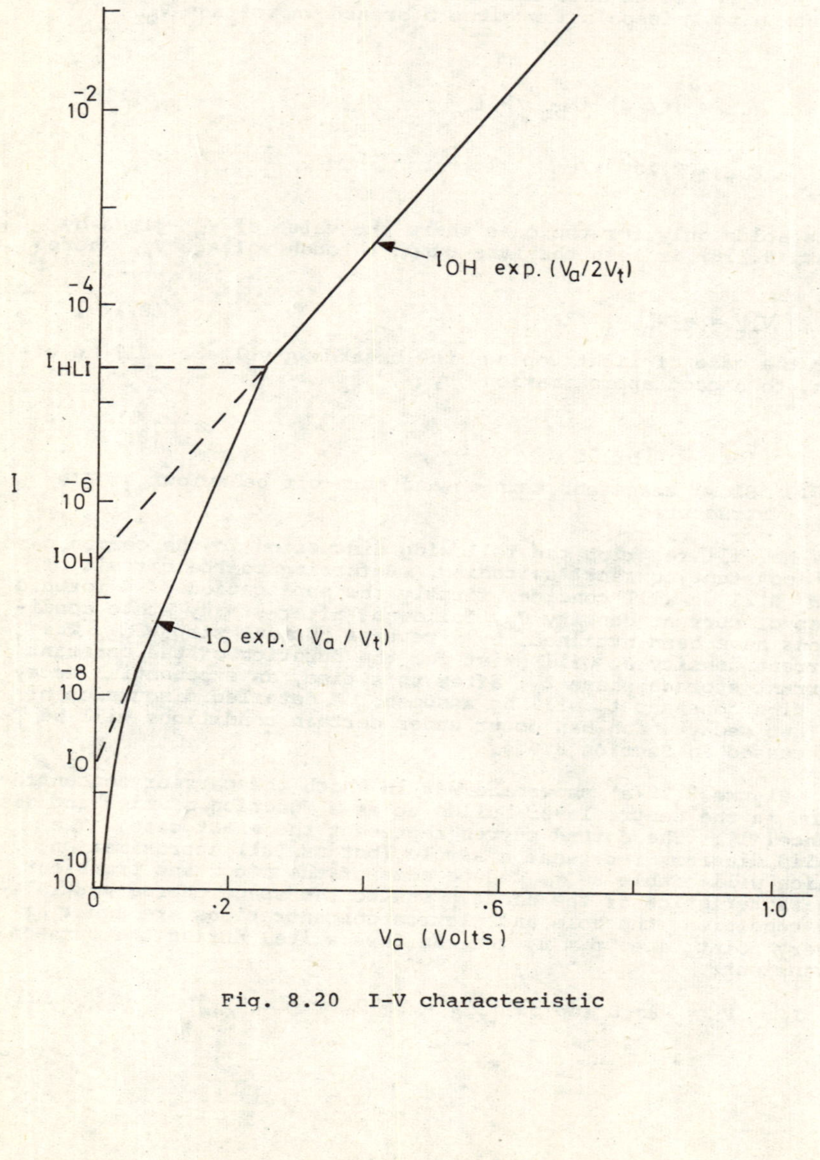

Fig. 8.20 I-V characteristic

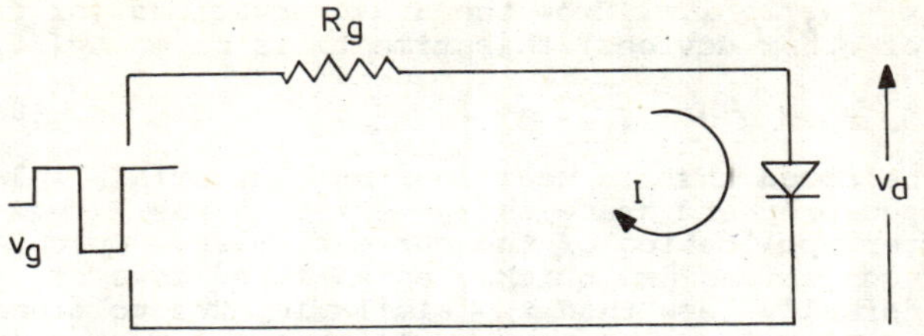

Fig. 8.21 "Current switched" circuit

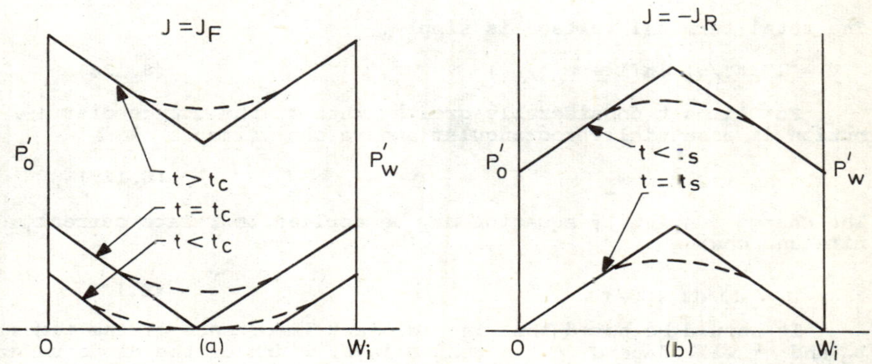

Fig. 8.22 Slow "turn-on"and "turn-off" transients

From the areas we have:

$$Q \simeq q\, p_o' x \tag{8.123}$$

Since the slope dp/dx is also given by p_o'/x, we have:

$$x = J_f t / q p_o' \tag{8.124}$$

It should be noted that these results are correct numerically to within a factor of about two. When the carrier concentration at $x = w_i/2$ starts to rise above zero, the behaviour will be different. Let us define this critical time as t_c. It is apparent

that t_c is simply the diffusion transit time of carriers from $x = 0$ to $x = w_i/2$. i.e. from the above equations (or from our knowledge of other devices) this time t_c is given by[36, 41].

$$t_c = w_i^2/8D \tag{8.125}$$

It should be noted that in near intrinsic material, a large transient voltage drop and space charge exist in the I region immediately after application of the current pulse. Space charge neutrality conditions are quickly established however, for times which are normally less than t_c; similarly, the voltage drop across the centre layer becomes small compared to the junction voltage drop in a time which is normally small compared to t_c. Apart from this initial perturbation, we can write, by combining eqns. (8.122) to (8.124):

$$p_o' = \sqrt{t}\ \sqrt{1/2D_p}\ J_f/q \quad \propto \sqrt{t} \tag{8.126}$$

$$= \sqrt{(QJ_f \ / \ 2D_p)}/q \quad \propto \sqrt{Q} \tag{8.127}$$

The total terminal voltage is simply:

$$V = (2\ KT/q)\ \ln(1 + p_o'/n_i) \tag{8.128}$$

For times t considerably greater than t_c the charge distribution is essentially rectangular and we can write:

$$Q \simeq q\ P_o'\ w_i \tag{8.129}$$

The charge continuity equation may be applied to relate current, time and charge:

$$J = dQ/dt + Q/\tau \tag{8.130}$$

It should be noted that the relevant importance of the times t_c and τ will depend on the centre layer width of the diode under consideration. It is useful to express the time t_c in terms of the lifetime τ and the carrier diffusion length L. From eqn. (8.129) we obtain:

$$t_c = \tau(w_i/L)^2/8 \tag{8.131}$$

Thus for a wide centre layer device where $L \simeq w_i$ eqns.(8.126) and (8.127) will describe the operation over most of the turn-on transient. For narrow centre layer devices (e.g. step-recovery diodes) on the other hand t_c will be much less than the lifetime τ and eqns.(8.129) and (8.130) may be applied over most of the turn-on transient. Note that in all cases the steady state forward current is reached for times t greater than the lifetime τ:

$$J_f = qp_o'(t\to\infty)\ w_i/\tau \tag{8.132}$$

For the reverse current step[37], for a complete treatment applied after steady state conditions have been attained, we replace J_f by $-J_r$ in (8.127). After a time t_s (measured from the instant at which the current is reversed) defined as the time at which the concentration becomes zero at either junction and given to a good approximation by solving the charge continuity equation with the condition Q equal to 0 at $t = t_s$.

$$t_s = \tau \ln (1 + I_f/I_r) \tag{8.133}$$

Equation (8.121) now becomes:

$$-J_n = -2qD_p(dp/dx)|_{x=0} = 2qD_n(dn/dx)|_{x=w_i} \tag{8.134}$$

The residual charge. approximated by the triangular distribution shown in Fig. 8.22(b) is given by:

$$Q_r \approx q\, p_{w/2}\, w_i/2 \tag{8.135}$$

With the assumption that all of this charge is evacuated during an exponential decay of the form:

$$J(t) = -J_r\, e^{-t/t_r} \tag{8.136}$$

it is apparent that the value of t_r is given by:

$$t_r = w_i^2/8D = t_c \tag{8.137}$$

To conclude this discussion, we see that the total turn-on and turn-off times are determined by the carrier lifetime τ. The charge continuity eqn. (8.130) may be combined with the voltage/carrier concentration relation (8.128) and (8.129):

$$Q = I_o\, \tau\, [e^{qV/mKT}-1] \tag{8.138}$$

This provides a useful model for the transient behaviour of PIN structures. For accurate results with wide centre layer widths, however, the more complicated square law behaviour described by (8.126) and (8.127) must be taken into account. Also, as mentioned above, the turn off transient of duration t_r can be followed by a very fast transient; this will be discussed in detail in the next section.

8.6.4 Fast turn-off transient

The PIN diode covers a very wide range of applications, as we have already mentioned. In nearly all of these applications the transient turn-off is of some importance. The PIN structure exhibits under certain conditions a rather complicated turn off behaviour towards the end of the "normal" decay time discussed previously (and equal to the diffusion transit time). In order to understand this phenomenon let us consider again the normal "resistive" switching circuit and waveforms shown in Fig.8.23.

Figure 8.23(c) splits the voltage waveform into several regions. The step 1-2 is due to the reversal of current in the finite diode series resistance. Region 2-3 corresponds to the constant current phase (note that point 2 and point 3 may be positive or negative depending on the magnitude of the reverse current and diode resistance). Region 3 to 4 corresponds to the

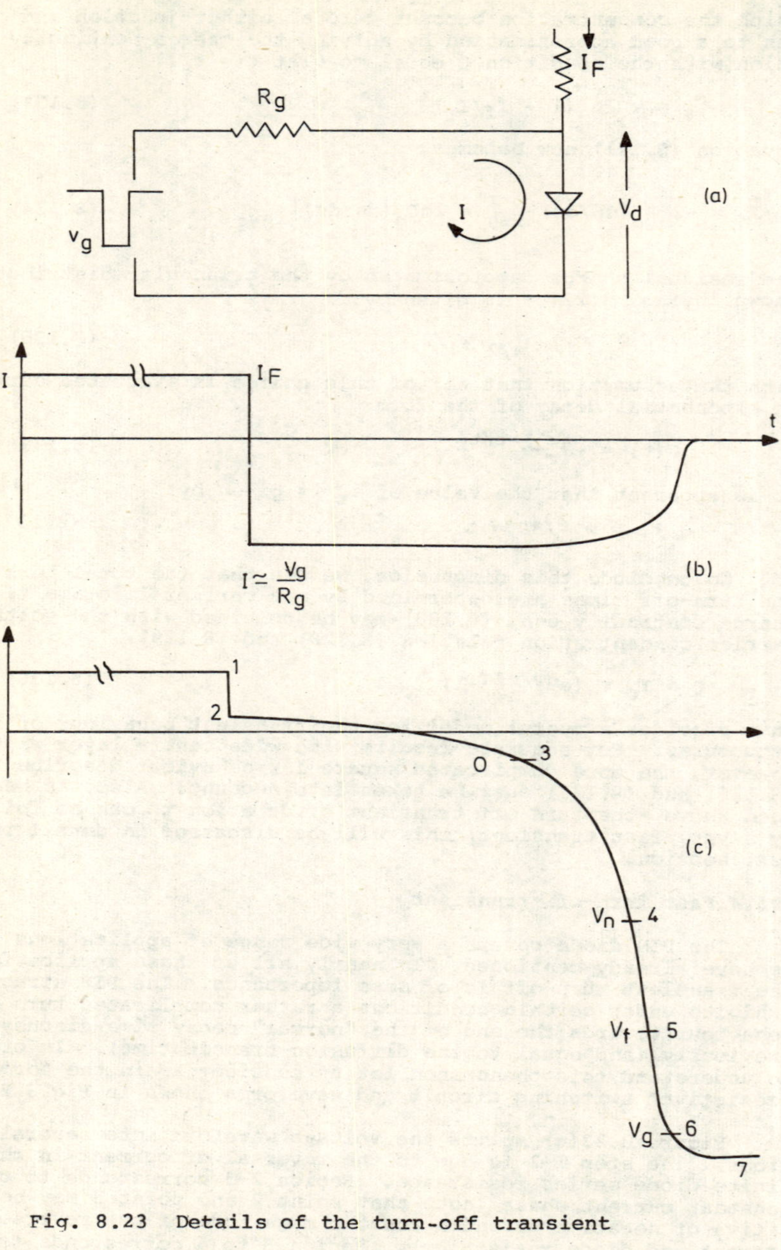

Fig. 8.23 Details of the turn-off transient

to the diffusion transit time, i.e. the evacuation of the residual charge. Let us examine this phase more closely than we did in sec. 8.6.3.

Figure 8.24(a) shows the excess carrier concentration diagram for the case of approximately constant current during this part of the decay phase. Whether or not the current can be maintained approximately constant will depend on the circuit and diode parameters (see below). Figure 8.24(b) shows the way in which the charge must build up in the space charge regions, which advance from each junction toward the centre, as time advances. Assuming that the field is sufficiently high for saturated drift velocity (v) conditions to exist, the space charge hole concentration must be given by[39].

$$p = I/(qAv) \tag{8.139}$$

At the end of the phase corresponding to the evacuation of the residual charge, i.e. after a time t_c:

$$t_c = w^2/8\,D \tag{8.140}$$

we have an approximately rectangular distribution of free carriers in the space charge region as shown in Fig. 24(b), '4'.

During this "slow" transient, Poisson's law in the space charge region in the left half is simply:

$$dE/dx = qp/\varepsilon \tag{8.141}$$

Integrating this twice gives the voltage across each space charge layer: Multiplying by two gives the total diode voltage:

$$V_d = x^2 I/(\varepsilon v A) \tag{8.142}$$

Notice that x and hence V_d are time dependent in eqn.(8.142). At the end of this phase (condition 4 in Fig.8.24), i.e. at a time $t = t_c$, we have

$$x = w/2 \tag{8.143}$$

and hence the voltage V_{dn} is given by

$$V_{dn} = w^2 I/(2\varepsilon v A) \tag{8.144}$$

This may be written in the form:

$$V_{dn} = wI/(4Cv) \tag{8.145}$$

where C is the capacitance of the diode for large reverse bias, i.e.

$$C = \varepsilon A/w \tag{8.146}$$

Notice that the current through the diode will be approximately constant during this "slow" transient if the generator voltage V_g is much greater than V_{dn}.

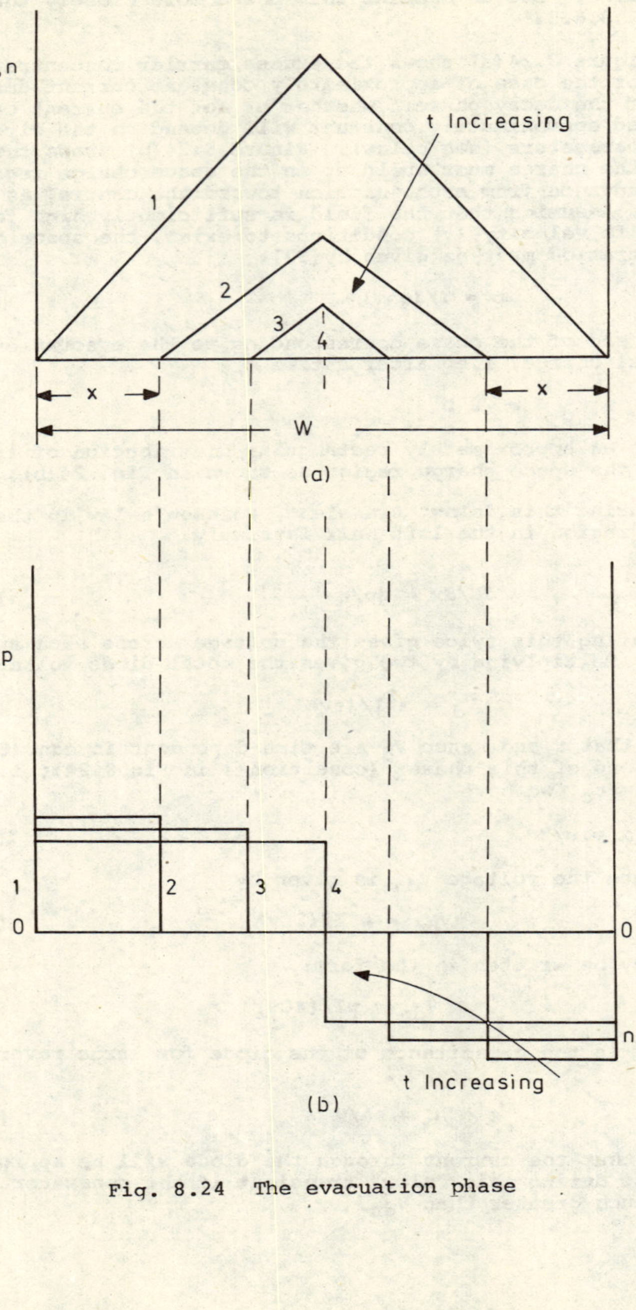

Fig. 8.24 The evacuation phase

The situation existing between points 4 and 5 of Fig. 8.23, is shown in Fig. 8.25. The free carriers in the space charge regions are extracted and the field builds up to a uniform value. The current during this phase at plane B' is a combination of particle current, I_{part}

$$I_{part} = qApv \tag{8.147}$$

and displacement current I_{disp}:

$$I_{disp} = A\varepsilon \frac{dE}{dt} \tag{8.148}$$

$$I = I_{disp} + I_{part} \tag{8.149}$$

The algebraic sum of these two currents must equal the terminal current. At the end of a time t_v (beyond the end of the slow transient t_c) the situation depicted in Fig. 8.25(c) will exist. All the free carriers will have been evacuated. Since these carriers are (we have assumed high field conditions) moving with their saturated drift velocity, the time t_v required to evacuate them from the space charge layers will be simply the drift transit time, i.e.

$$t_v = w/2v \tag{8.150}$$

From this time onwards (i.e. beyond point 5 in Fig. 8.23), the current is entirely displacement current and the rate of decay will be determined by the time constant:

$$t_d = R_g\, C \tag{8.151}$$

where R_g is the external circuit resistance and C (eqn.(8.146)), the diode capacitance.

The exact shape of the current decay and voltage rise beyond point 4 of Fig. 8.23 is relatively easy to determine by combining the preceding equations with Kirchoff's law relating the total (time dependent) diode voltage and current to the fixed generator voltage and circuit resistance.

It is intuitively obvious (and may be rigorously shown) that for the special case:

$$R_g\, C = w/2v \tag{8.152}$$

the fastest overall current decay, point 4 to point 6, Fig. 8.23, is obtained.

Normally there will be a rounding off the "tail" of the transient due to the fact that at some time during the transient the free carrier concentrations will fall below the doping level and the diode will cease to function as a PIN device.

8.6.5 Practical PIN devices

The PIN microwave switch (or modulator)

This device is designed so that for reverse bias, a high

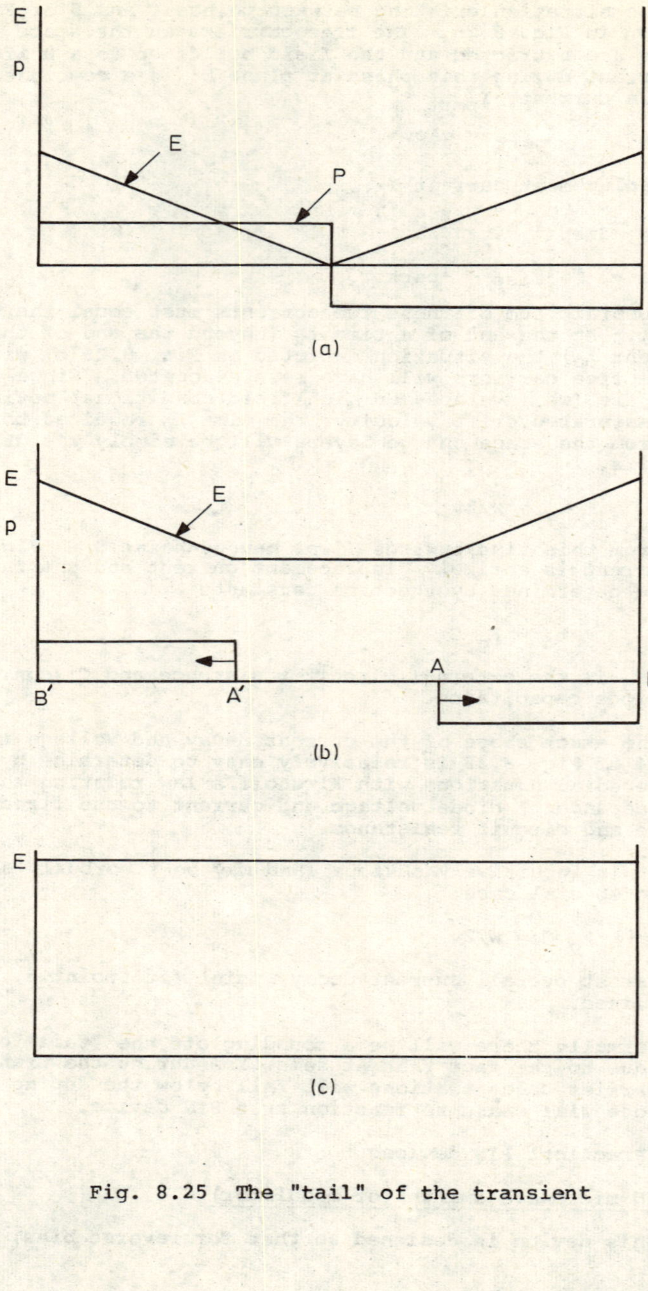

Fig. 8.25 The "tail" of the transient

(but capacitive) impedance is presented to the microwave signal and for forward bias, the diode represents essentially a very low resistance. Referring to Fig. 8.26(a) which shows a (microwave) signal superimposed on the d.c. bias, it is apparent that for the device to remain in the 'high impedance' state, in the presence of large microwave voltages, the d.c. reverse bias must be high. A high power PIN microwave switch is therefore characterised by a wide I region (which must not be much greater than a diffusion length L_d, however; otherwise the centre layer voltage drop will increase and the forward bias impedance will become high). Furthermore, in order that the reverse bias capacitance be low, the cross sectional area must be small; it must not be so small, however, as to increase the forward bias impedance.

The resistance of the I region with a d.c. current I flowing is determined by the conductivity ($q\mu_n n$) and width w_i of the centre layer. Refering to Fig.8.19, it is apparent that this can be written as:

$$R = (V_t/I)\ (w_i/L_d)^2 \qquad (8.153)$$

(since $R = w_i/q\mu_n n$) and $I = Q/\tau = qAw_i n/\tau$).

It can be shown that in the case of a high frequency signal, the impedance becomes modified due to the presence of capacitive and inductive terms. Note that this resistance is essentially in series with the 'conventional' parallel RC circuit consisting of the diffusion resistance (slope of the V-I curve) and diffusion capacitance.

For most practical purposes, at microwave frequencies eqn. (8.153) gives a sufficiently accurate representation and we see that the resistance is inversely proportional to the d.c. current flowing.

In the reverse bias condition, for lightly doped or intrinsic centre layers, the impedance is almost entirely capacitive with:

$$C = \varepsilon\ A/w_i \qquad (8.154)$$

Figure 8.26(b) shows the complete microwave equivalent circuit of a PIN diode switch. The additional small series resistance is due to finite substrate and contact resistance; it can, however, play an important part in determining the forward bias transfer characteristics of a circuit.

The turn-on and turn-off times are important parameters since they limit the ultimate switching speed of the diode. From the previous section, it can be seen that the turn-on time is given by:

$$t_c = w_i^2/8D \qquad (8.155)$$

and the turn-off time is given by:

$$t_{off} = \tau\ \ln(1+I_f/I_r) + t_c \qquad (8.156)$$

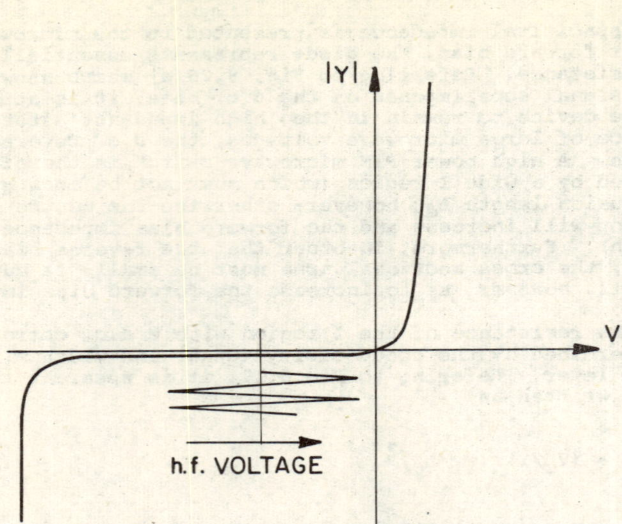

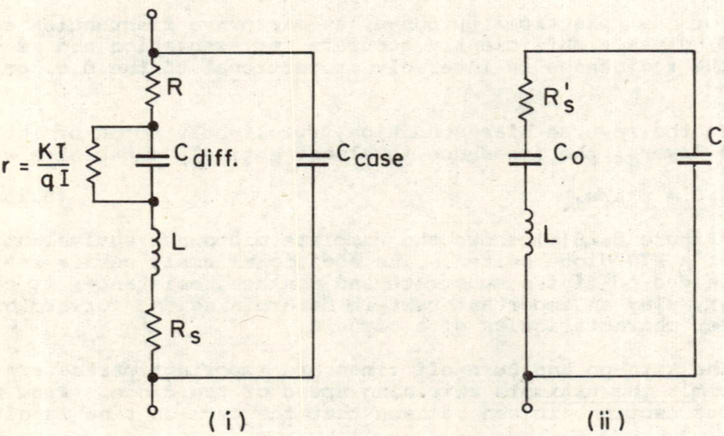

Fig. 8.26 PIN diode used as a microwave switch

In some applications, only the last part of this transient is of importance.

The PIN Diode limiter

The limiter diode is very similar to the switching diode. In the case of a limiter, it is the microwave signal itself which causes the diode to pass into the forward bias condition. Refering to Fig. 8.26(a), when the peak r.f. amplitude exceeds the bias voltage some charge is injected into the centre I region. This charge recombines, giving rise to a d.c. component of current. Representing the r.f. current by I_1, it can be shown that the resistance and d.c. component of current are given by:

$$R \simeq V_t w/(I_1\sqrt{D/\omega}) \tag{8.157}$$

$$I \simeq I_1 w/(2L\sqrt{\omega\tau}) \tag{8.158}$$

Breakdown voltage

It should be noted that in both the switch and limiter applications, the breakdown voltage will be given by:

$$V_{br} \simeq E_{br}\, w_i \tag{8.159}$$

For the wide centre layer structure, where $w_i \simeq L_d$ we see that the breakdown voltage is not independent of the carrier lifetime τ, since:

$$V_{br} \simeq E_c \sqrt{D\tau} \tag{8.160}$$

The PIN step-recovery diode

The step-recovery or snap-off diode is designed to give the fastest possible value of 'fast recovery time'. Since most harmonic generator applications require large power handling capabilities, a large reverse breakdown voltage is desirable. Since, from eqn. (8.150) $t_r = w_i/2v_{lim}$ and since eqn. (8.159) is fairly accurately valid, we see that there is a very direct relation between breakdown voltage and fast recovery time:

$$t_r = V_{br}/2\, E_c v \tag{8.161}$$

$$\simeq V_{br}/2.10^{12}$$

or, in other words, a fast recovery time of about 0.5 picoseconds per volt.

Power handling capabilities dictate a large cross section A, but this is limited by the fact that to obtain the value of t_r given above, the $R_g C$ time constant must be kept small eqn. (8.151).

The input frequency of a multiplier using a PIN step-recovery

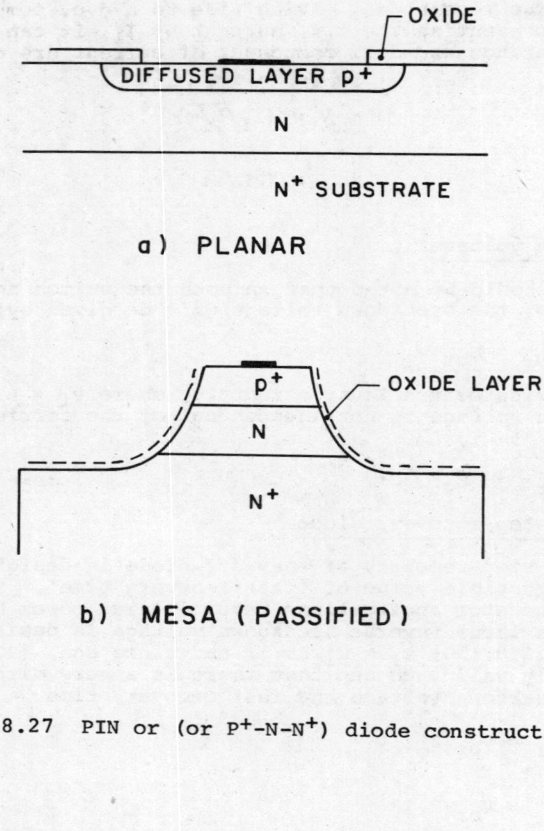

Fig. 8.27 PIN or (or P^+-N-N^+) diode construction

diode must be greater than $1/2\pi\tau$, otherwise a significant amount of stored charge will be lost during each input cycle and a detected current will result (with consequent reduction in power conversion efficiency). The maximum frequency multiplication ratio for efficient operation is thus given by:

$$n < \tau/t_r = 2\, v_{lim}\, \tau/w_i \qquad (8.162)$$

Construction of PIN diodes

Figure 8.27(a) shows the conventional planar construction, and should be self-explicit. The planar structure has the disadvantage however of adding "unwanted" capacitance at the periphery of the diffused region. High frequency current passing through this 'edge' capacitance will have to pass through the relatively high resistance of the surrounding I region. This series R-C circuit causes losses at high frequencies, which can be severe in low area diodes.

The mesa structure shown in Fig. 8.27(b) eliminates these deleterious edge effects completely. In its passivated form, surface recombination is reduced to a small value. This appears to be the best small area structure available.

Very high power PIN devices with breakdown voltages in the 1000 Volt range are made by starting with near intrinsic material. An initial diffusion of for example phosphorous is then carried out on both sides. One side is then etched and polished until the required thickness is obtained. The second diffusion e.g. boron, is then carried out (the other, n-type, side being protected by an oxide layer). By this means PIN structures with very wide high resistivity (e.g. 4000 Ωcm.) centre layers can be fabricated.

References

1. Yu, A.Y.C., The Metal-Semiconductor Contact : an old Device with a new Future, IEEE Spectrum, Vol.7, March 1970, pp.83-89.

2. Anand, Y. and W.J. Moroney, Microwave Mixer and Detector Diodes, Proc. IEEE, Vol.59, No.8, Aug. 1971, pp.1182-1190.

3. Kahng, D. and M.P. Lepselter, Planar-epitaxial silicon Schottky Barrier Diodes, Bell System Tech. Jr., Vol.44, 1965, p.1525.

4. Watson, H.A. (Ed.), Microwave Semiconductor Devices and their circuit applications, McGraw-Hill, New York, 1969, Chapters 11,12.

5. Sze, S.M., Physics of Semiconductor Devices, Wiley-Interscience, New York, 1969, Ch. 8.

6. Cowley, A.M. and S.M. Sze, Surface States and Barrier height of metal-semiconductor systems, J. Appl. Phy. Vol.36, 1965, pp. 3212-20.

7. Crowell, C.R. and S.M. Sze, Current transport in Metal-Semiconductor Barriers, Solid-State Electronics, Vol.9, 1966, pp. 1035-1048.

8. Mead, C.A., Metal Semiconductor Surface Barriers, Solid-State Electronics, Vol.9, 1966, pp. 1023-1034.

9. Padovani, F.A. and R. Stratton, Field and Thermionic-Field Emission in Schottky Barriers, Solid-State Electronics, Vol.9, 1966, pp. 695-708.

10. Cooke, H.F., Microwave Transistors: Theory and Design, Proc. IEEE, Vol.59, Aug. 1971, pp. 1163-1181.

11. Archer, J., State-of-the-Art Small Signal Microwave Transistors, Microwave Jr., Vol.15, Feb. 1972, pp. 48-50.

12. Watson, H.A.(Ed), Microwave Semiconductor Devices and Their Circuit Applications, McGraw-Hill Book Co., New York, 1969.

13. Sze, S.M., Physics of semiconductor devices, Wiley Interscience, New York, 1969, Ch. 6, pp. 279-294.

14. Hasan, M.M. and S. Venkateswaran, Cut-off frequency of a hetrojunction transistor, Int. J. Electronics, Vol.28, May 1970, pp. 407-415.

15. Bealle, W.E., K.E. Daburlas, and W.H. Eckton, Jr., Design, Fabrication and Characterisation of Microwave Transistor, IEEE Trans. Electron Devices, Vol.ED-16, Jan.1969, pp.125-138.

16. Johnson, E.O., Physical Limitations of Frequency and Power Parameters of Transistors, RCA Rev., Vol.26, 1965, pp.163-177.

17. Lange, J., A survey of the present state of microwave transistor Modeling and simulation as applied to circuit design, IEEE Trans. Electron Devices, Vol.ED-18, Dec. 1971, pp.1168-1174.

18. Lange, J. and W.N. Carr, An application of device modeling to microwave power transistors, IEEE J. Solid State Circuits, Vol. SC-7, Feb. 1972, pp. 71-80.

19. White, M.H. and M.O. Thurston, Characterisation of Microwave Transistors, Solid State Electronics, Vol.13, May 1970, pp. 523-542.

20. Roulston, D.J. et al., Simplified computer-aided analysis of double diffused transistors including two dimensional high level effects, IEEE Trans. ED-19, pp. 809-820, June 1972.

21. Shockley, W., A unipolar field effect transistor, Proc.I.R.E. Vol.40, p. 1365, 1952.

22. Hauser, J.R., Characteristics of Junction Field-effect devices with small channel length to width ratios, Solid State Electronics, Vol. 10, p. 577, 1967.

23. Drangeid, K.C., R. Sommerhalder, Dynamic performance of Schottky barrier field effect transistors, IBM Jour., pp.82-94, March 1970.

24. Roulston, D.J., Frequence maximum d'oscillation d'un transistor a effect de champ Internal C.S.F. report No. 109-M.L., 16 March 1963.

25. Penfield, P., Jr. and R.P. Rafuse, Varactor Applications, MIT Press (1962)

26. Roulston, D.J., Frequency multiplier using charge storage diode in an inductive circuit, Proc. IEEE, Vol.55, p.1220, July 1967.

27. Boctor, S., and D.J. Roulston, Design and performance of series mode frequency multipliers using the step-recovery diode, Intl. Jour. Electronics, Vol.31, pp 333-351, Oct.1971.

28. Roulston, D.J. and A.R. Boothroyd, A large signal analysis and design approach for frequency multipliers using varactor diodes, IEEE Trans. Circuit Theory, Vol.CT-12, pp 194-205, June, 1965.

29. Johnston, R.H., and A.R. Boothroyd, Charge storage frequency multipliers, Proc. IEEE, Vol. 56, pp 167-176, Feb. 1968.

30. Granchamp, G., Etude Assiste par ordinateur d'un multiplicateur de frequence a diode snap-off, Internal report Thomson-CSF, No. 629, March 1972.

31. Penfield, P., Jr., and R.P. Rafuse, Varactor Applications, MIT Press (1962).

32. Roulston, D.J., Low noise photoparametric up-converter, IEEE Jnl. Solid State Circuits, Vol.SC-3, pp. 431-440, Dec. 1968.

33. Obregon, J., Amplificateur parametrique integre a 13 GHz, Internal report Thomson-CSF, No. 540, March 1971.

34. Moll, J.L., Krakauer, S., and R. Shen, P-N junction charge-storage diodes, Proc. I.R.E. 1962, Vol.50, pp. 43-53.

35. Benda, H. and E. Spenke, Reserve recovery processes in Silicon Power Rectifiers, Proc. IEEE, 1967, Vol.55, pp.1331-1354.

36. Varshney, R.C. and D.J. Roulston, Transient behaviour of a range of p^+-n-n^+ diodes with narrow centre regions, Solid State Electronics, 1970, Vol.13, pp.1081-1095.

37. Venkateswaran, K. and D.J. Roulston, Recombination dependent characteristics of p-n-n epitaxial diodes, Solid State Electronics, Vol.15, pp.311-323, 1972.

38. Benda, H., A. Hoffmann and E. Spenke, Solid State Electron., Vol.8, p 887, 1965.

39. Varshney, R.C. and D.J. Roulston, Turn-off transient behaviour of PIN diodes, Solid State Electronics, Vol.14, pp.735-745.

40. Moll, J.L., and S. A.Hamilton, Proc. IEEE, 57, 1250 (1969).

41. Roulston, D.J., and R.C. Varshney, Semiquantitative treatment of transients in silicon epitaxial diodes, electronic letters, Vol.5, pp. 548-550, Oct. 1969.

9

ACTIVE DEVICES FOR MICs - II

by

R. Sharan* and Amarjit Singh**

* Sections 9.1 and 9.2
** Sections 9.3 to 9.5

Chapter Nine

This chapter describes some of the semiconductor devices for microwave integrated circuits. Transfered electron devices, impatt diodes, double drift impatts, barrier injected transit time devices and surface wave amplifiers are discussed briefly.

9.1 TRANSFERRED ELECTRON DEVICES

Bulk semiconductor devices that can generate or amplify signals at microwave frequencies have obvious appeal. Transferred electron device (TED), including Gunn effect devices, belong to this class and have been the subject of some excellent review publications [1-4]. Here we would briefly describe the principle of operation of these devices and mention some of the recent advances in the area.

N-type samples of some semiconducting materials (e.g. GaAs, InP) behave in an unusual manner when an electric field above a certain threshold value (E_T) is applied across them. In this case one observes that the current passing through the sample (say GaAs), in response to a d.c. voltage, becomes oscillatory and does not remain constant with time as would normally be expected. This is shown in Figure 9.1, where a GaAs sample of length 10 μm is chosen. The period of socillation T_t of the current, as will be shown later, is approximately 0.1 nanosecond. Thus the fundamental component of oscillation frequency is 10 GHz i.e. in the microwave range. This effect was experimentally observed by Gunn[5] in 1963 and is named after him. It was earlier predicted by Ridley and Watkins[6], and by Hilsum[7], who called it 'transferred electron effect'.

The origin of the transferred electron effect is related to the band structure of the materials in which it is observed. This is discussed in detail in Sec. 9.1.1. Here, we briefly note that there are two features of transferred electron devices which

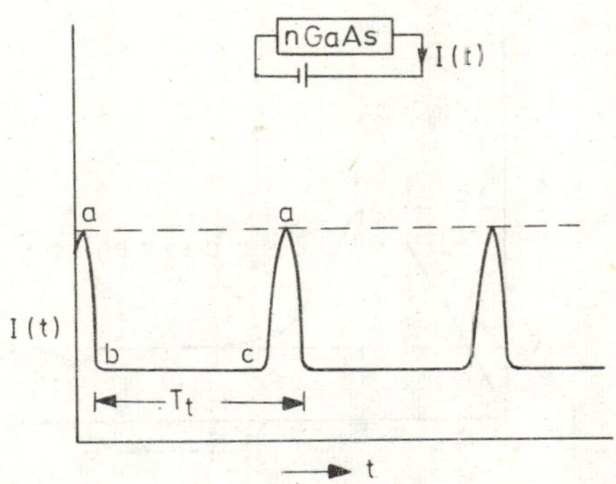

Fig. 9.1 Current waveform across a GaAs sample

are particularly remarkable. One is that Ohm's law is violated in these crystals to the extent that negative differential mobility (n.d.m.) develops. This is shown in Figure 9.2, where the drift velocity (v) of electrons is plotted against the electric field (E). The region where the velocity decreases with increasing electric field is the n.d.m. region. The second feature, which is related to the n.d.m., is that the crystal develops space charge instabilities. These instabilities are responsible for the oscillatory current waveform shown in Fig. 9.1 and are discussed in Sec. 9.1.2.

9.1.1 v-E characteristic

Most of the commercial transferred electron devices are fabricated from GaAs,. It is of particular interest to study that part of the band structure of this material, which gives rise to the n.d.m. region. A simplified band diagram of GaAs is shown in Fig. 9.3. This is a plot of the energy of the electrons (ε) versus their wave vector (K). One notes that the conduction band has two types of valleys. One is with its minimum at $K = 0$ and is called the Γ valley. Other valleys are along <100> directions and are centered at the boundaries of the Brillouin zone. Because of the zinc-blende symmetry of GaAs, there will be six such symmetrical valleys which are called X valleys. However, since the six X valleys are centered at the boundaries of the Brillouin zone, these are effectively counted as $6/2 = 3$ valleys for purposes of density-of-states calculation which is needed to explain the transfer of electrons from Γ to X valleys. The important parameters of Γ and X valleys are summarized in Table 9.1. To understand it consider an electric field (E) applied across an n-type GaAs. At the instant that the electric field is applied there are a certain number n_o of light mass, highly mobile electrons in valley Γ and virtually no electrons in valleys X i.e. $n_o = n_\Gamma$. The field supplies energy to the electrons and causes them to drift with a velocity $v_\Gamma = \mu_\Gamma E$, which sets up a current

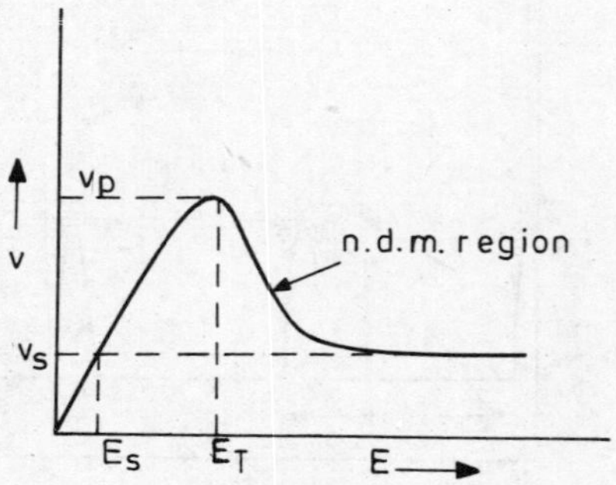

Fig. 9.2 v-E characteristic showing negative differential mobility region

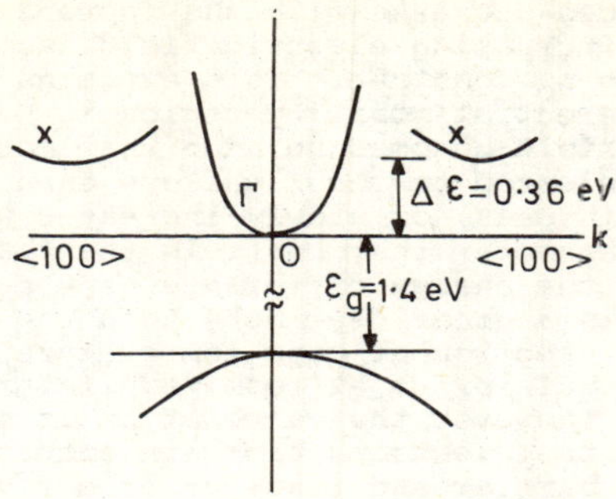

Fig. 9.3 ε-K diagram of GaAs

Table 9.1 Characteristics of Γ and X valleys of GaAs at 300°K

Electron mobility (cm^2/v sec)	μ_Γ μ_X	8000 150
Effective mass	m_Γ m_X	0.072 m_e 0.35 m_e (x 6/2)
Energy gap (ev)		1.4
$\Delta\varepsilon = \varepsilon_X - \varepsilon_\Gamma$ (ev)		0.36

in accordance with Ohm's law. If the field intensity is raised, the average energy of the electrons in valley Γ increases roughly as the square of the field. When the field becomes high enough, some of the electrons have enough energy to transfer to the valleys X. This process is further aided by the fact that the cumulative density-of-available-states in valleys X is very large compared to those of Γ valley. This electron transfer process, however, is constrained by the law of conservation of particles and one obtains $n_o = n_\Gamma + \Sigma n_X$. The electrons which have transferred to the valleys X move with a much smaller velocity $v_X = \mu_X E$ and the average drift velocity is given as:

$$v = (n_\Gamma v_\Gamma + \Sigma n_X v_X)/(n_\Gamma + \Sigma n_X) \qquad (9.1)$$

Expressing the velocities in terms of mobilities and electric field one obtains:

$$v = (n_\Gamma \mu_\Gamma + \Sigma n_X \mu_X) E/n_o \qquad (9.2)$$

In GaAs, since $\mu_\Gamma > \mu_X$ from Table 9.1, eqn. (9.2) shows that average drift velocity can decrease with the increasing electric field, provided the increasing electric field increases n_X and decreases n_Γ keeping n_o constant. This explains the appearance of the negative differential mobility region in GaAs as shown in Fig. 9.2. When the field becomes so high that most of the electrons have been transferred to the X valleys than the velocity either saturates as in GaAs, or slowly increases with field. An important critical parameter that controls the transfer of electrons from Γ to X is the energy difference $\Delta\varepsilon = \varepsilon_X - \varepsilon_\Gamma$. In GaAs $\Delta\varepsilon = 0.36$ ev, which is a great deal more than the thermal energy kT ($\simeq 0.026$ ev) of an electron at room temperature. Therefore under conditions of equilibrium at room temperature, all electrons are in valley Γ. However the value of Δε is such that at high electric fields the electrons can gain enough energy to surmount this energy barrier and transfer from Γ valley to X valleys.

It is also important to note that Δε in GaAs is much smaller than the band gap ε_g, so that the electric field needed for electron transfer between Γ and X valleys is much less than the electric field needed for impact ionization.

In summary, there are three requirements that should be met by any material which would exhibit negative differential mobility, due to transferred electron effect:

(a) The mobility of carriers in the lower valleys should be much greater than that of the carriers in upper valleys.

(b) At room temperature, and at low electric fields, all the charge carriers should be in the lower valley (i.e. $\Delta\varepsilon >> kT$).

(c) With moderately high electric field, charge carriers should be in a position to transfer from the lower to the higher valleys. The applied electric field required for this transfer of charge-carriers, should be small enough so that avalanche breakdown does not take place, (i.e. $\Delta\varepsilon < \varepsilon_g$).

Besides GaAs there are other materials (e.g. CdTe) which meet the above three requirements. However either because the fabrication technology of other materials is not as well developed as that of GaAs or because the threshold fields needed for electron transfers are higher, the commercial exploitation of GaAs for fabrication of transferred electron devices has predominated over the exploitation of other materials.

Detailed calculations of the v - E characteristic of GaAs are available in published literature [1-4]. Basically this requires the solution of coupled Boltzmann transport equations, but the details tend to be involved. For engineering design purposes, however, use is made of the following empirical relation:

$$v(E) = [\mu E + v_s(E/E_o)^4]/[1+(E/E_o)^4] \tag{9.3}$$

where μ is the low field mobility, v_s is the saturation velocity and E_o is a constant electric field. For GaAs, some typical values are: μ = 8000 cm^2/v.sec, v_s = 8.5 x 10^6 cm/sec and E_o = 4 kV/cm. The threshold electric field, E_T, for onset of the n.d.m. region in GaAs is about 3.6 to 3.8 kV/cm.

9.1.2 Domain Formation

If we apply a field greater than the threshold electric field E_T to a suitable sample of n-type GaAs, we will bring some part of the crystal under n.d.m. region. At this point some remarkable behaviour occurs due to onset of space charge instabilities. To get an accurate picture of what happens one would need the solution of a set of coupled nonlinear differential equations for which computer calculations will be required. Here we present a simplified, though necessarily crude, picture of what happens when space charge instabilities set in.

Let us consider a bulk crystal of length L across which a voltage V_a has been applied. If the crystal is not in the n.d.m. region then the voltage profile would be linear with constant slope as shown by the curve I in Figure 9.4. However, in a crystal biased in the n.d.m. region, one finds the voltage profile to be a function of time due to the onset of space charge instabilities. At an instant t_1, the voltage profile will deviate from I and will look like curve II in Figure 9.4. The linear profiles shown in curve II are a gross simplification, in an actual case the profile will be nonlinear. In curve II, there is a short middle region across which a large fraction of the applied voltage drops. This region is called the high electric field domain. The domain nucleates near the cathode. With passage of time, the domain starts growing and it also propagates towards the anode end under the influence of applied electric field. The reasons for growth are discussed next. The growth of the domain is limited by the nonlinearities of v-E characteristic and in a long sample it matures before reaching the anode. When the do-

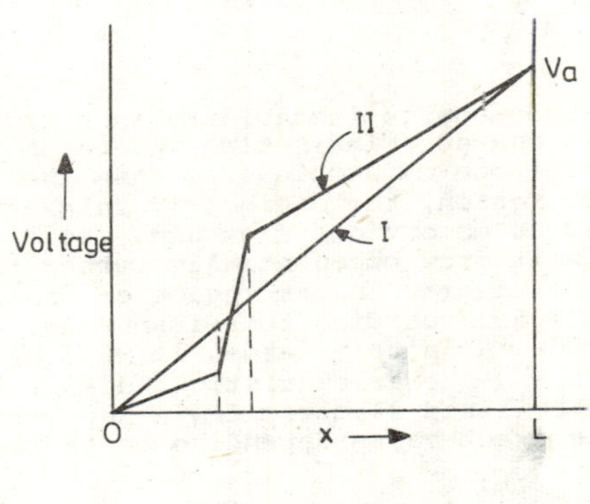

Fig. 9.4 Voltage profiles showing high field domain

main reaches the anode it is annihilated and the voltage profile returns to curve I. Shortly after this, a new domain is nucleated near the cathode and the whole process repeats itself. This periodic nucleation; propagation and annihilation of domains can explain the current profile of Fig. 9.1, which can be subdivided into three time intervals a-b, b-c and c-a. During the interval a-b the domain nucleates and continues to grow while propagating towards the anode. At time b the domain has matured and hereafter the current remains constant till the domain reaches the anode at time c. Time interval c-a is required for the annihilation of the domain after which the process repeats itself. The period of oscillation (T_t) for a 10 μm long sample is 0.1 nanosecond if the average drift velocity of domain in GaAs is taken to be 10^7 cm/sec.

From the above description the first question that arises is how does a high field domain grow? The answer can be found by a study of dielectric relaxations in solids. The continuity equation gives

$$\partial \rho_n/\partial t = -\nabla . J_n \qquad (9.4)$$

where ρ_n is the charge density and J_n is the current density given by:

$$J_n = \sigma E + eD_n \nabla n \qquad (9.5)$$

Here σ is the conductivity, e is charge on carrier and D_n is the diffusion constant. Neglecting the diffusion component of the current, for the time being, and substituting eqn. (9.5) in eqn. (9.4) and using the relation $\nabla . E = \rho_n/\varepsilon$ where ε is the dielectric constant, one obtains:

$$\partial \rho_n/\partial t = -\sigma \rho_n/\varepsilon = -\rho_n/T_D \qquad (9.6)$$

where T_D is defined to be ε/σ and is called the dielectric relaxation time. The solution of eqn. (9.6) is given as:

$$\rho_n = \rho_n(0)\ e^{-t/T_D} \qquad (9.7)$$

It shows that any space charge instability in a crystal dies down exponentially with a characteristic time T_D, which is of the order of 10^{-12} sec for 1 ohm-cm sample. In case the crystal is biased in the n.d.m. region, the dielectric relaxation time itself is negative and as is obvious from eqn. (9.7), any space charge instability will grow exponentially rather than decay. The electric field will also grow in the region of instability with the growing space charge according to Poisson law. The field, however, will not grow infinitely large, it will be limited by the nonlinear shape of the v-E characteristic and will thus stabilize. Typical values of stabilized electric field may perhaps be from 10 to 100 KV/cm; the exact value depending on the circuit conditions.

Finally let us try to find the constraints that should be satisfied before domains can form in a sample. A domain can form only if the negative dielectric relaxation time (T_{DN}) is smaller than the transit time (T_t) given by L/v_D where L is the length of the sample and v_D is the velocity of propagation of the domain i.e. $T_t > T_{DN}$. Therefore one obtains the relation that

$$n_o L > \varepsilon v_D / (e|\mu_N|)$$

where $|\mu_N|$ is the differential mobility of electrons in the n.d.m. region. Substitution of values for the parameters in the case of GaAs gives $n_o L > 4 \times 10^{10}$ cm^{-2} and a more exact analysis gives $n_o L > 2 \times 10^{11}$ cm^{-2}.

To conclude, stable dipole domains can form and propagate if the doping and length of the GaAs are suitably adjusted to obey the restriction on their product $n_o L$.

9.1.3 Domain Shape

The actual shape of the domain depends on the electron concentration (n_o), the length of the sample (L) and also the external circuit. To study the propagation of this domain, one needs to solve the following coupled, nonlinear, partial differential equations [8]:

$$\partial E(x,t)/\partial x = e[n(x,t)-n_o]/E \qquad (9.8)$$

and $$\partial I(t)/\partial x = 0 \qquad (9.9)$$

where $$I(t) = en\,v(E) - e\partial[D(E)n]/\partial x + \varepsilon\,\partial E/\partial t \qquad (9.10)$$

Here I is the current density, D is the diffusion coefficient and n is the carrier distribution. n_o depends upon the doping profile.

Initially, to simplify matters we will consider a resistive load. Also let us assume that a stable high field domain has formed and it is uniformly propagating with velocity v_D without change in shape. Since it has been assumed that the domain has already matured, it would be advantageous to transform the frame of reference as $y = x - v_D t$. Thus one obtains [8]:

$$\partial E/\partial x = \partial E/\partial y \text{ and } \partial E/\partial t = -v_D \partial E/\partial y \qquad (9.11)$$

Using eqn. (9.11), eqns. (9.8) and (9.10) can be written as:

$$dE/dy = e(n-n_o)/\varepsilon \qquad (9.12)$$

$$d[D(E)n]/dy = n[v(E) - v_D] - n_o[v_R - v_D] \qquad (9.13)$$

In eqn. (9.13), I has been replaced by $en_o v_R$ where v_R is the velocity of the electrons which are outside the high field domain.

From eqns. (9.12) and (9.13) one obtains:

$$(e/E)\, d[D(E)n]/dE = \{n[v(E)-v_D] - n_o[v_R-v_D]\}/(n-n_o) \qquad (9.14)$$

The form of a typical solution of this equation is given in [8] and reproduced in Fig. 9.5, where the carrier concentration is such that there is a leading depletion region and a trailing accumulation region. The resulting electric field has a peak value E_D and the field outside the domain is E_R. The excess domain voltage V_d is given by the shaded area of the electric field profile in Fig.9.5, A V_d versus E_R plot characterizes the Gunn effect device. This plot is given in Figure 9.6.

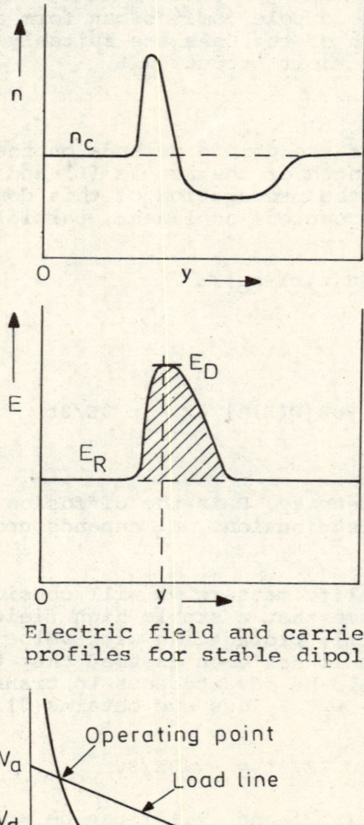

Fig. 9.5 Electric field and carrier concentration profiles for stable dipole domain

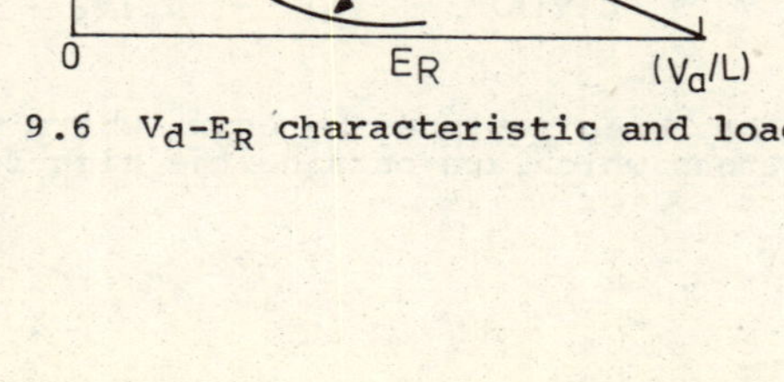

Fig. 9.6 V_d-E_R characteristic and load line

If the bias voltage (V_a) and the length of the sample (L) are known, a load line can be constructed in Fig. 9.6 using $V_a = V_d + E_R L$. It is assumed that the domain width is very small compared to L. The intersection of the load line and the characteristic curve of the device gives the domain voltage (V_d) and outside field (E_R) at the operating point. The knowledge of E_R allows the determination of current by Ohm's law if the conductivity and area of cross section of the sample are known. Thus the constant current level in Fig. 9.1 for a given applied voltage can be determined.

9.1.4 Gunn effect devices

In order to make an oscillator based on Gunn effect all that is necessary in principle, is to take a thin layer of n-GaAs with suitable value of n_oL product, put contacts on it, and apply a large enough d.c. or pulse voltage; the formation of domains does the rest. Although the structure is very simple, there are complications in building an operational device.

The oscillatory current waveform is dependent, apart from other factors, on the loading circuit. The current waveform for a resistive loading is shown in Fig. 9.1 and is such that the efficiency of oscillation is very low, as the current waveform consists of short duration of spikes. Another drawback of the resistive loading is that the frequency of oscillation is controlled only by the length of the sample and not by any externally controllable circuit parameter. Due to its disadvantages the resistive loading is hardly ever employed in practical oscillator circuits; it is mostly used for physical investigation of the transferred electron effect and is denoted as transit time mode of oscillation.

The increase in the efficiency of oscillation as well as an external control of the frequency of oscillation can be obtained by operating a Gunn device in a tunable resonant circuit as shown in Figure 9.7. In this case there will be an r.f. field which is superimposed on the d.c. bias field across the device. The relative amplitude of the r.f. and d.c. fields determines the mode of operation. The amplitude of r.f. field is controlled by the Q of the cavity and depending on this and some other factors the following modes of operation can be obtained:

(a) Delayed domain mode

(b) Quenched domain mode

(c) Limited space-charge accumulation (LSA) mode, and

(d) Hybrid mode.

A good discussion of these modes is given by Bott and Fawcett[4], who describe the behaviour of these modes in terms of the threshold electric field E_T and sustaining electric field E_s which are shown in Fig. 9.2. If the average electric field falls below E_T no new domains can be nucleated, whereas if the electric field falls below E_s even a nucleated domain cannot be sustained.

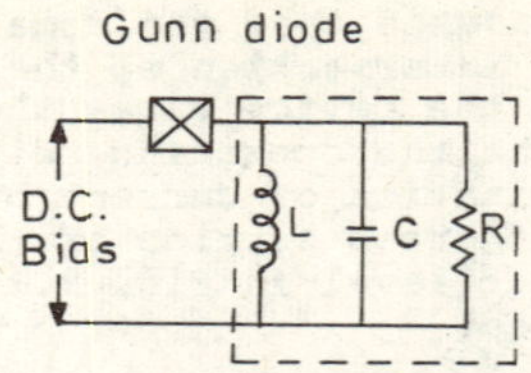

Fig. 9.7 Gunn device in a resonant circuit

(a) **Delayed domain mode: Let $2E_{r.f.}$ be the peak to peak value of the sinusoidal r.f. field and $E_{d.c.}$ be the d.c. bias across the crystal. The lowest value of the total field is $(E_{d.c.}-E_{r.f.})$. If $E_S < (E_{d.c.} - E_{r.f.}) < E_t$, then the mode of oscillation is called delayed domain mode. Under these conditions** the domains travel the full length of the crystal but at the time they are annihilated at the anode, the value of the total instantaneous field is less than E_T. No new domains can be nucleated till the total field becomes larger than E_T. In this case the period of r.f. oscillation is larger than the domain **transit time and the theoretical efficiency has been calculated to be about 27 percent[2].**

(b) Quenched domain mode: In this case $(E_{d.c.} - E_{r.f.}) < E_S$, thus a fully formed domain is quenched before it reaches the anode. Here the period of r.f. oscillation can be smaller than a full domain transit time and the efficiency of oscillation has been calculated to be near 13 percent[2].

(c) LSA mode: In this mode of operation the build-up of any space charge accumulation layer in the crystal is prevented by a suitable choice of r.f. amplitude and frequency. The d.c. and r.f. electric fields along with the v - E characteristic are shown in Figure 9.8. The crystal is biased in the n.d.m. region. The r.f. amplitude is large enough to drive the crystal below E_T during every r.f. cycle, thereby quenching any accumulation layer that might have formed. The frequency of r.f. field is chosen to be so high that there is not enough time for any space charge accumulation to grow during that part of r.f.cycle when the composite field across the device is greater than E_T. Since the frequency of operation is not limited by transit time this mode of operation holds substantial promise for oscillation in higher frequency ranges e.g. milimeter wave region.

(d) Hybrid mode: Domain modes (delayed or quenched) as well as L.S.A. mode have some advantages and disadvantages. In hybrid mode operation a compromise is made in as much as in this mode there is significant space charge accumulation, but $E > E_T$ for so short a time that domains cannot fully mature. As compared to L.S.A. mode the hybrid mode operation is less sensitive to the r.f. circuit conditions.

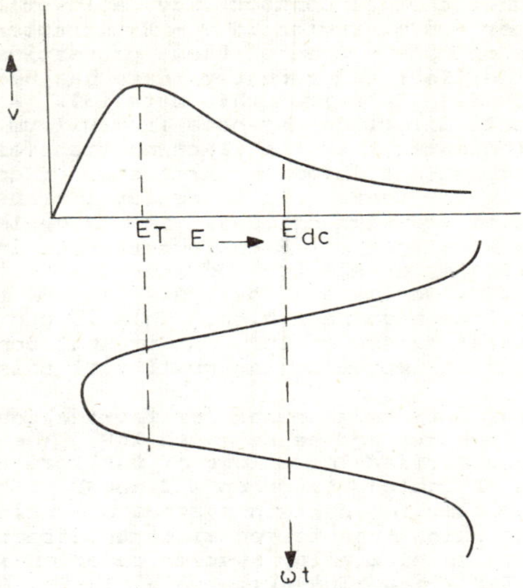

Fig. 9.8 D.C. bias and r.f. fields for LSA modes

Practical Gunn effect and L.S.A. oscillators can be fabricated in a waveguide or co-axial cavity, or in a microstrip configuration. Many types of circuits are available in published literature, e.g. an LSA mode oscillator for microstrip has been designed and fabricated by Quine[9]. The design of oscillators is considerably simplified if an equivalent circuit for the device is developed. A generalized equivalent circuit to represent a Gunn diode with a steadily propagating domain has been recently reported by Gunshor and Kak[10].

Amplifiers based on transferred electron effect have been fabricated from n-GaAs for both $n_oL > 2 \times 10^{11}$ cm^{-2} and $n_oL < 2 \times 10^{11}$ cm^{-2}. The details are available in [2] and [3].

9.1.5 Recent developments

Gunn oscillators, made from GaAs, are now commercially available. However GaAs suffers from one serious limitation namely its low thermal conductivity. Since the efficiency of operation of Gunn oscillators is rather low, there is a substantial amount of heat dissipation in the crystal, which, due to poor thermal conductivity of the substrate, raises the temperature of the crystal. At higher temperature the n.d.m. region of v - E characteristic becomes shallower and the efficiency of oscillation decreases further, thus giving rise to still higher temperatures of operation. It is obvious then that a more desirable material for fabrication of the transferred electron devices will be one with a higher thermal conductivity, a lower threshold field and a sharper n.d.m. region in v - E characteristic as compared to those of GaAs. Some of these properties are present in Indium Phosphide (InP) and, recently, there has been a large interest in fabricating TEDs from this material. An important contribution in this direction has come from Hilsum and Rees in 1970[11]. They indicated that the electron transfer mechanism in InP is more complex and involves three sets of conduction band states i.e. in InP there is a three-level transfer as compared to the two level transfer of GaAs. The three-level electron transfer provides a sharper v - E characteristic in the negative slope region. Transferred electron devices fabricated from InP have been reported by Taylor and Colliver[12], but the failure rate of these devices have been high. This is due to poor quality of the epitaxial layers of InP. At present considerable effort is being put in improving the quality of this material.

The search for optimum material for fabrication of transferred electron devices has not ceased with InP. Theoretical calculations have been carried for alloys of the form $A_xA'_{1-x}B$ (or $A\,B_xB'_{1-x}$) where A, A' belong to group III and B, B' to group V. It has been found that for certain systems a continuous range of x is available for which transferred electron effect can occur. One can choose from several alloy systems based on InSb or InAs; e.g. $Ga_xIn_{1-x}As$, $InSb_xP_{1-x}$, $InAs_xP_{1-x}$, $Al_xIn_{1-x}Sb$, $Al_xIn_{1-x}As$ or $Ga_xIn_{1-x}Sb$. It is noted that in first three cases, one of the components exhibits transferred electron effect whereas in the last three cases, neither of the components exhibits this effect at room temperature. Of all these compounds, the preparation of $Ga_xIn_{1-x}\,Sb$ is easiest and for this compound Gunn oscillations in the range $0.3 < x < 0.54$ have been observed[13]. Hilsum and Rees[11] have performed theoretical calculations for $x = 0.8$ and have obtained the threshold field to be 550 volts/cm and the peak to valley ratio of the velocity curve to be 3.8, which is large compared to that of GaAs. There are many problems, however, with the metallurgy of these compound semiconductors and it would take some time before commercial TEDs are fabricated from these materials.

9.2 IMPATT DIODES

Semiconductor diodes when operated in the Impatt mode can generate microwave power. These devices are now commercially available. Impatt is an acronym for 'impact avalanche transit time', which basically mentions the mechanisms on which the operation of this device depends. This mode of operation of semiconductor diodes was proposed by Read[1] in 1958 using a p-n-i-n or n-p-i-p structure. The basic idea of Read is very simple and is based on the fact that when a charge carrier moves opposite to the direction that an electric field tends to push it in, the charge carrier gives energy to the electric field. The procedure involved in the Read diode is that of obtaining a situation in which a d.c. field pushes the carriers in a direction opposite to that of an a.c. field at that instant. This leads to a transfer of energy from the d.c. field to the a.c. field. However, the situation gets complicated due to the simple fact that an a.c. field alternates whereas the d.c. field does not and this will lead to zero net transfer of energy. The way to obtain a net transfer of energy, then, is to provide a means of shutting off the carrier flow for half a cycle, when the a.c. field is in the same direction as the d.c. field and then turning on the carrier flow at just the right time. This is obtained in a Read diode by extracting carriers due to impact avalanche in a narrow region of space charge of a p^+-n junction and then letting them drift in an almost intrinsic region. The details are described later. However, one should note that a p-n-i-n structure is difficult to fabricate for high frequency oscillation. Hence Read's proposal remained only as an idea till 1965.

In 1965, Johnston et al.[15], obtained oscillation in the microwave region from a p-n junction diode of Si. This diode was fabricated by diffusing p-type impurity to a depth of 25 μm in an n-type Si of thickness 175 μm and of 0.15 ohm-cm resistivity. The area of diode was 125 μm by 125 μm. Microwave power was generated when the reverse potential in such diodes became larger than a certain critical value. Johnston et al. had obtained a pulsed power output of 80 mw (efficiency 0.5%) at 12 GHz. The state-of-the-art has changed dramatically since then and more efficient devices with higher power outputs at higher frequencies are now available [16-19].

9.2.1 Structure of Read diode

Before considering the operation of the Read diode let us consider a much simpler problem[19]. The problem is to calculate the output power and efficiency corresponding to the voltage and current waveforms shown in Fig.9.9. Here the voltage waveform consists of a sinusoidal r.f. voltage superimposed on a d.c. voltage.

The average power dissipated for these waveforms which are shown in Fig. 9.9, is:

$$P = (1/2\pi) \int_{\pi y}^{\pi x} I_{max} V_{RF} \sin \omega t \, d(\omega t) \qquad (9.15)$$

$$= (1/2\pi) V_{RF} I_{max} (\cos \pi y - \cos x\pi) \qquad (9.16)$$

The maximum current I_{max} is related to the d.c. current by:

$$I_{d.c.} = (1/2\pi) \int_{\pi y}^{\pi x} I_{max} \, d(\omega t) \qquad (9.17)$$

$$= I_{max} (x-y)/2 \qquad (9.18)$$

Using eqns. (9.18) and (9.16), one obtains:

$$P = (V_{RF} I_{d.c.} /\pi) (\cos \pi y - \cos \pi x)/(x-y) \qquad (9.19)$$

The efficiency η is then given by:

$$\eta = (V_{RF}/\pi V_{d.c.}) (\cos \pi y - \cos \pi x)/(x-y) \qquad (9.20)$$

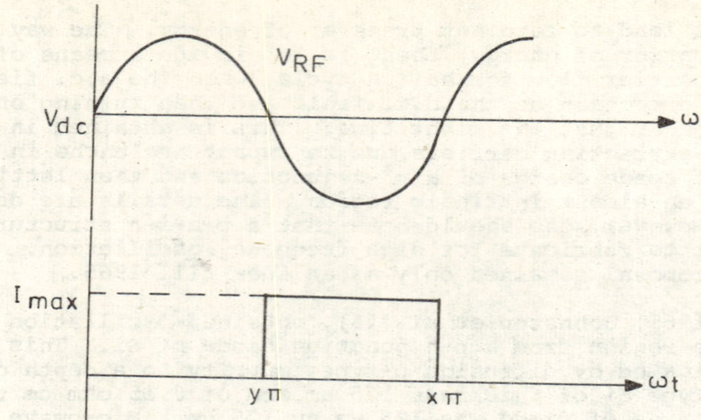

Fig. 9.9 Impatt voltage and current waveforms

A consideration of eqns. (9.19) and (9.20) shows that for $0 < y \leq 1$, maximum power output and efficiency will result for $y \cong 1$ and are given by

$$P = (-V_{RF} I_{d.c.}/\pi)\ (1+\cos \pi x)/(x-1) \tag{9.21}$$

$$\text{and}\quad \eta = (-V_{RF}/\pi V_{d.c.})\ (1+ \cos \pi x)/(x-1) \tag{9.22}$$

The negative signs in eqns. (9.21) and (9.22) indicate that power is delivered by a device which has the wave forms shown in Fig.9.9.

The voltage and current waveforms across a Read diode, as will be shown later, are similar to corresponding waveforms in Fig. 9.9. and it follows from earlier considerations that a.c. power can be obtained from this device.

A p^+-n-i-n^+ Read structure, alongwith its d.c. electric field profile, is shown in Fig.9.10. The diode can be subdivided into the following functional parts.

a) A source of carriers: This is a narrow region in the p^+-n depletion layer where electric field is higher than the threshold value which will give rise to impact ionization. Both electrons and holes are produced by the impact ionization and their numbers depend on their ionization coefficients, which in turn are sensitive functions of the electric field.

b) A region where the carrier transport current is modulated by the r.f. signal. Since the ionization coefficient and hence the multiplication of carriers are quite sensitive to the magnitude of electric field present, the avalanche region is also the place where the carrier transport current is modulated by the r.f. field.

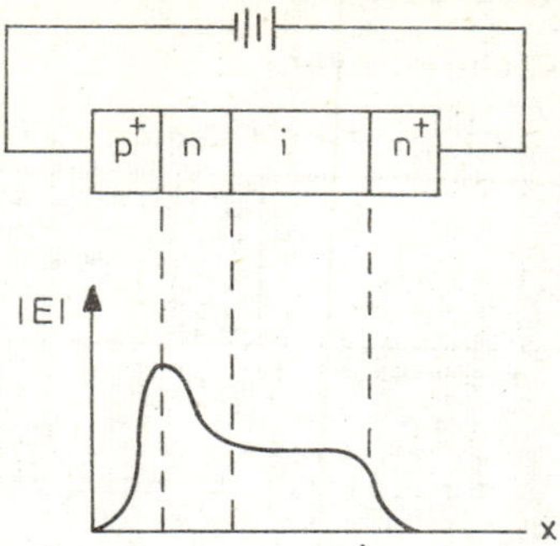

Fig. 9.10 Read diode structure and d.c. electric field

c) A drift region: This is the intrinsic region in which the carriers drift with saturated velocity and where no impact ionization occurs. This region alongwith the avalanche region can also be looked upon as the working region where the carriers produce useful power output by interacting with the r.f. field.

d) The end contacts: These act as collectors of carriers. Both electrons and holes are produced by impact avalanche and due to the reverse bias on the diode the holes are collected by the p^+ contact on the left. The electrons which drift through the intrinsic region are collected by the n^+ contact on the right.

9.2.2 Mechanism of operation of Read diode

To study the mechanism of operation of Read diode we will first consider the characteristics of the propagation in the intrinsic region [16] and then the characteristics of the avalanching region [19].

Propagation in intrinsic region [16]: Consider a uniform region of n-type semiconductor shown in Fig. 9.11. The electric field across the semiconductor is high enough so that the velocity of the carriers in the drift region is a constant, say v_s. We postulate a control layer on the left side of the semiconductor bar which injects an electron current of density J_n. If V is the voltage across the whole device then $J_n = Y_i V$ specifies the control layer. The admittance Y_i is presumed to be known.

Now we can consider two cases, one that J_n is independent of time and the other that J_n is time dependent. In the first case the current at the terminals of the device will also be J_n. However, if J_n is time dependent then the current at the terminals J_e will be different from J_n. Let us try to find the relation between J_n and J_e. The difference between J_n and J_e is due to induced current[16]. This happens because each bit of charge travelling down the device induces its own current contribution into the external circuit. One can see that the particle current density at a point x arises from the current emitted at a time x/v_s beforehand at the control layer. Thus the terminal current

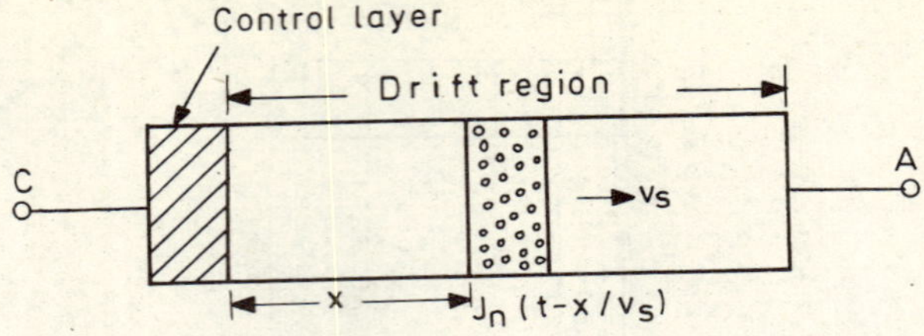

Fig. 9.11 A generalised transit time device

at time $t = t_o$ is induced from carriers generated at time $t = t_o - \tau$, where τ is the total transit time of carriers. The average delay between the injected particle current (J_n) and the induced current (J_e) is therefore, $\tau/2$. Assuming that both currents vary sinusoidally at a frequency ω, the delay may be written as:

$$J_e = J_n \, A \exp(-j\omega\tau/2) = V \, A \, Y_i \exp(-j\omega\tau/2) \tag{9.23}$$

where A is a constant available from detailed calculation. The admittance Y_i has some phase associated with it, and one can now see that the phase difference between J_e and V will be $(\theta-\omega\tau/2)$. If the total phase lag is greater than 90° but less than 270° then (J_e/V) will have a component of negative conductance across the terminals. When this phase difference is 180°, then there will be a pure negative conductance at the terminals of the device. In a Read diode a situation is obtained where $\theta = -\pi/2$ and $\omega\tau = \pi$, thus a pure negative conductance is available and the device can be used either for generation or amplification of microwave power. We will now consider the availability of a phase lag of $\theta=\pi/2$ from the avalanching region of Read diode.

Inductive Avalanche [19]: Let us assume that a d.c. and an r.f. voltage are applied across the diode, and the d.c. voltage is approximately equal to the breakdown voltage. We want to show that the current generated by the avalanche process will have its maximum when the r.f. field goes to zero. This is shown in Fig. 9.12. To justify the current waveform in Fig. 9.12 we note that [19]:

(i) The rate of generation of electron-hole pairs in the avalanche region is proportional to both the electric field and the density of electrons and holes that are already present.

(ii) Because of the small r.f. voltage of sufficiently high frequency, superimposed on d.c. voltage, the rate of generation of electron-hole pairs will exceed the rate at which the pairs leave the avalanche region.

(iii) Thus both the density of carriers and the current will grow exponentially with time as long as the r.f. and d.c. voltages add to give a total field above breakdown.

(iv) When the r.f. voltage changes sign and subtracts from the d.c. voltage, causing the field to fall below breakdown, the generation rate starts decreasing causing the current to decrease.

(v) Thus the current generated by the avalanche processes will have its maximum when the r.f. field goes to zero. This is what we had set out to show.

We conclude that the avalanche process contributes a 90° inductive phase lag to the current generated in the avalanche region as shown in Fig. 9.12. The external current through the device is shown in Fig. 9.12(c) and it is interesting to note the similarity between Fig. 9.12(a) and (c) and the voltage and current waveforms in Fig. 9.9. In the next section we summarise the operation of Read diode and then proceed to describe some more practical Impatt structures.

9.2.3 Structure of Impatt diodes

A. Read diode: The Read diode (p^+-n-i-n^+) structure and the electric field profile under reverse bias are already shown in Fig.9.10. This diode has a thin avalanche region (p^+-n) at one end, providing the modulated avalanche current and phase shift nearly $\pi/2$. The remaining phase shift ($\approx \pi/2$) is provided by adjoining drift space which is about one-half cycle long. The field in the drift region is substantially lower than breakdown values, so that impact ionization is negligible in the drift space, and is high enough so that drift velocity of carriers is

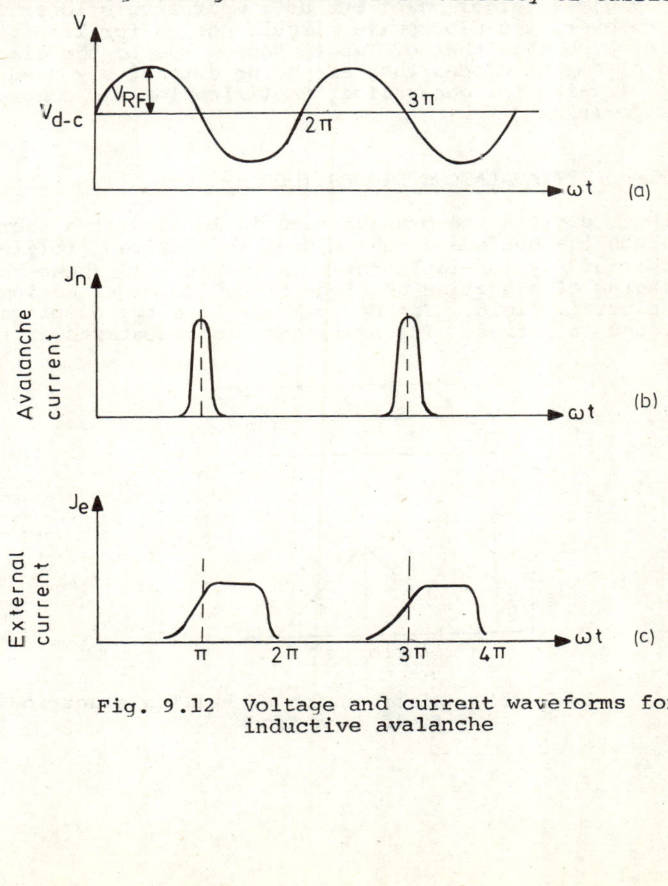

Fig. 9.12 Voltage and current waveforms for inductive avalanche

saturated. The negative conductance of the Read diode is apparent from Fig. 9.12(a) and (c). It should be emphasized that though the operation of Read diode is easy to visualize, its fabrication is more complicated and it is not a favoured structure for commercial production.

B. PIN and p^+-n structure: Misawa[20] has shown that a negative conductance of Impatt nature can be possessed by a junction diode with almost any doping profile. To do this he used a structure in which the avalanche region and the electron bunch transit region are common, and the whole phenomenon is rather complex and not easy to visualise as in the case of Read diode. A PIN diode best represents the structure suitable for Misawa theory and is known as Misawa diode. Since the effective avalanche region of a p^+-n junction is also narrow, an abrupt p^+-n junction can be classified as a Read-type diode. Commercial Impatt diodes are either Misawa diode or p^+-n diodes.

9.2.4 Conclusion

In principle Impatt diodes can be made from any semiconductor from which p-n junctions can be fabricated. However, commercial Impatts are mostly made from silicon. Germanium and gallium arsenide devices are also available and have some advantages over silicon diodes. Recently another mode of operation called TRAPATT (trapped plasma transit time) mode in structures similar to Impatt devices has given larger power but at comparatively lower frequencies. However, the miocrowave circuit design for Trapatt mode is more involved than that of Impatt mode. Due to the simplicity of design of Impatt diodes they are being extensively used in microwave circuits for generation, amplification and control of microwave power.

9.3 DOUBLE DRIFT AVALANCHE DIODES (DDR's)

In these devices the p-n junction is located at a short distance beneath the surface of the diode, the desired profile being usually obtained by ion-implantation. Figure 9.13 shows the schematic diagram of a section of diode alongwith distribution of built-in electric field. The device acts like two complementary Impatt diodes in series. The avalanche process starts at the p-n

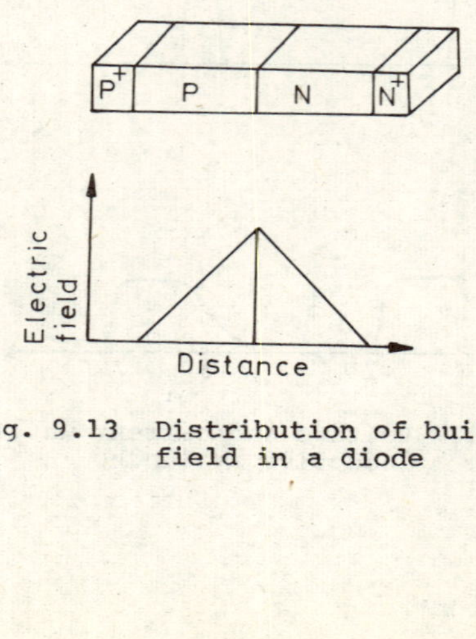

Fig. 9.13 Distribution of builtin electric field in a diode

junction. The holes and electrons drift in opposite directions.

For negative resistance, the transit angle for each type of carrier is close to π. To a first approximation it is given by $\omega w/u$, where w is the width of the drift region, and u is the drift velocity assumed to be constant. In the case of DDRs, the total width that can be used is twice the width for a single drift region avalanche diode. The impedance of the device and the power per unit area are approximately double. In case it is coupled into a circuit with a given impedance, then in comparison with a single drift device, the area of cross-section can be doubled, and on this basis the power output could be quadrupled [21]. Also under these conditions, the volume of the active material is quadrupled in comparison with a single drift diode. Higher efficiency is achieved due to (a) smaller percentage loss from parasitic series resistance and (b) avoidance of storage of minority carriers near the avalanche region, which can spoil the phase relationship between voltage and current at the next cycle.

Techniques of fabrication: By using this approach, a power output of 1 watt at 50 GHz has been obtained with an efficiency of 13.3% [22]. The starting material is a n-type epitaxial layer on a n^+ substrate. The typical layer has a thickness of 1.4 microns and a doping density of $6 \times 10^{16}/cm^3$. By multiple ion implantation with Boron ions, a p-n junction is created at a depth of 0.6 microns, the p region also acquiring a net donor density of $6 \times 10^{16}/cm^3$.

A shallow diffusion to a depth of 0.15 microns forms the p^+ contact. This operation is performed at 875°C for 30 minutes during which annealing takes place and the Boron ions which had been implanted become substitutional [23].

By an adjustment of profile using ion-implantation techniques it has also been possible to fabricate an Impatt diode back-to-back with a Trapatt diode [24]. In this case the Impatt diode triggers the operation of the Trapatt mode which operates between 4 to 6 GHz.

9.4 BARRIER INJECTION TRANSIT TIME DEVICES (BARRITTS)

In these devices, there is a transit of carriers through a semiconductor region, the transit angle being such as to give negative conductance. Unlike Impatt diodes, where the carriers originate in a process of avalanching, in Barritts, the carriers are injected through a p-n junction or a Schottky barrier. At the operating voltage a "punch through" or a "reach through" condition is established, so that the current is a highly sensitive function of the voltage. The injected current is thus in phase with the r.f. voltage.

The negative conductance is proportional to $\sin(\omega t)/\omega t$, the maximum negative conductance being obtained when ωt is $3\pi/2$ radians. This contrasts with the avalanche diode where the maximum negative conductance is obtained when the transit angle is π radians, the extra $\pi/2$ phase difference having been introduced between voltage and current by the avalanching process. In the case of Barritts for a quarter of the cycle the charge carriers are absorbing energy from the r.f. field, while for half a cycle the energy is being given to it. Because of the fact that for part of the cycle energy is absorbed, Barritts have lower efficiency as compared with avalanche diodes.

However, because of the fact that a noisy process like avalanche multiplication is not involved, these devices have a much lower noise figure.

Two kinds of embodiments have been experimented upon: In the MSM diode [25], there are two metal layers on the opposite sides of a thin layer of semiconductor. In one embodiment reported, the semiconductor material was silicon with a doping density of $4 \times 10^{14}/cm^3$. The contacts consist of platinum silicide, on the two sides of a silicon layer 10 microns thick. Active areas of 5×10^{-4} cm^2 were defined by etching. Such devices gave a power of 50 milliwatts at 5 GHz, at an efficiency of 1.8 per cent. As amplifiers, such devices gave a noise figure of 15 db and gain of 19 db.

In Barritts with semiconductor junctions, experiments have been conducted with a configuration of n^+-p-n^+ as well as p^+-n-p^+. In this case the maximum E-field at the injecting junction, at the time of "punch through", has to be less than the breakdown voltage, which is of the order of 3.2×10^5 V/cm in silicon. Gain of the order of 15 db at 6.35 GHz, with a noise figure of 9 db; and a power output of 45 mW with an efficiency of 2.5% have been reported [26].

9.5 SURFACE WAVE AMPLIFIERS

Amplification at microwave frequencies can be obtained by an interaction between charge carriers drifting under the influence of a d.c. field, and an elastic wave in a semiconductor. In particular, surface waves, offer certain advantages, as follows:

a. The velocity of propagation is typically 5 orders of magnitude smaller than the velocity of light. Correspondingly, the guide wavelengths for microwave frequencies are in the micron range and result in miniaturisation.

b. Surface waves give accessibility for transducers, in which cumulative interaction with the surface waves is possible.

c. Advancement in the fabrication techniques and materials technology in the field of integrated circuits have made surface wave devices possible which would otherwise not be feasible.

The disadvantage of surface waves in comparison with bulk waves is that they are more sensitive to any irregularities or surface contamination. Transducers commonly used consist of inter-digital arrays with several variations involving change of pitch, etc [27].

Coplanar guides have been used for coupling to the transducers for surface wave amplifiers. Because of their inherent properties mentioned above, they are likely to find use in microwave integrated circuits.

A broad survey of modern trends and techniques has been given in reference[28]. Broad bandwidths, of the order of 2 to 12 GHz,has been reported by experimenters at RCA.

References

1. Levinstein, M.E. and M.S. Shur, The gunn effect (review), Soviet Physics - semiconductors, Vol.5, p.1561, 1972.

2. Sterzer, F., Transferred electron (Gunn) amplifiers and oscillators for microwave amplification, Proc. IEEE, Vol.59, p. 1155, 1971.

3. Carrol, J.E., Hot electron microwave generators, Edward Arnold, London, 1970.

4. Bott, I.B. and W. Facett, The Gunn effect in Gallium Arsenide, Advances in Microwaves, Vol.3, Ed.,Lee Young, Academic Press, 1968.

5. Gunn, J.B., Microwave Oscillations of current in III-V semiconductors, Solid State Comm., Vol.1, p.88, 1963.

6. Ridley, B.K. and T.E. Watkins, The possibility of negative resistance in semiconductors, Proc. Phys. Soc., London, Vol. 78, p.293, 1961.

7. Hilsum, C., Transferred electron amplifiers and oscillators, Proc. IRE, Vol.50, p.185, 1962.

8. Butcher, P.N., W. Fawcett and C. Hilsum, A simple analysis of stable domain propagation in the Gunn effect, Brit. J. Appl. Phys., Vol.17, p.841, 1966.

9. Quine, J.P., An LSA-mode transferred electron diode oscillator for microstrip, Proc. IEEE, Vol.58, p.1291, 1970.

10. Gunshor, R.L., and A.C. Kak, Lumped circuit representation of Gunn diodes in domain mode, IEEE Trans. Electron Devices, Vol. ED-19, p. 765, 1972.

11. (a) Hilsum, C., and H.D. Rees, Three-level oscillator: a new form of transferred electron devices, Electronics letters, Vol.6, p.277, Apr. 30, 1970.

11. (b) Hilsum, C. et al., Instabilities of InP 3-level transferred electron oscillators, Electronics letters, Vol.6, May 14, 1970.

12. Taylor, B.C., and D.J. Colliver, InP microwave oscillators, IEEE Trans. Electron Devices, Vol.ED-18, p.835, 1971.

13. McGroddy, J.C., M.R. Lorenz and T.S. Plaskett, Solid State Communication, Vol.1, p.901, 1969.

14. Read, W.T., A proposed high frequency negative resistance diode, Bell Syst. Tech. J., Vol.37, p.401, 1958.

15. Johnston, R.L., B.C. DeLoach, Jr. and B.G. Cohen, A silicon diode microwave oscillator. Bell Syst. Tech. J., Vol.44, p. 369, 1965.

16. Carroll, J., Hot-electron microwave generators, Edward Arnold, London, 1970.

17. Watson, H.A., (Ed.), Microwave semiconductor devices and their circuit applications, McGraw-Hill, New York, 1969, (Chap. 15).

18. Sze, S.M. and R.M. Ryder, Microwave Avalanche Diodes, Proc. IEEE, Vol.59, p.1140, 1971 (August).

19. Haddad, G.I., P.T. Greiling, and W.E. Schroeder, Basic principles and Properties of Avalanche Transit-Time Devices, IEEE Trans. Microwave Theory Tech., Vol.MTT-18, p. 752, 1970 (November)

20. Misawa, T., Negative resistance in p-n junctions under avalanche breakdown conditions, IEEE Trans. on Electron Devices, Vol.ED-13, p.137, 1966 (Jan.)

21. Scharfetter, D.L., W.J. Evans, and R.L. Johnston, Double drift region (p^+-p-n-n^+) avalanche diode oscillators, Proc. IEEE(Lett.), Vol.58, July 1970, pp.1131 - 1133.

22. Seidel, T. and D.L. Scharfetter, High power millimeter wave IMPATT Oscillators with both hole and electron drift space made by ion-implantation, Proc.IEEE (Lett.), Vol.58, July 1970, pp. 1135-1136.

23. Seidel, T.E., R.E. Davis, and D.E. Iglesias, Double-Drift-Region ion-implanted multimeter-wave IMPATT diodes, Proc.IEEE 59, pp.1222-1228, Aug. 1971.

24. Evans, W.J., T.E. Seidel and D.L. Scharfetter, A novel TRAPATT oscillator design, Proc. IEEE, Vol.58, Aug.1970, pp. 1294-1295.

25. Coleman, D.J. and S.Z.E. Say, A low noise metal-semiconductor-metal (MSM) Microwave Oscillator, Bell Syst. Tech., Vol.50, pp. 1095-1099.

26. Liu, at Device Research Conference, Edmonton, Canada, June 21 - 23, 1972.

27. White, R.M. Surface Elastic waves, Proc. IEEE, Vol.58, (1970), p. 1238 - 1276.

28. Nergaard, L.S., Amplification - modern trends, techniques and problems II, RCA Rev., Dec.1971, Vol.32, pp. 519-566.

10

DESIGN OF MICROSTRIP CIRCUITS

by

D.J. Roulston* and P.K. Wahi**

* Sections 10.1 to 10.3

** Section 10.4

Chapter Ten

10.1 INTRODUCTION

Hybrid MIC techniques lend themselves to the construction of an extremely wide range of active and passive circuits. In Section 10.2 we shall discuss a few examples of MIC power sources. In section 10.3 some low power circuits will be dealt with. The circuits to be considered have all been realised practically; it should be borne in mind, however, that there exist many different ways of performing the same function and it is the design engineer's job to seek the best way of solving a particular problem. In this chapter we shall see how some of the basic circuits and devices dealt with in the previous chapters may be put to practical use.

10.2 HIGH POWER CIRCUITS

10.2.1 Transistor Oscillator

In designing MICs which use active devices operating under large signal conditions it is almost essential to perform large signal measurements on the active element or elements before attempting a prototype design. Unfortunately, measurements are not possible without introducing stray reactive elements (series L and shunt C) when connecting the active chip to the measurement circuit. The ultimate 'measurement' circuit will be the prototype MIC itself. Where possible this prototype should therefore be designed to allow for adjustments after initial fabrication.

Figure 10.1 shows a 2.2 GHz transistor oscillator in microstrip form together with the corresponding lumped equivalent circuit. The output resonator in this case is tuned to 2.2 GHz although the input resonators are tuned to 1.1 GHz. The transistor is, in this example, operating as an oscillator at 1.1 GHz and the collector depletion layer capacitance is being used as a 'varactor' frequency doubler. This enables useful power to be extracted at a frequency which is higher than that obtainable otherwise.

The operation of the circuit is self explanatory. Note the use of high impedance strip line ($\lambda/4$) for the collector bias and for grounding the emitters for d.c. The output circuit is a variation on the standard $\lambda/4$ coupled strip discussed earlier.

10.2.2 Step-recovery diode frequency multiplier

The principle of operation of the step-recovery diode frequency multiplier has been outlined in Chapter 8. The basic i.f. circuit is reproduced in Fig. 10.2(a). Note that there exist two possible circuits which are essentially duals, one driven by a low frequency current source, the other driven by a low frequency voltage source. In both cases the bias may be either externally applied or obtained by choice of a suitable value of resistance (which allows the bias to be developed due to the small detected current). A microstrip version of the multiplier is shown in Figure 10.2(b). This is a 2.2 to 8.8 GHz multiplier (i.e. X4) and the experimental version gave a power conversion loss of about 5dB.

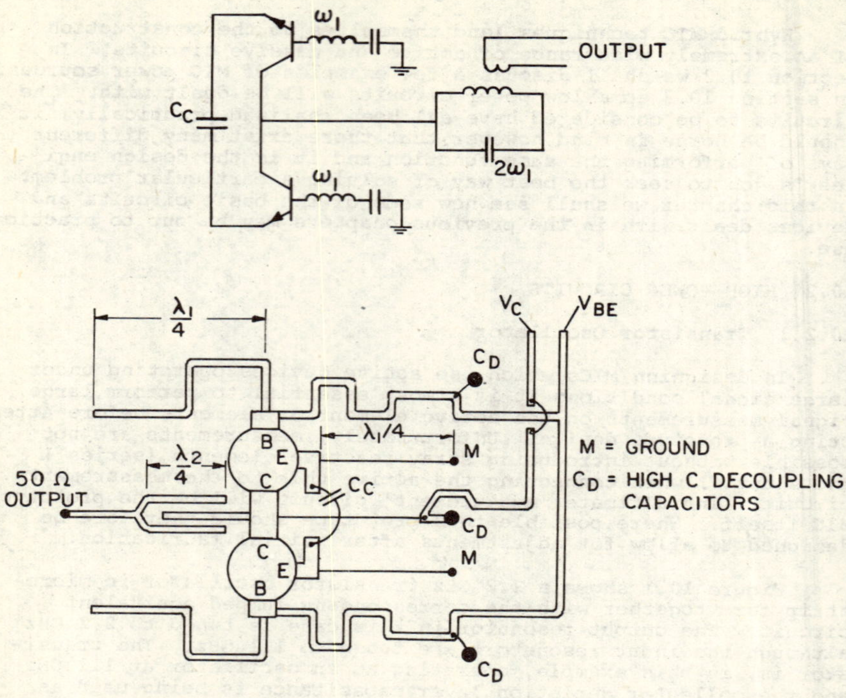

Fig. 10.1 Microstrip transistor oscillator.

As explained in Sec. 8.5 there is no simple analytic theory which enables one to predict the efficiency, power output and optimum input and output impedances of such a multiplier[1]. Low frequency measurements and approximate theory do, however, enable one to estimate the approximate orders of magnitude of these quantities. The circuit shown gives an effective source impedance at point 'A' of about 10 ohms at 2.2 GHz. The effective load impedance at point 'B' is of the same order of magnitude. Note that the pulse forming circuit in this case is formed by lumped constant elements; a MOS capacitor and a short length of 30 micron dia. wire. The output filter is of the parallel quarter wave coupled resonator type described in Section 2.3. Note, however, that impedance transformers are used at both ends of the filter. The bias resistance is added to the complete circuit and provides a very simple form of final adjustment. The pulse formation is very sensitive to bias voltage; since the detected current is a function of the carrier lifetime of the diode, varying this bias resistance allows both compensation for diodes with different lifetimes (remember that the

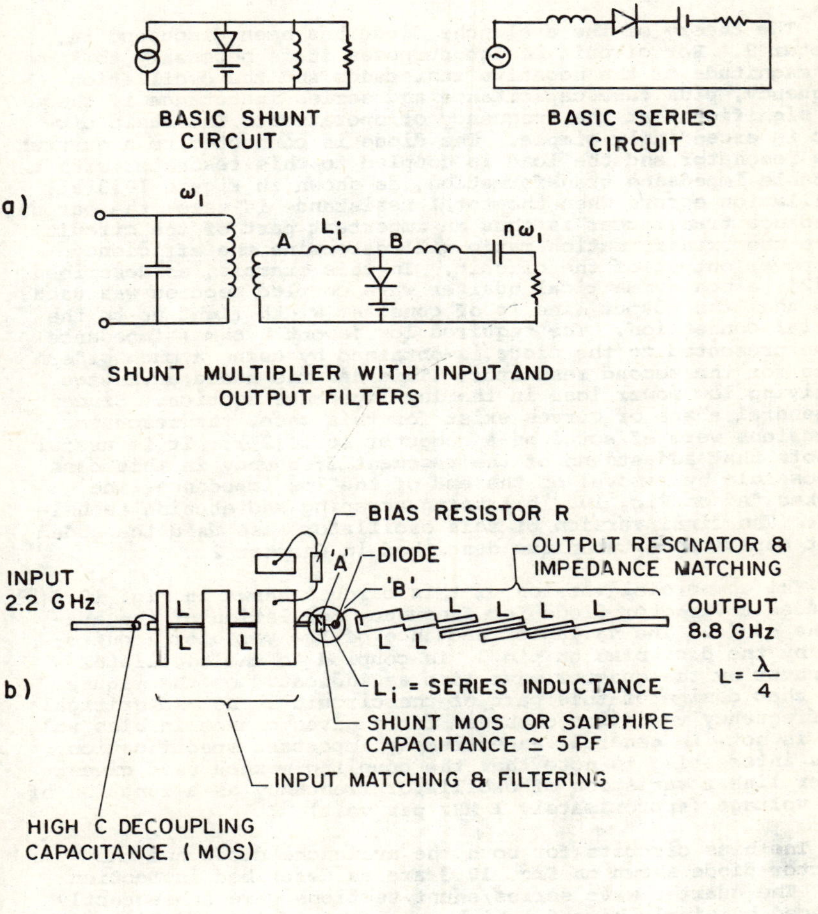

Fig. 10.2 Step recovery diode frequency multiplier [Ref. 3].

(a) Basic circuit
(b) Microstrip version

carrier lifetime is subject to quite wide variations from one diode to another of the same type) and also allows for minor variations in input and output impedances. The input filter and impedance transformer is of the series L, shunt C type discussed in Chapter 2.

10.2.3 Avalanche diode oscillator (with varactor frequency modulator)

The theory of the avalanche diode has been discussed in Chapter 9. For circuit design purposes it is necessary to know the magnitude of the negative resistance and the oscillation frequency, plus case capacitance and series inductance if these are significant at the frequency of operation. The basic circuit is exceedingly simple. The diode is connected to a quarter wave resonator and the load is coupled to this resonator with a suitable impedance transformation, as shown in Figure 10.3(a). Oscillation occurs when the total resistance is zero; the output impedance transformer is thus an important part of the circuit, since the transformation ratio will determine the efficiency and power output of the circuit. In this example, as described in [2], a non symmetrical quarter wave coupled section was used. Note that the output line is of constant width right up to the coaxial connection. The required low (about 6 ohms) impedance to be presented to the diode is obtained by using a much wider strip for the second resonator. This has the added advantage of giving low power loss in the low impedance section. Since no general abacs or curves exist for this case, the resonator dimensions were effected on a computer (c.f.[2]). It is useful to note that adjustment of the resonant frequency in this case is possible by removal of the end of the low impedance line (marked "A" on Fig. 10.3(b)) using scraping and etching techniques. The final version of this oscillator also used the added shunt capacitance technique described in Chapter 2.

The commercial version of this circuit shown in Fig. 10.3(b) includes a varactor diode for frequency modulation of the avalanche diode. The variable reactance of the varactor (controlled by the d.c. bias on pin 3 is coupled to the oscillator resonator by the quarter wave line as indicated on the figure. Note that design of this part of the circuit is not so critical; the frequency variation obtained for a given change in bias voltage is not, in general, an extremely important specification. It is interesting to note that the coupling method used gives a near linear variation of oscillator frequency as a function of bias voltage (approximately 1 MHz per volt).

The bias circuits for both the avalanche diode and the varactor diode shown on Fig. 10.3 are as described in Section 2.5. The quarter wave series/shunt sections were subsequently replaced by single high C - high impedance series $\lambda/4$ bias networks. The highest power obtained to date with this circuit at X-band is 1 watt with an efficiency of abcut 6%[3].

It should be noted that if the avalanche diode characteristics are not available from the manufacturer it is possible to measure the small signal equivalent circuit by mounting the diode in a 50 ohm microstrip line using a microwave S parameter net-

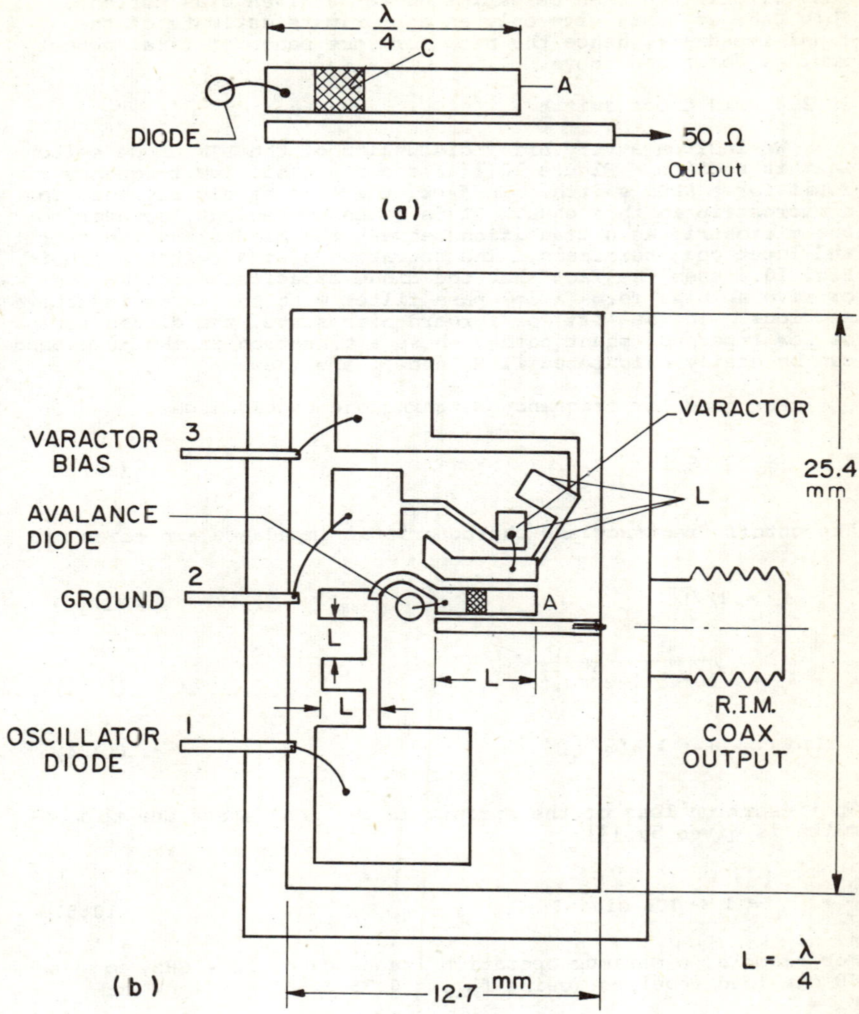

Fig. 10.3 Avalanche diode oscillator (with varactor diode modulator) [Ref.3].
(a) Resonator details
(b) Microstrip circuit layout

work analyzer (e.g. Hewlett Packard). The large signal equivalent circuit can then be estimated for a given bias current. This can, at best, give only an approximate estimate of the final impedance, hence the need for some means of final adjustment as described above.

10.2.4 PIN diode switch

We include a very brief discussion of the PIN diode switch in this section. Figure 10.4(a) shows a basic low frequency circuit for a SPST switch. In fact, there is hardly any need for a a microstrip in this case. It is often convenient, however, to use microstrip as a transition between the diodes and the output and input coax terminals. The operation of the switch shown in Fig. 10.4 uses the fact that the diode capacitance in the 'on' or reverse bias forms a low pass filter with the three inductive sections. In the 'off' or forward bias state, the diodes act as low impedance shunt paths, whose attenuation in the pass-band can be easily calculated if R_i and R_s are known.

For good low frequency matching one should have:

$$R_o = \sqrt{L/C} \tag{10.1}$$

The cutoff frequency and characteristic impedance are given by:

$$\omega_c = 2/\sqrt{LC} \tag{10.2}$$

$$Z_o = \sqrt{(L/C)(1-\omega^2/\omega_c^2)} \tag{10.3}$$

$$\text{Also } \cos\theta = 1-2(\omega/\omega_c)^2 \tag{10.4}$$

The insertion loss of the circuit in the 'on' state due to mismatch is given by [4]:

$$L_n = 1 + 0.5\sin^2 n\theta\,(R_o/Z_o - Z_o/R_o)^2 \tag{10.5}$$

For example, a maximum operation frequency of 12.4 GHz, with a 50 ohm load requires (using $f_{max} \approx 0.75\ f_c$) C = 0.4 pF and L_s = 1 nH.

It can be shown that the insertion loss due to finite series diode resistance is given by[4]:

$$L_n = |\cos 2\theta\ [1+(Z_o/R_o + R_o/Z_o)\{Z_o/R_1(1+\cos\theta)\}] - (Z_o/R_o-R_o/Z_o)\{Z_o/R_1(1+\cos\theta)\} + j\sin 2\theta\ [(Z_o/R_o+R_o/Z_o)/2 + 2Z_o/\{R_1(1+\cos\theta)\}]|^2 \quad (10.6)$$

where $R_1 = (\omega c)^{-2}\ R_s^{\ -1}$.

At the maximum frequency of operation $\omega = \omega_c/\sqrt{2}$, this gives:

$$L_{max} = (1 + 2R_s/R_o)^2 \simeq 1 + 4R_s/R_o \quad (10.7)$$

To have an overall loss at the highest frequency less than 1.5 dB, it is therefore necessary to have a series diode resistance less than 3 ohms (for R_o equal to 50 ohms), allowing for a maximum loss due to mismatch of 0.5 dB. The 'microstrip' version of the circuit is shown in Fig. 10.4(b).

SPMT switches require more sophisticated circuitry. An example of a multi-throw circuit is shown in Figure 10.5. In this case the diode is used as a series element, presenting a high series impedance in a given output path when reverse biased and a low series impedance when forward biased. It must be remembered that the 'conducting port' will have in parallel the reverse bias capacitance (with the shunt inductances) of the other ports. The calculation of cut-off frequency and loss must take all these parameters into account.

10.3 LOW POWER CIRCUITS

The design of low power circuits is, in general, a more precise affair than for large signal circuits, the reason being that it is often possible to measure small signal impedances with a high degree of precision on the active element before the final microstrip circuit is built. We shall consider three examples of common 'receiver' circuits, although the third example brings us back to the domain of non-linear operation (limiter).

10.3.1 Schottky diode balanced mixer

The balanced mixer is one of the best known microwave circuits; it lends itself readily to fabrication in microstrip form.

Figure 10.6(a) shows the basic circuit: the signal and local oscillator are fed to a pair of diodes via the ring coupler made up of $\lambda/8$ sections. An additional quarter wavelength of line is added before diode D1 so that the signal has the same phase at both diodes. After mixing, the two outputs are combined at the i.f. after suitable low-pass filtering.

Figure 10.6(b) shows the complete microstrip version. The

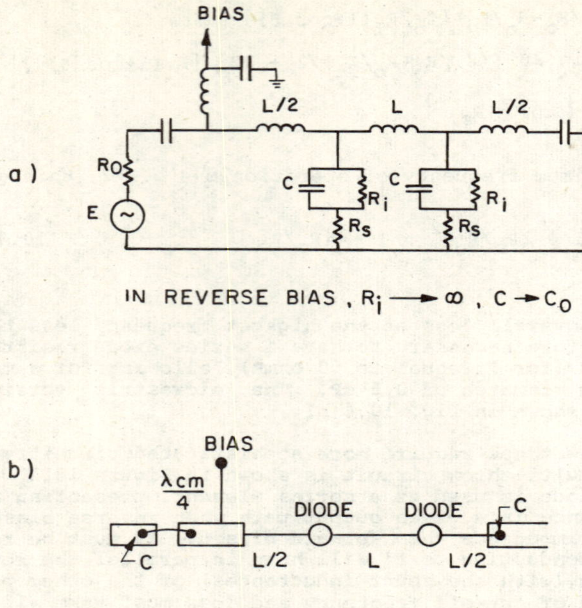

Fig. 10.4 PIN diode SPST switch.

(a) Basic circuit
(b) 'Microstrip' version

ring is grounded to d.c. signals by the series $\lambda/4$, shunt $\lambda/4$ sections shown on the right. Note the mounting of the diodes in opposition. The i.f. filters consist of high impedance series sections and low impedance shunt sections.

The balanced mixer has the advantage of being practically immune to local oscillator noise. The circuit shown has been realised [3] with noise figures of 7.5 dB in X-band; outside case dimensions are 3.1 cm by 3.1 cm.

10.3.2 Parametric amplifier

The theory of operation of the parametric amplifier is discussed in Chapter 8. For design purposes we require two basic equations:

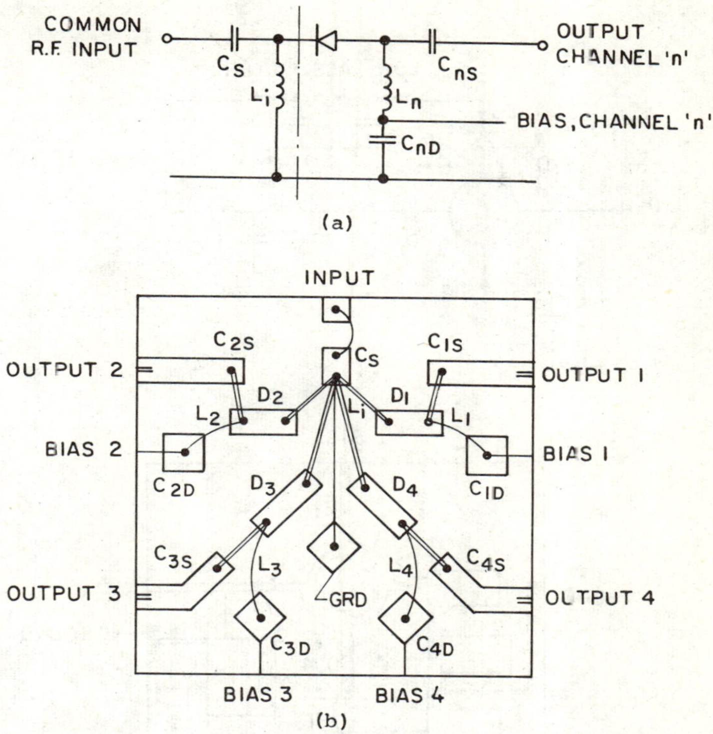

Fig. 10.5 PIN diode one way and four way switches.
(a) Circuit of one channel of SPMT switch
(b) Microstrip version of 4 way PIN switch

$$G_{eq} = K^2\ (\omega_s c_o)^2\ \ (\omega_3/\omega_s)\ G_3 \tag{10.8}$$

$$F = 1 + (\omega_s\omega_3/\omega_c{}^2K^2) + \omega_1/\omega_3 \tag{10.9}$$

Here, G_{eq} is the value of the negative conductance which appears at the signal frequency ω_s, ω_3 is the idler frequency $\omega_p - \omega_s$, where ω_p is the pump frequency, G_3 is the conductance in parallel

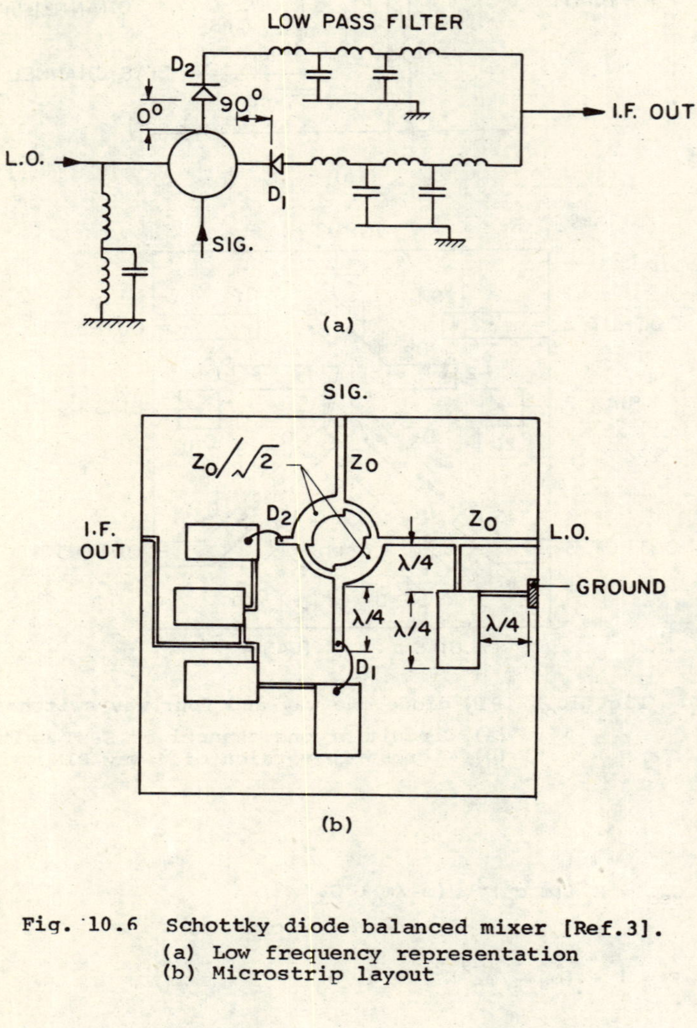

Fig. 10.6 Schottky diode balanced mixer [Ref.3].
(a) Low frequency representation
(b) Microstrip layout

with the diode at the idler frequency and $K = \gamma/2$ where $\gamma = 1/2$ for an abrupt junction and 1/3 for a linearly graded junction.

It is important to note that the value of G_3 and hence G_{eq} is determined by the diode (and possibly circuit) losses at the idler frequency. We can rearrange eqn.(10.8) in the form:

$$R_{eq} = (1/\omega_1 C_o)(\omega_c/\omega_1)\gamma^2/4 \qquad (10.10)$$

where R_{eq} is the series negative resistance at the signal frequency and ω_c is the cut-off frequency $(2\pi R_s C_o)^{-1}$ of the diode at the operating bias.

Since the power gain is determined by the magnitude of the load at the signal port, referred to the diode terminals, it is quite clear that for a given gain we must arrange the input (signal frequency) circuit to give the required impedance transformation ratio. At microwave frequencies it is usually necessary to take the case capacitance into account, together with the diode lead inductance, as shown in Fig. 10.7(c). It can be shown that the resistance R_{eq}' at the input may be expressed as:

$$R_{eq}' = (-R_{eq} + R_s)/(1 + n^2 C_a/C_o)^2 \qquad (10.11)$$

$$\text{where } n^2 = 1 - (\omega_1/\omega_r)^2, \qquad \omega_r = 1/L_s C_o \qquad (10.12)$$

Since the gain and negative resistance are related by:

$$G = 4/(1 - R_{eq}/R_g)^2 \qquad (10.13)$$

It is a simple matter to design the required input microstrip circuit.

Figure 10.7(d) shows one example of a microstrip parametric amplifier designed to operate at 13 GHz with a gain of 10 dB and a noise figure of 3.5 dB[5]. The diode has a cut-off frequency of 500 GHz, $C_o = 0.3$ pF, $C_a = 0.18$ pF. The idler frequency is chosen to be the self resonant frequency of the diode. The idler stub therefore consists of a quarter wavelength of line (at the idler frequency, in this case 27 GHz). The idler stub is tapered to give a wide bandwidth. At the signal frequency the 50 ohm input is converted to the desired resistance value by use of, in this case, two quarter wave sections. As shown in the Figure, this enables a larger transformation ratio to be obtained without unduly wide or narrow strips. The capacitive component of Z_{in} was compensated for by making the first $\lambda/4$ section slightly longer than $\lambda/4$ (c.f. eqn. 2.7).

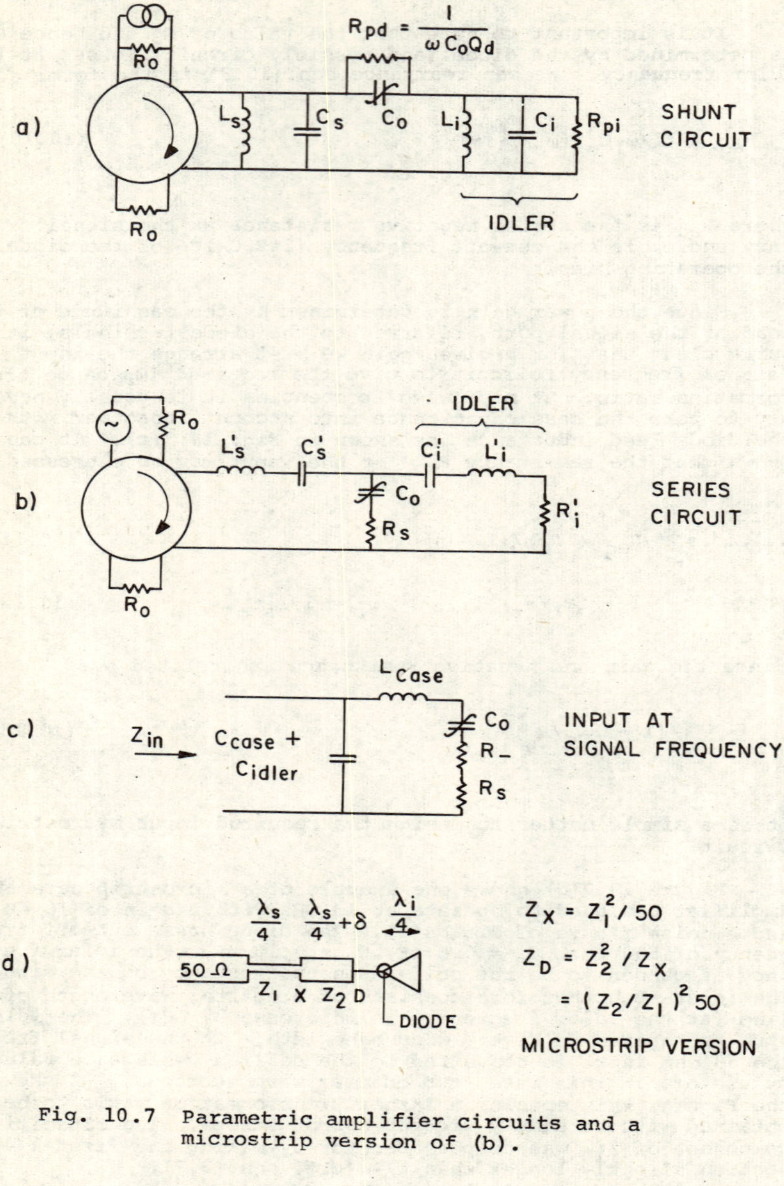

Fig. 10.7 Parametric amplifier circuits and a microstrip version of (b).

The pump in this circuit was introduced at 40 GHz via a waveguide to the right of the diode. For a diode operating at 0 Volt bias, no bias circuit is required. In this case the pump frequency was the limiting factor in determining the noise figure. The 'optimum' frequency relation $\omega_3 = \omega_c\gamma/2$ would imply a pump at about 125 GHz. It should be noted, however, that the loss in noise figure is not so important as might be expected. Equation (10.9) yields F = 1.58, or 2.0 dB. The circulator (0.8 dB), input circuit (0.4 dB) and idler (0.4 dB) bring the overall noise figure to 3.6 dB.

The above described amplifier [5] was made by first measuring the diode parameters with the diode connected directly to a 50 ohm length of microstrip, via an OSM connector to the Hewlett-Packard network analyser. The same measurement was then effected with the idler stub in position. By this means, a very accurate estimate of the input impedance was obtained before construction of the input transformer.

10.3.3 PIN diode limiter

The PIN diode limiter is very similar to the switch [6]. The main difference in the circuit, Fig. 10.8, is that a d.c. return path is provided in the circuit and the diode is operated at zero bias in the 'switch-on' state. At low input signal powers the diode impedance is high (essentially the depletion layer capacitance) and the losses are given by the formulae discussed for the switch in the previous section. When the input power increases, the diode impedance decreases and the signal is thus attenuated. At very high powers, the diode impedance reaches a near constant value R_s and the output power again rises linearly with input power, but with a large, fixed, attenuation. Figure 10.8(c) shows a typical set of characteristics. Notice that, like the PIN diode switch, the losses increase at high frequencies.

10.4 PIN DIODE PHASE SHIFTERS

10.4.1 Introduction

A major application of phase shifters is in phased array radar (PAR) systems where these are used in quantities of thousands. A phased array consists of a number of antenna elements. The relative phase of radiation from each of these elements is adjusted electronically by using phase shifters.

There are two different classes of variable phase shifters: those using switching diodes (such as PIN diodes) and those using ferrites. Both the types are possible in MIC form. Phase shifters using ferrites have been discussed in Section 7.6. Design considerations for a PIN diode phase shifter for a phased array radar are discussed in this section.

The phased array antennas in PAR are used both for transmitting and receiving the signal. Therefore, the phase shifters have to be reciprocal. Since the array is switched digitally with the help of an electronic computer, digital phase shifters are used. For quick movements of the beam, switching times less than

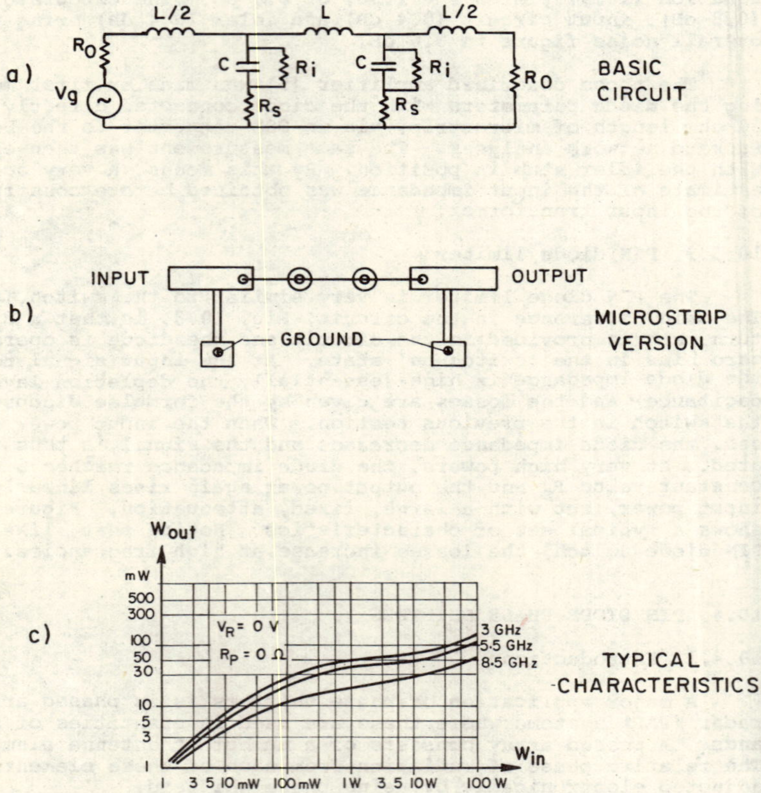

Fig. 10.8 PIN diode limiter [Ref.3].

1 microsecond are necessary.

The above conditions are fulfilled by phase shifters using PIN diodes. Usually the design philosophy is to construct 1 bit phase shift modules and to cascade four or five of these to get the digital phase shifters of the desired quantization. Different 1 bit sections can have different circuit configurations also.

In a phased array, the individual elements are spaced at distances of about half wavelength ($\lambda/2$) and therefore small size phase shifters (less than $\lambda/2 \times \lambda/2$) are desirable.

The design example described in this section is suitable for a one bit phase shifter in hybrid integrated circuit form. The selection of circuit configuration out of various alternatives available is discussed. Design of 3 dB hybrid which forms a part of the circuit used is described. Circuit layouts for circuits on alumina and polystyrene substrates are given. The considerations for selecting the PIN diodes required are also discussed.

10.4.2 Selection of circuit

There are mainly two types of semiconductor diode phase shifters. Firstly, the transmission type and secondly the reflection type. In the transmission type phase shifters, the phase shift is obtained by perturbing the parameters of the transmission line. On the other hand in reflection type phase shifters, the phase shift is obtained by changing either the value or the location of the terminating impedance. A hybrid coupler or a circulator is used to separate the input and the output in the reflection type phase shifter.

Transmission type phase shifters are either of periodically loaded line type or of switched line type. Reflection type phase shifters are also of two types namely of switched reactance type or switched length type.

Both the reflection type and the transmission type phase shifters can be fabricated in hybrid integrated circuit form. Figure 10.9 shows the different types of phase shifters.

The choice of a circuit configuration for the phase shifter depends upon factors such as the number of diodes required, the insertion loss, the power handling capacity, the ease of fabrication and the cost. These factors have been analysed in literature [8, 9, 10] for different types of phase shifters. Table 10.1 summarises some of these results.

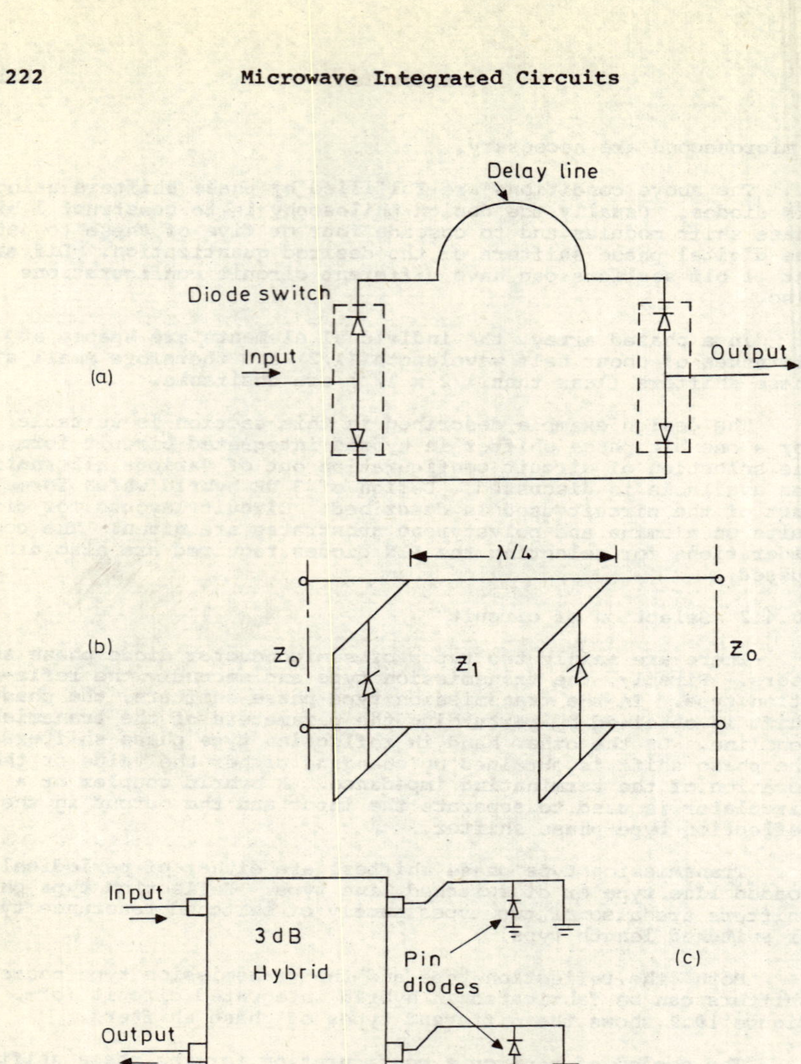

Fig. 10.9 Different type of phase shifters
(a) Switched line transmission type one bit phase shifter.
(b) Loaded line transmission type one bit phase shifter.
(c) Hybrid coupled reflection type one bit phase shifter.

Table 10.1 Comparison of diode phase shifters for peak power, loss and number of diodes required

(V_d = peak rf voltage rating of the diode, ϕ = phase shift)

	Loaded line transmission type: Stub mounted	Loaded line transmission type: Shunt mounted	hybrid coupled reflection type	Switched line transmission type
Peak power capacity	$\frac{V_d^2}{8Z_o \sin^2\phi/2}$	$\frac{V_d^2}{8Z_o}$	$\frac{V_d^2}{4Z_o \sin^2\phi/2}$	$\frac{V_d^2}{2Z_o}$
Loss (dB)	$17.4f/f_c$ tan $\phi/2$	$17.4f/f_c$ tan $\phi/2$	$17.4f/f_c$ sin $\phi/2$	$17.4f/f_c$ (per bit)
Total diode loss for 4 bits	$52\ f/f_c$*	$52\ f/f_c$*	$40\ f/f_c$	$69\ f/f_c$
Number of diodes	30*	30*	8	16

* Assuming 15 sections of 22.5° phase shift each.

From these results, one finds that a hybrid coupled (reflection type) phase shifter has the least insertion loss and uses the minimum number of diodes. For this type of the phase shifter, the breakdown voltage requirement for the diodes is highest for 180° 1-bit phase shifter but is reduced by a factor of $\sqrt{\sin(\phi/2)}$ for smaller values of phase shift [9].

As mentioned earlier, there are two types of hybrid coupled reflection type phase shifters, switched reactance or switched length type. In the case of switched reactance type phase shifter, lumped inductances and capacitances are generally used for line terminations.

Based on the above considerations a switched length, hybrid coupled, reflection type phase shifter has been selected as an example for detailed considerations. This phase shifter uses a 3-db hybrid ring with PIN diode switches and shorted lines connected to the two coupled ports of the hybrid as shown in Figure 10.9(c).

10.4.3 Design of the Hybrid

Most commonly used hybrid (for narrow band operation) is a rat-race ring. Layout for the rat-race ring is given in Figure 10.10. A detailed analysis of this hybrid ring is given in [11]. From this analysis, it is found that the value of impedances of all the sections of the ring is required to be $\sqrt{2}Z_o$ for a 3 dB hybrid, as shown in Fig.10.10, where Z_o is the characteristic impedance of the connecting arms. For an input at port I, the outputs at ports III and IV, are 180° out of phase.

Another type of hybrid ring which is based on the principle of branch-line coupler is shown in Fig. 10.11. Unlike the rat-race ring, it has branches of equal lengths but of different impedances. This hybrid is slightly smaller in size as the circumference is only one wavelength, λ, compared to 3λ/2 in the rate-race type. Coupled-line 3 db directional couplers may also be used in place of hybrid rings, but, because of small spacing between the lines, their fabrication in hybrid MIC form is not very convenient.

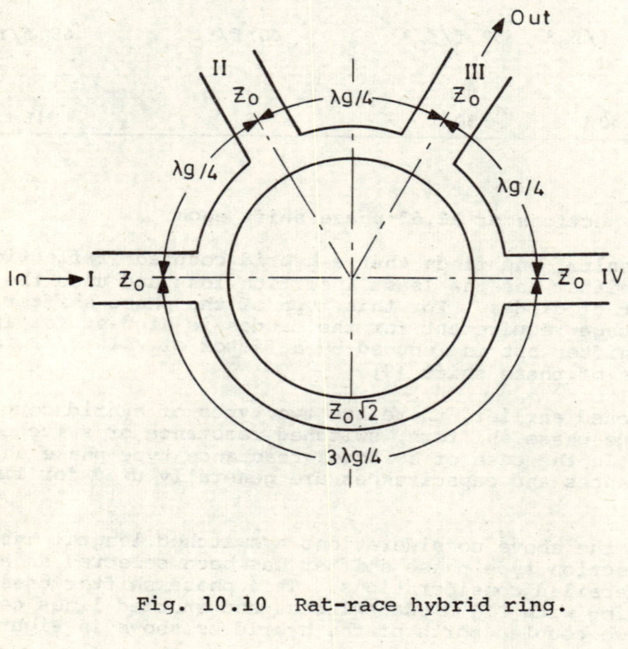

Fig. 10.10 Rat-race hybrid ring.

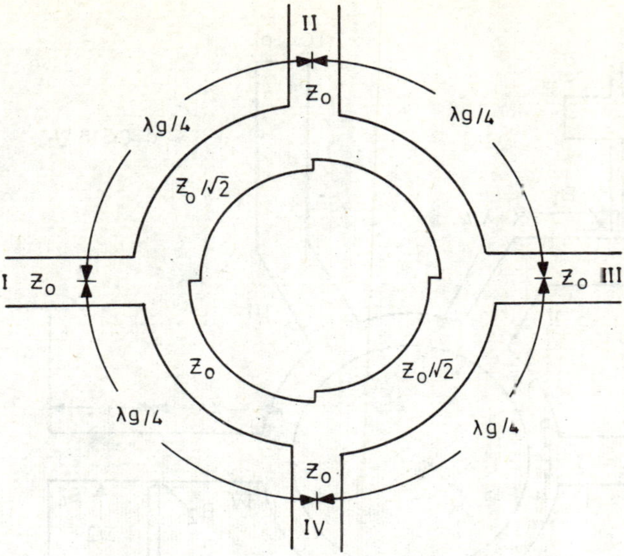

Fig. 10.11 Branch line hybrid ring

10.4.4 Circuit layouts

Figure 10.12 shows the circuit layout of a hybrid coupled one bit phase shifter using 3-db rat-race ring. The input is given at port I and output is obtained at port III. Ports II and IV are terminated in short circuits at the ends of the transmission line sections connected to ports. B_1 and B_2 are the diode positions.

The input applied at port I divides in two parts, one travelling in direction 'a' and other in direction 'b' as shown in Fig. 10.12. These are reflected back from ports II and IV either at B_1 and B_2 or at B_3 and B_4 depending upon the state of the diode switch (we have assumed an ideal diode switch). Thus the length Y, between B_3 and B_1 or B_2 and B_4 determines the phase shift. The phase shift is equal to twice the electrical length Y. This assumes that the PIN diodes are ideal and effects of package parameters are not taken into account. An extra length $\lambda/4$ is added to the line in arm II so that outputs coming from ports II and IV add up in phase at port III, the output port.

The widths of the microstrip lines of different impedances have been calculated, for alumina and polystyrene substrates.

The formula used is as given by

$$Z_o = \{87/(\varepsilon_r + 1.41)^{1/2}\} \ln \{5.98d/(0.8W + t)\} \qquad (10.14)$$

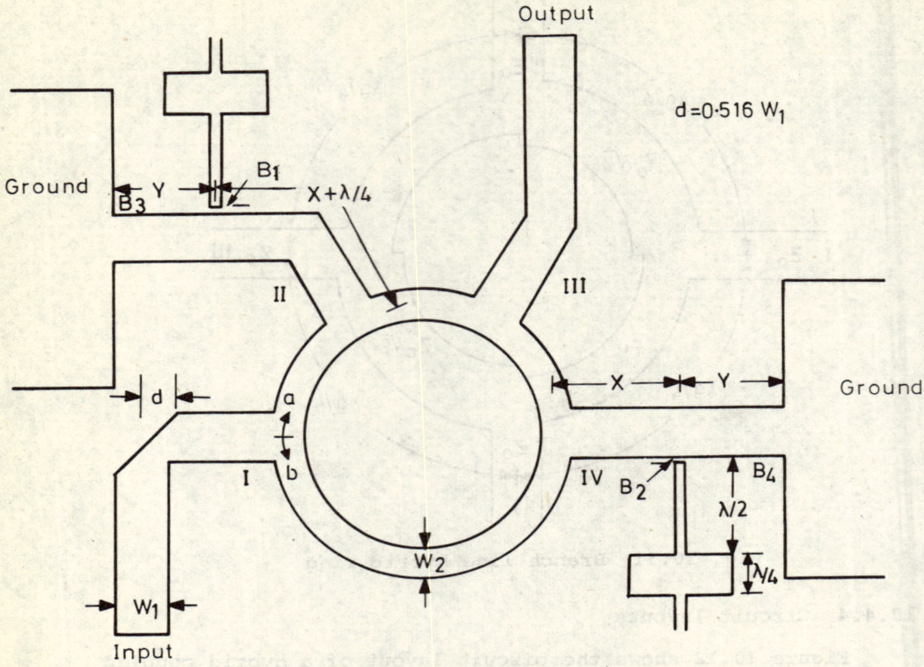

Fig. 10.12 One bit phase shifter using rat-race hybrid ring.

where

Z_o = Characteristic impedance

ε_r = Relative dielectric constant of the substrate

d = Thickness of the substrate

W = Width of the line

t = Thickness of the strip

This formula has limited accuracy for high W/d ratios. For such cases formulas given in [13] can be used.

The guide wavelength was calculated using the values of effective dielectric constant.

$$\lambda_g = \lambda_o \sqrt{\varepsilon_{reff}} \qquad (10.15)$$

where ε_{reff} is the effective dielectric constant. Among the many formulas available for calculating ε_{reff} the one given in [12] is used. This takes into account the dispersion in microstrip lines. Values of the line widths and guide wavelengths for different substrates are given in Table 10.2. A circuit used for biasing the diodes is also shown in Fig. 10.12. It consists of a half wavelength line of high impedance (≥ 100 ohms) followed by a quarter wavelength line of low impedance. This gives a low impedance at the diode location. Other type of biasing circuits, discussed in section 2.5 can also be used.

10.4.5 Selection of the diodes

As mentioned earlier, the diode switches have been assumed to be ideal and thus the diode characteristics have not been introduced in the design so far. But in general, the performance of a phase shifter is described in terms of loss, power handling ability, band width and switching time. These parameters can be related directly to the basic diode switch characteristics.

Technical data sheets for different types of switching diodes specify mainly the following parameters.

(a) Total power dissipation (depending upon thermal resistance of the diode)

(b) Peak power handling capacity

(c) Reverse breakdown voltage

(d) Switching time

(e) Operating and storage temperature

and (f) the different package styles available.

These parameters are choosen according to the requirements of the phase shifters. An analysis of a hybrid coupled 180° one bit phase shifter gives the peak power capacity[10].

$$P = V_d^2/4Z \tag{10.16}$$

where

V_d = breakdown voltage and,

Z = optimum impedance level for equal loss in both switching states.

Thus for high peak power capacity, a PIN diode with large breakdown voltage is desirable. Most of the PIN diodes available, usually satisfy the switching time requirements. Diodes in beam-lead packages or in chip form are well suited for the circuit discussed here.

Table 10.2: Dimensions for phase shifter (Fig.10.12) on different substrates.

Substrate Material	Polystyrene	Polystyrene	Alumina
Relative dielectric constant (ε_r)	2.55	2.55	9.7
Thickness of Substrate (d)	62.5 mils	125.0 mils	25.0 mils
Thickness of conductor (t)	5 mils	5 mils	2 mils
Frequency (f)	3 GHz	3 GHz	9 GHz
Width of line, W_1, having impedance = 50 ohms	142.619 mils	291.49 mils	25.02 mils
Width of line, W_2, having impedance = 50 $\sqrt{2}$ ohms	86.896 mils	180.04 mils	10.04 mils
Effective dielectric constant	2.059	2.084	6.336
Guide wavelength	2.743 in.	2.726 in.	0.521 in.

References

1. Roulston, D.J. "Frequency multipliers using charge storage diode in an inductive circuit", Proc. IEEE, Vol.55,pp.1220-1221, July 1967.

2. Obregon, J., Zamansky, P., "Calcul des lignes microstrips et application aux oscillateurs a avalanche", Revue Technique Thomson-CSF, Vol.3, pp.5-18, March 1971.

3. Thomson-CSF catalogue for DMH (MICs) - 1972.

4. Funck, R., "Etude et realisation d'une preserie de modulateurs limiteurs en bandes S-C et X" Internal Thomson-CSF report No. 1825, May 1970.

5. Obregon, J., "Amplificateur parametrique integre a 13 GHz", Thomson-CSF internal report no. 540, March 1971.

6. Funck, R., "Realisation de dispositifs integres microondes: limiteur" internal Thomson-CSF report no. 1619, Aug. 1967.

7. Kaupp, H.R., "Characteristics of Microstrip transmission lines", IEEE Trans. on Electronic Computers, Vol.EC-16, No.2, pp 185-193, April 1967.

8. Burns, R.W. and L. Stark, "PIN diodes advance high power phase shifting", Microwaves, Vol. 4, pp 38-48, November 1965.

9. Stark, L., R.W. Burns and W.P. Clark, "Phase shifter for arrays" Radar Handbook, Ed. M. Skolnik, McGraw-Hill, N.Y., 1970, pp 12-45 - 12.65.

10. White, J.F., "High power PIN diode controlled microwave transmission phase shifter", IEEE Trans. on Microwave Theo. and Tech., Vol. MTT-13, pp 233-242, March 1965.

11. Chuck, Y.P., "Hybrid ring directional coupler for arbitrary power divisions", IEEE Trans. on Microwave Theo. and Tech., Vol. MTT-9, p 529, Nov. 1961.

12. Jain, O.P., V. Makios, "Coupled model of dispersion in microstrip", Electronics Letters, Vol. 7, No. 14, 15th July, 1971.

13. Schneider, M.V., Microstrip lines for microwave integrated circuits, B.S.T.J., Vol. 48, No.5, May, 1969.

11

TECHNOLOGY OF HYBRID MICs

by

Amarjit Singh* and K.C. Gupta**

* Sections 11.1 to 11.5
** Sections 11.6 and 11.7

Chapter Eleven

11.1 INTRODUCTION

Advancement in the state of the art of microwave integrated circuits has been dependent on the development of suitable materials and processes. In the case of monolithic integrated circuits, the materials and technology for passive and active components are very closely interlinked. This is not so in the case of hybrid technology which has been mainly used for MICs so far.

The materials can in this case be broadly divided into the categories of substrates and thick or thin films which are deposited on them. Semiconductor substrates are used essentially in monolithic MICs and are treated separately. In the following, we take up first dielectric substrates. The materials used for films, the techniques of deposition on substrates, and the subsequent processing are different for the cases of thick and thin films. They are, therefore, discussed separately, first for thick and then for thin films, followed by a discussion on methods of testing as well as encapsulation and mounting of devices.

11.2 DIELECTRIC SUBSTRATES

In choice of substrates for different applications, one has to optimise different characteristics, keeping in mind the general requirements. The most important parameters are as follows:

a. Loss tangent: In order to use the substrate at microwave frequencies, the losses must be kept low. These are frequency dependent. The requirements on purity and nature of material become more stringent at higher frequencies.

b. Dielectric constant: The higher the dielectric constant, the larger the reduction of guide wave length, and the greater is the degree of miniaturisation in relation to the wave length in free space for the operating frequencies. Too high a dielectric constant can create problems of propagation of spurious modes in the case of striplines or of excessive miniaturisation in the case of lumped elements. Typically, the dielectric constant is near about 10 for alumina substrates which are the ones most commonly used.

c. Surface finish capability: Losses at microwave frequencies are heavily dependent on the degree of surface finish[1], the fineness of finish required increasing with increasing frequency. Typically for use at X-band the r.m.s. roughness must not exceed 0.1 micron.

d. Adherence qualities of deposited films: This parameter is also dependent on the degree of surface finish. Adherence becomes more difficult as the surface becomes more highly polished. Higher degree of polish is also necessary in the case of thin films so as to avoid a break in the film due to surface roughness. In order to get over this problem, in thin film deposition, a very thin layer (a few hundred A° thick) of an active metal such as chromium or titanium is used as an intermediate layer.

e. Temperature coefficient of parameters, such as dielectric constant: This has to be low in order to ensure stability of the circuit over a range of operating temperatures.

f. Dielectric field strength: Insulation properties and the ability to withstand voltages applied, decide the power levels that can be obtained. It may be mentioned here that in hybrid circuits considerably higher levels are obtainable than in monolithic ones, because of the insulation properties of the dielectric.

g. Mechanical strength and uniformity: In processing, specially of thick films, uniformity of thickness in the substrates and from substrate to substrate is very essential in order that the deposited thick film may be uniform. A typical degree of uniformity is 0.5 per cent.

h. Thermal conductivity: This is particularly important where high heat dissipation rates are encountered.

i. Cost: This has to be optimised in relation to the application in view.

Table 11.1 summarises the properties of some typical substrate materials as regards surface roughness, loss tangent, dielectric constant and thermal conductivity[2]. High alumina ceramics are the most commonly used material, because of a desirable range of parameters for most applications at frequencies upto X-band. An alumina content of the order of 99.5 per cent becomes necessary in order to ensure surface smoothness and low loss tangent. At still higher frequencies sapphire has to be used. Fused quartz can also give a high degree of surface polish as well as low loss tangent. However, its dielectric constant being lower, it is more suitable for use with lumped elements. Beryllia has the highest coefficient of thermal conductivity. It may be used as an insert in a combined substrate where the rate of heat transfer is high and a metal cannot be used. Glass can be used only at the lower frequency end, because of the high loss tangent. Ferrites and garnets are used as substrates or inserts where non-reciprocal elements are required. Silicon and Gallium Arsenide are used in monolithic integrated circuits. In the case of silicon it is difficult to maintain a high resistivity in various processes of heat treatment.

11.3 THICK FILM TECHNOLOGY AND MATERIALS

The basic process in this case involves the deposition of certain kinds of pastes (also called "inks") on substrates through a screen printing process followed by firing at high temperatures. Figure 11.1 shows a generalised flow diagram. Since different alternatives are possible, these have been indicated as steps that may be taken if necessary. For instance, where different types of pastes are used, more than one cycle of screen printing, drying and firing would be required. Etching would be introduced where further improvement in delineation is necessary. In those cases where several circuits are made on the same substrate, (e.g. for lumped element circuits), they would be separated

Table 11.1 Properties of substrates [2]

Material	Surface Roughness Δ (μm)	$\tan \delta \times 10^4$ (10 GHz)	ε_r	K W/cm^2/°C
Alumina (99.5 per cent)	2-8	1-2	10	0.3
(96 per cent)	20	6	9	0.28
(85 per cent)	50	15	8	0.20
Sapphire	1	1	9.3 - 11.7	0.4
Glass	1	20	5	0.01
Quartz (fused)	1	1	3.8	0.01
Beryllia	2-50	1	6.6	2.5
Rutile	10 - 100	4	100	0.02
Ferrite/garnet	10	2	13 - 16	0.03
GaAs (high resistivity)	1	16	13	0.3
Si (high resistivity)	1	10 - 100	12	0.9

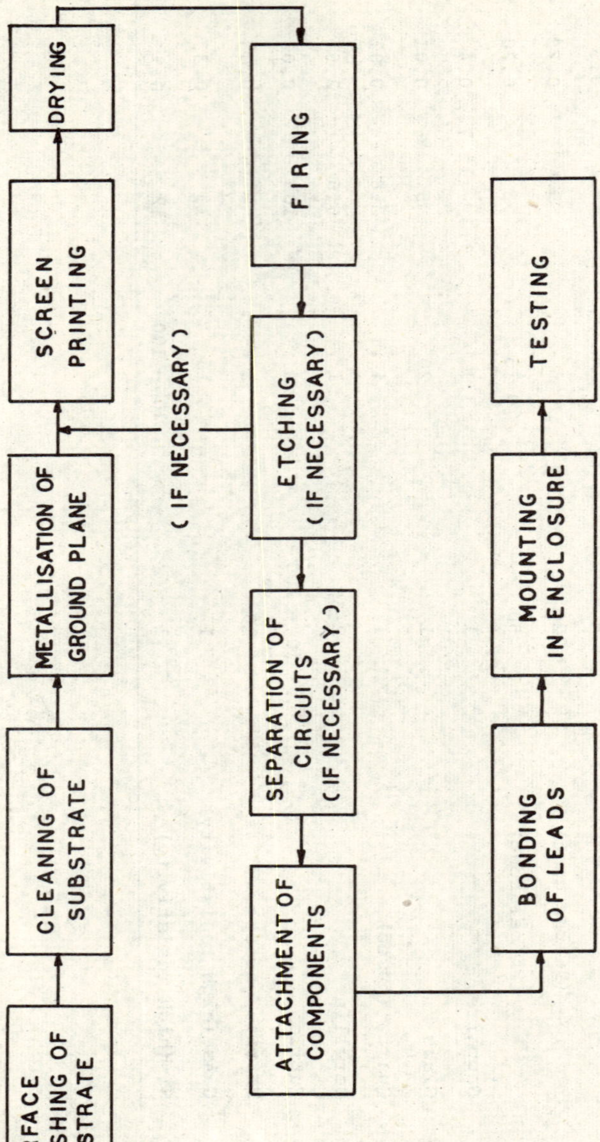

Fig. 11.1 Possible sequences of steps for fabrication of thick film MICs

before further processing.

Different pastes have been developed to give conductors, resistors and insulating layers. The pastes contain a suspension of noble metals or metal-oxides with glass frit and organic binder. The least expensive of the metal combinations is palladium-silver. However, from the point of view of strength and stability of soldered joints, palladium-gold is preferable[3]. Platinum-gold and fine gold have also been used in the paste for conductors. For insulating layers a mixture of barium titanate and glass frit is employed where a high dielectric constant is required. The pastes have to be suitably stirred. Some pastes have to be continuously stirred or rolled and others have to be stirred only before application. Vigorous stirring can also cause harm, because of a change of viscosity and by trapping air in the paste.

The "silk" screen commonly used is a metal screen using stainless steel wire with a mesh of the order of 105 to 500. It has a photo-resist layer on which the pattern is transferred from the original art work.

In the process of printing, accurate positioning of the pattern with reference to the substrate is essential so that the sequences of printing processes are in proper register. The printing machine is so adjusted as to ensure the separation of 10 to 60 mil between the screen and the substrate. The sqeegee travels at a rate of about 2 - 10 inch per second and has an angle of attach of the order of 45°. After the sqeegee passes over the screen surface, the latter should lift up from the substrate. Proper gap between the two in the normal position, is necessary in order to give sharp lines and to avoid smearing.

After printing,the substrate is allowed to lie for a few minutes, to permit levelling of the paste to take place, thus removing the marks of the mask wires. The substrate is dried for about 15 minutes at about 100°C to 175°C. The temperature has to be raised slowly, so as to avoid formation of bubbles. Drying is followed by firing in which the binder is burnt off and the glass frit reflows.

This firing is a rather critical operation. The final electrical parameters of the ink patterns depend not only on the temperature to which the substrate was fired, but also on the rate of rise of temperature. This rate usually lies between 30 to 70°C per minute. In order to ensure a correct rate of rise and desired final temperature, the furnace normally has about 4 zones whose temperatures are carefully controlled. The rate of movement of the substrates on the conveyor belt has also to be controlled in order to get good reproducibility of the properties of thick film components. The final firing temperature lies between 500 - 1000°C depending upon the nature of paste used. Subsequent operations have to be at successively lower temperatures. It is normally not necessary to have a protective atmosphere, a draft of the clear fresh air at about 2-5" per second being adequate. However, it is necessary to use a non-oxidising atmosphere, if copper is used in the conductive paint.

Since the conducting layers consist of metal particles in a glassy medium, the conductivity of the films is considerably less than that of the bulk metal, the percentage being as low as 10 per cent.

The delineation can be improved by etching. However, even so, the boundaries are not as straight and sharply defined as is possible in the case of thin films, which consist of simpler materials[4]. A comparison of the conductor thickness and width commonly achievable in the case of thick and thin film technology, is given in Table 11.2. However, in more recent literature [10], line thickness down to 4 mil (100μ) has been reported, by use of 500 mesh screen.

The thick film process has the advantage of lower capital cost and absence of stringent requirements regarding clean environment. However, because of the limitations regarding conductivity and delineation of the conducting paths, this process is useful only in the lower part of the microwave spectrum. One of the problems is that the materials technology of the pastes and the processes which go on inside the paste as it is fired, are not fully understood.

11.4 THIN FILM TECHNOLOGY AND MATERIAL

Thin films are generally less than 5 microns in thickness. However, the term now applies more to the technology than the thickness itself, being usually associated with deposition of films by vacuum evaporation or sputtering: The films are often of a composite nature, so that more than one source material is required in the same vacuum envelope.

Chemical vapour deposition and plating can be used to add layers of differnt materials to those deposited by evaporation or sputtering.

Figure 11.2 shows possible sequences of steps for fabrication of thin films MICs, with different alternative routes indicated. These depend upon techniques used for adding on layers of different materials. Some of the initial and final steps are common with thick film circuits.

Vacuum evaporation: This process is applicable to simple substances with vapour pressures in the range of 1 to 10 x $10^{(-3)}$ Torr[5, 6]. The source is maintained at the required temperature either by tungsten heaters or by RF heating of boats of molybdenum, carbon or boron nitride. Focussed electron beams are also used to evaporate the materials. This eliminates contamination of source but is a more costly and complex process. Deposition of alloys is more difficult to handle by this process because of differential evaporation leading to change of composition. This problem can be mitigated by flash evaporation.

The equipment used often has more than one filament and source available, as well as shutters and masks. The latter have been found successful only for broad delineation. The most commonly employed means of delineation of pattern is by etching

Table 11.2 Comparison of conductor thickness and width in thick and thin-film technologies[4]

	Thin films	Thick films
Conductors		
Line width, μm	Etched : 12	Etched : 50
	Stencil mask : 127	Screened : 254
Film thickness, μm	2 to 10	25
Sheet resistivity, ohms/square	0.005 - 0.01	0.01 - 0.1

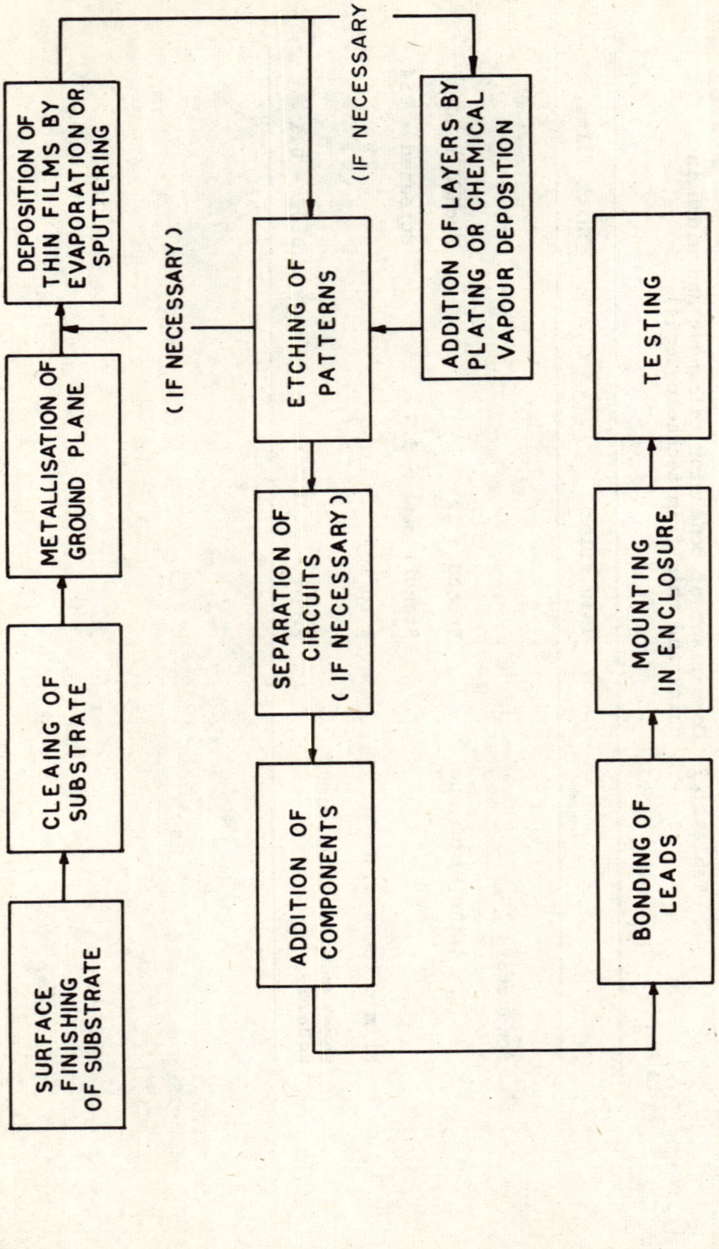

Fig. 11.2 Possible sequences of steps for fabrication of thin film MICs

after deposition of the film.

Sputtering: In this case the source material is placed as the cathode in a gas discharge. The impact of positive ions leads to evaporation of material and diffusion to the nearby substrates where deposition of thin films is required. The process can be used with dc voltage applied to source (a target) if it is a conducting material. If an insulator were to be used as a source, with a dc applied voltage, its surface would get positively charged and the process of ion bombardment would stop. This problem is overcome in RF sputtering. In this case, the positive charge is neutralised by electrons in the negative half of the RF cycle, In fact, because of the higher mobility of the electrons, the surface acquires a high negative potential, in order to reach equilibrium with the electrons in the plasma, during the negative swing.

An inert gas such as Argon is used where the composition of the film is desired to be the same as that of the source. However, if the gas present can react chemically with the material, a layer which is chemically different from the source can be deposited. This process is called 'reactive sputtering' and is used in the deposition of films such as Si_3N_4.

As mentioned already, when the substrates are highly polished, it is difficult to obtain a proper mechanical bond between the thin film and the substrate. In such a case, a chemical bond becomes necessary for proper adhesion. Accordingly, a composite layer is used, in which a thin layer of an active metal is first deposited on the substrate which forms a good bond with the latter. Chromium and titanium are typical examples of such a first layer. The bond between a substrate and the film is formed by the oxide of the metal. This is followed by deposition of a conductive film such as copper or gold, or copper followed by gold. The thickness is built up to several skin-depths so as to minimise losses. In this case the conductivity of the film can be as much as 50 per cent of the value for the bulk metal. Chromium-gold and titanium-palladium-gold are combinations commonly employed. The latter combination is more suitable where the subsequent operations are likely to lead to high temperatures. Table 11.3 shows the properties of the conducting materials used in thin films[7].

Dielectric films can also be deposited by a process called chemical vapour deposition. A typical case is the deposition of silicon dioxide (SiO_2) by a reaction between silane (SiH_4) and oxygen. The properties of the films are improved by subsequent heating. Tantalum pentoxide is another material used for low loss dielectric thin films. The properties of various dielectric films are given in Table 11.4.

Etching: As indicated already, the final definition of patterns in the case of thin films is done by a process of etching. The etchants suitable for different materials are summarised in Table 11.5

Table 11.3 Conductor characteristics [7]

Material	DC Resistivity relative to copper	Skin Depth at 2 GHz (μm)	Surface Resistivity ohms/square ($\times 10^{-7}\sqrt{f}$)	Coeff. of Thermal Expansion (α_T/°C $\times 10^{-6}$)	Adherence to Dielectric Film or Substrate
Ag	0.95	1.4	2.5	21	Poor
Cu	1.0	1.5	2.6	18	Very poor
Au	1.36	1.7	3.0	15	Very poor
Al	1.60	1.9	3.3	26	Very good
W	3.2	2.6	4.7	4.6	Good
Mo	3.3	2.7	4.7	6.0	Good
Cr	7.6	4.0	7.2	9.0	Good
Ta	9.1	4.4	7.9	6.6	Very good
Mo-Mg	16.5	-	-	-	Very good

Table 11.4 Properties of dielectric films[7]

Material	ε_r	Dielectric strength (V/cm)	Microwave Q
SiO (Evaporated)	6-8	4×10^5	30
SiO_2 (Deposited)	4	10^7	100 - 1000
Al_2O_3 (Anodized or Evaporated)	7-10	4×10^6	-
Ta_2O_5 (Anodized or Sputtered)	22-25	6×10^6	100
Si_3N_4 (Vapor-phase)	7.6	10^7	-
(Sputtered)	6.5	10^7	-

Table 11.5 Etchants for selected deposited films[4]

Deposited Film	Possible etchant	Comments
Aluminium	Ferric chloride	Tantalum, titanium, gold, and platinum unaffected by these etchants
Chromium	Sodium hydroxide, Hydrochloric acid, Ferric chloride, concentrated hydrochloric acid	Sulfuric acid also will work but is less desirable.
Copper	Ferric chloride, Ammonium persulfate	Also for high-copper alloys
Gold	Aqua regia	
Molybdenum	Concentrated sulfuric acid, diluted nitric acid	Can also be etched anodically in chromic acid.
Nichrome	Concentrated hydrochloric acid and water	Suitable for nickel and nickel-based magnetic alloys
Platinum	Aqua regia	Also for platinum alloys
Silicon	Ferric chloride, concentrated nitric acid, and concentrated hydrofluoric acid	
Silicon dioxide	Saturated aqueous solution of ammonium fluoride	Removes about 100 nm of oxide per minute
Silver	Ferric nitrate	Dilute nitric acid may also be used but is less desirable
Tantalum	Concentrated hydrofluoric acid, diluted nitric acid	Gold and nichrome unaffected by this etchant, but on glass these may slough off if not protected.
Titanium	Diluted hydrofluoric acid	Gold, nichrome, and tantalum unaffected by this etchant

11.5 METHODS OF TESTING

Bulk parameters such as thickness can be tested by conventional methods. For examination of the film on a microscopic scale, scanning electron microscopes have been found extremely useful. These instruments have a relatively large depth of field and can give resolution of the order of 100A°. They make it possible to study various phenomena such as nature of the film at the boundaries, or cracks appearing in processing or in subsequent use. Of particular interest are those areas where the surface underneath the film is not perfectly level or smooth [8].

In order to evaluate the overall performance of a given type of film deposited on a substrate a standardised resonant circuit can be used[9]. The performance of the system is judged from the value of Q obtained in the resonator. In this manner, a large number of combinations of materials, substrates, thin and thick films and methods of processing can be evaluated. Some illustrative data are given in Table 11.6.

11.6 ENCAPSULATION OF DEVICES

11.6.1 Diode packages

Diode packages normally used for other types of transmission lines are cylindrical in geometry and use a ceramic tubing to separate two metallic studs. Typical geometry of such a package is shown in Fig. 11.3. The diode chip is mounted on one of studs. The contact to other stud is through a fine wire which is bonded to the diode. These diodes may be shunt mounted in strip line circuits but they are not suitable for mounting in microstrip circuits. A modified form of this package called the 'Pill' or 'Micropill' form shown in Fig. 11.4 has been used for shunt mounting in microstrip circuits on alumina substrates also. Shunt mounting of the 'micropill' package requires a hole to be drilled in the alumina substrate (usually with an ultrasonic impact grinder).

Diode packages that can be mounted in the plane of the microstrip are conveniently suited for use in hybrid MICs. Several type of packages are available for this purpose[11, 12]. The most commonly used ones are:

(i) Passivated chips

(ii) Beam lead devices

(iii) LID's (Leadless inverted device package)

11.6.2 Passivated chips

Typical dimensions and outline for the passivated chips normally available are shown in Fig. 11.5. Chips are available with moly-gold contact for cathode and anode ends. This contact system permits temperatures as high as 400°C for a maximum of 10 minutes without degradation of characteristics[13].

Table 11.6 Comparison of thin-film and thick-film processing methods of eight different substrate materials[9] (numbers indicate average O values at 7.9 to 12.4 GHz)

Substrate	Vacuum Deposition thin films with photoetched patterns			Screened and fired screened patterns			Thick films etched pattern
	Cr-Au film Au plate	Cr-Au film Cu plate	Cr-Au film Ag plate	Pt-Au	Au	Ag	Ag
Sapphire	-	382	-	-	-	-	198
Alumina (99 per cent pure)	251	195	148	63	138	151	171
Alumina (96 per cent pure)	137	140	128	58	120	141	145
Alumina (Glazed)	75	67	60	-	-	-	-
Beryllia (99 per cent pure)	83	72	65	47	56	90	112
Beryllia (Glazed)	56	54	58	-	-	-	-
Quartz	51	63	27	-	-	-	-
Glass (Borosilicate)	49	55	45	-	-	-	-

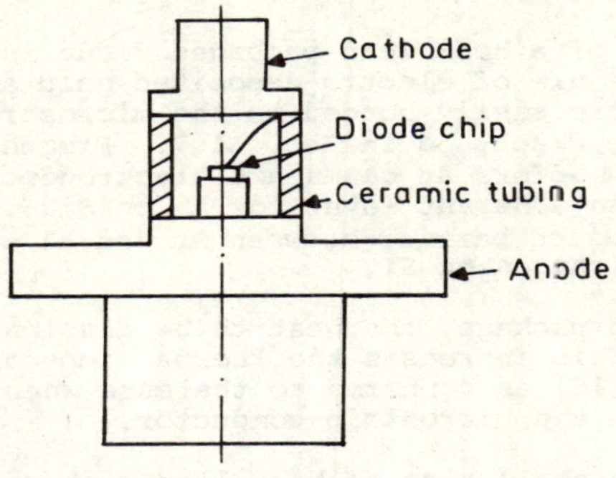

Fig. 11.3 Cylindrical stud package

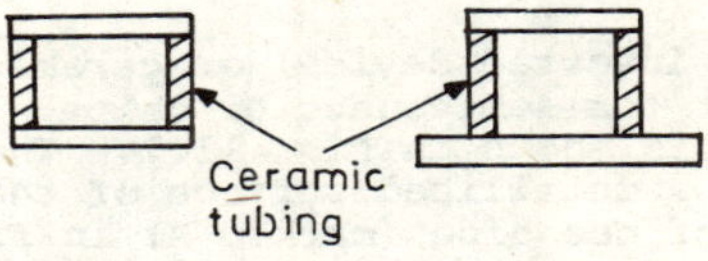

Fig. 11.4 'Pill' or 'Micropill' package

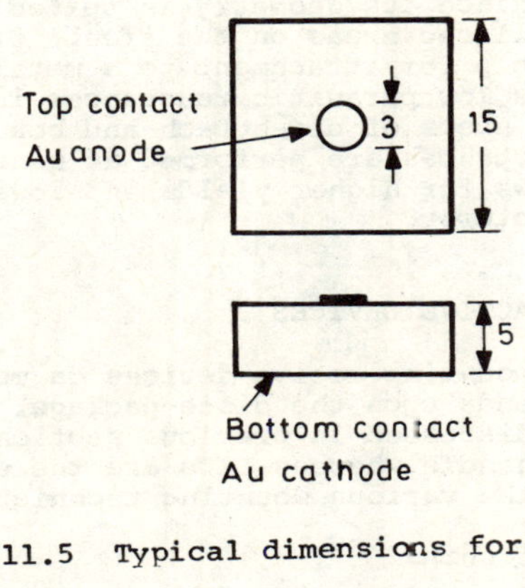

Fig. 11.5 Typical dimensions for chips

The use of passivated chips in hybrid MICs eliminates parasitic reactances associated with the packages. This can lead to improvement in the circuit performance. Also chips are less costly. It is claimed the cost of MICs can be brought down considerably by using diode chips and lumped version of passive elements[14].

11.6.3 Beam lead devices

The structure of a beam lead packaged diode is shown in Figure 11.6. Leads are of electro-deposited gold approximately 0.6 mil thick and are easily bonded to the microstrip conductor. Bonding methods are discussed in Sec. 11.7. Frequently Ti-Au or Mo-Au is evaporated before Au beams are electrodeposited[15]. Mo or Ti provides an adherant layer for Au on SiO_2. Also Mo or Ti provides a diffusion barrier between Au and Al when the latter is used for metallisation on Si.

In a beam lead package, the heat to be dissipated flows through the leads. This increases the thermal impedance of the beam lead packages[16] as compared to the case when the chip is mounted directly on the microstrip conductor.

Because of the small size of beam lead packaged diodes (as well as passivated chips) handling and positioning presents considerable difficulties.

11.6.4 LID package

LID (Leadless inverted device) or ceramic carrier package is a rugged carrier for semiconductor chips. Typical geometry of the LID package is shown in Fig. 11.7. The chip is mounted or die-attached on a metallised surface of the ceramic which extends to the foot of one side (marked A) in Fig. 11.7. A lead is bonded from the top terminal of the chip to the other side (marked B). Shaded portions are metallised and make contact with microstrip conductor when the device is mounted in MICs.

The LID package is easy to handle and is ideal for high volume applications since its geometry is suited for automated production. The metallised areas on the 'feet' of the package allow a single solder cycle for attachment to a metallised substrate. This package offers low parasitic reactances in a rugged structure. The critical steps of die-attach and bond as well as the measurement on reactances are performed at manufacturing stage itself. This allows for higher yields and lower assembly costs for the MIC manufacturer.

11.7 MOUNTING OF ACTIVE DEVICES

Technique of mounting active devices on metallised substrates in hybrid MICs depends upon the diode package. Out of the three types of packages discussed in previous section, chips are the most difficult to handle whereas LIDs are the easiest. This section discusses the various mounting techniques used.

11.7.1 Mounting of chips

Techniques of bonding the connecting leads to semiconductor chips have been discussed widely in literature[3, 7, 8].

There are two different processes involved in mounting the chips in hybrid MICs. The first operation is 'die bonding' so

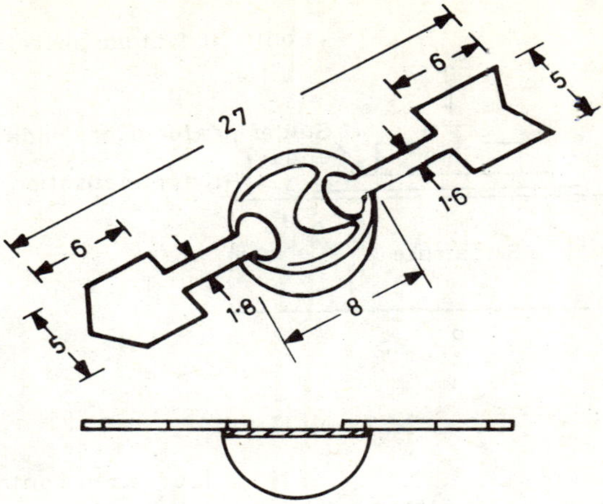

Fig. 11.6 Beam lead package

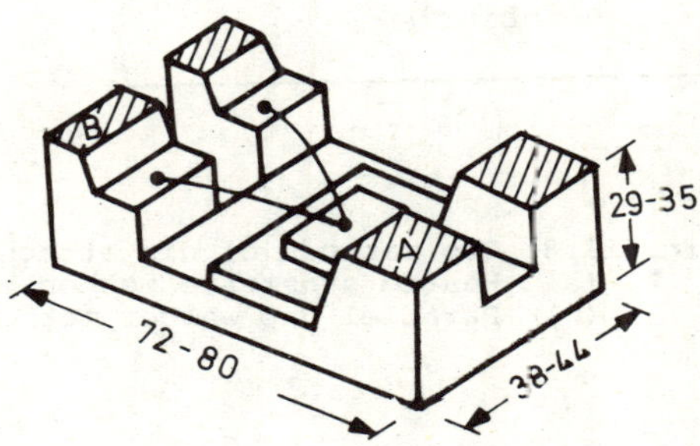

Fig. 11.7 LID or ceramic carrier package

that the bottom electrode makes contact with metallisation on the ceramic substrate. A solder preform or conductive epoxy is used for adhesion. Since a temperature of 300 - 400°C is used, a controlled(inert) atmosphere or at least a localised controlled atmosphere is needed. Usually, flush with nitrogen gas is used for this purpose. Two methods commonly used for die attaching are shown schematically in Figs. 11.8(a) and (b). In the scheme

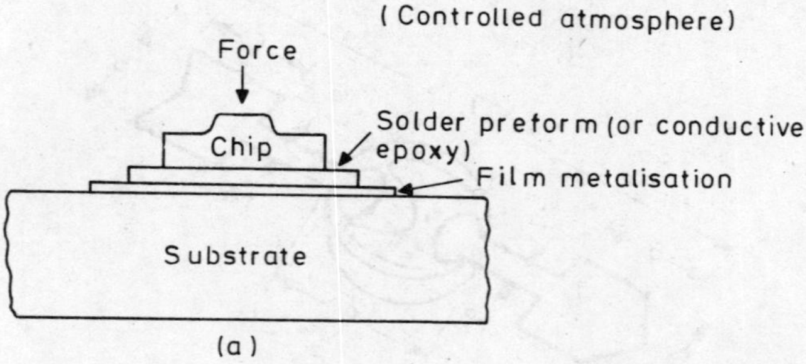

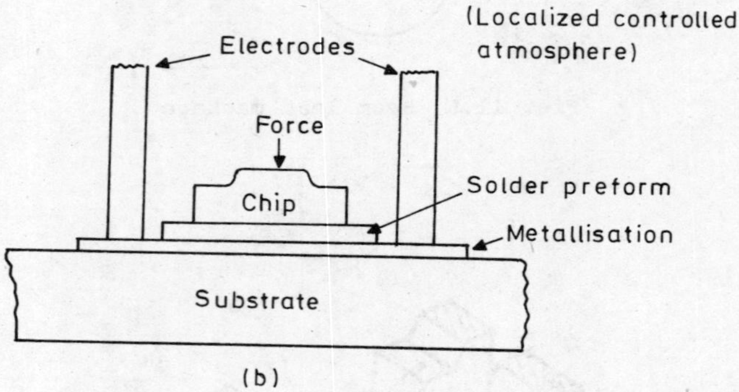

Fig. 11.8 Two methods of die attach
(a) Heated substrate method
(b) Parallel gap welder method

depicted in Fig.11.8(a) the substrate is heated by holding it in a suitable electrically heated holder. In Fig. 11(b) a parallel gap welder is used.

Substrate heating, as in Fig. 11.8(a), requires a smaller pad area for mounting the chip than that needed for localised heating technique of Fig.11.8(b). Generally a 5 mil border around the chip is adequate.

In the localised heating method (using parallel gap resistance heating technique) controlled current flows between the electrodes as shown. This heats the area under the chip so that the solder preform melts and wets both the surface metallisation and the chip. This method provides ease of handling as the chip pick up can be integrated with the weld electrodes. This method also offers a speedy melt cycle, the use of localised atmosphere controls and the use of room temperature ambient for the substrate.

The second step involved in mounting of chips is more critical and delicate. This involves attaching a thin wire or strip form the top surface of the chip to a nearby metallisation on the alumina substrate. Various techniques used for this purpose are

1. Thermocompression bonding

2. Ultrasonic bonding

3. Parallel gap or step welding

Whereas the first two bonding methods (thermocompression and ultrasonic) are used for connecting wires or ribbons to the top surface of the chip, parallel gap or step welding is more suitable for connecting the other end of the wire or ribbon to the metallisation on the substrate.

Thermocompression bonding[18]

Thermocompression bonding is defined as a joining process for two metals, whereby the metals being joined are brought into intimate contact by pressure using a shaped smooth bonding tool at a temperature below that required for interface melting. In the region to which pressure is applied, diffusion occurs between the metals being joined and a metallurgical bond results. A variety of methods for making this bond are available. Application of this bonding technique to high frequency transistors has been discussed by Chirstensen[19]. Two variations of the thermocompression technique are shown schematically in Fig. 11.9 (a) & (b). In the first one a heated wedge is used. This special shaped wedge is made of micropolished tungsten carbide or sapphire. The ductile lead material is gold wire or ribbon with typical dimensions 0.003 in. to 0.0003 in. The second method uses a heated capillary of special shape (made of tungsten carbide). Gold wire is cut off by using a flame or scissors. Typical wire diameter is 0.0007 in. to 0.0015 in.

Ultrasonic bonding

Ultrasonic bonding involves essentially the same mechanism as the thermocompression bonding except that the source of energy is mechanical rather than thermal. An elastic vibration is created by rapid expansion and contraction of a magnetostrictive transducer driven by a source of high frequency alternating current. As shown in Fig. 11.10 the bonding tip undergoes excursions in a direction parallel to the interface of the weld. Ultrasonic welding is very successful with aluminium because aluminium is soft and deformation occurs at relatively low pressures. A detailed description of the process is given in [20].

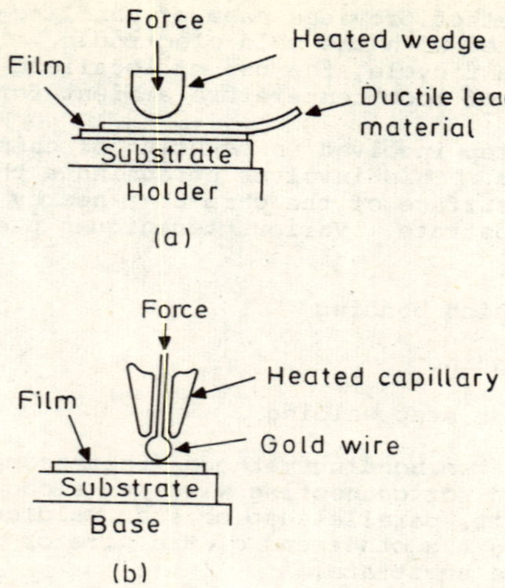

Fig. 11.9 Two variations of thermocompression bonding
(a) Heated wedge method
(b) Heated capillary method

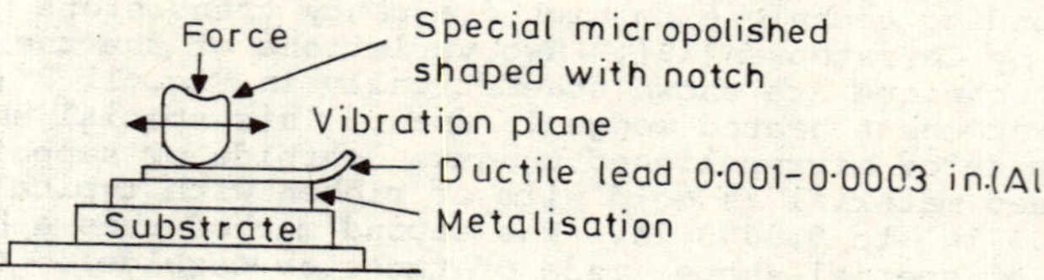

Fig. 11.10 Ultrasonic bonding

Welding

Two spot welding techniques which are used frequently are shown in Fig. 11.11. In parallel gap welding most of the current flows through the lead whereas in step welding the current flows through the metallisation on the substrate.

Comparison of various techniques

1. Temperature: Thermo-compression bonding requires a high temperature (200 - 400°C) and a long time (2 - 30 seconds) bond

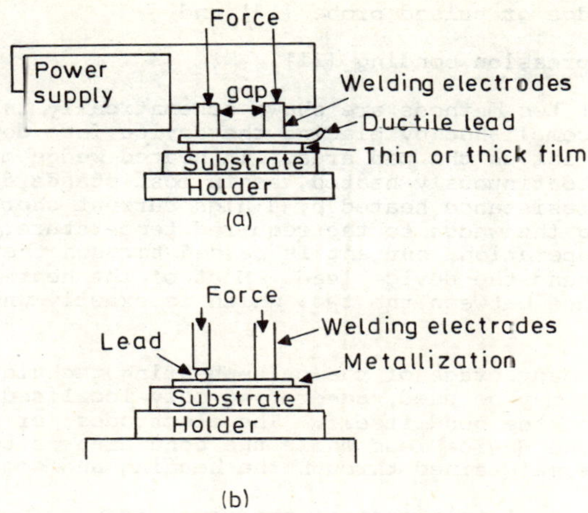

Fig. 11.11 Two spot welding techniques
(a) Parallel gap welding
(b) Step welding

cycle. Welding and ultrasonics bonding use a cold ambient and have extremely short time cycles (in the milli-second range).

2. Materials: Thermocompression bonding is generally applied to gold or silver wires and ribbons. Ultrasonic bonding is generally applied to aluminum wire. Parallel - gap welding usually requires gold or gold plated ribbon. When the welding process is used for larger size bonds, the limitations on materials are reduced and the machine can be used for soldering and welding as well as for bonding. There are certain advantages regarding reliability, in using aluminium wire when the chip has Al bonding pads.

3. Size: In terms of joint size, the thermocompression technique can bond to the smallest area (0.0005-in. dia spot). The smallest size of bonds obtained by ultrasonic technique is generally limited to a 0.002 in. dia spot and by the parallel gap welder to a 0.005 in. dia area.

11.7.2 Mounting of beam lead devices

Mounting of beam lead devices in hybrid MICs has been discussed in [13] and [21]. Various techniques suggested for this purpose are

i) parallel gap welding [13]

ii) heated wedge or pulsed probe [13] and

iii) thermocompression bonding [21].

The first two methods are shown schematically in Fig.11.12. Bonding is accomplished by placing the device face down with the tabs resting flat on the pad area. In heated wedge method, the wedge may be continuously heated, as in most standard equipment, or it may be resistance heated by a high current short duration pulse, raising the wedge to the required temperature. In the step-welding operation, current is passed through the substrate metallisation and the device lead. Most of the heat is generated at the interface between the two, which is exactly where it is needed.

The major advantage of the pulse heating technique is that a cold ambient may be used, generating only localised heating in the vicinity of the bond itself. The electrodes (or wedge) can be placed on the device lead while the bond area is cold. A constant force is maintained through the heating and cooling cycle.

Analysis and development of thermocompression bonding schedule for beam lead devices has been reported by Adams and Bonham [21]. A 'wobble tool' as shown in Fig.11.13, is used for this purpose. The critical parameters for this type of bonding have been established. The main results of this analysis are:

1. For small beam lead devices the bonding tool temperature has to be limited so as to prevent device degradation.

2. The bond interface temperature (T_i) can be expressed as a linear combination of tool (T_t) and substrate temperatures (T_s). The linear relation $T_i = 0.11\ T_t + 0.89\ T_s$ was obtained theoretically and verified experimentally.

3. Wobble speed is not critical provided that the bond time is greater than 0.4 sec. per side.

4. Thermocompression bond failure rate for uncleaned devices was three times greater than the bond failure rate for cleaned devices. Consequently devices and substrates should be cleaned prior to bonding.

5. A bond schedule for four beam transistors has been developed in terms of only two variables, bond pressure and interface temperature.

11.7.3 LID attachment

Since the LID is a rugged ceramic carrier and has metallised pads terminating on one side, it can be attached to a substrate with a single solder operation. The preferred method would be to pre-wet both the feet of the LID and the metallisation on the substrate and then use a flux or a controlled non-oxidising atmosphere. A weight can then be used to apply equal force to the legs of the LID. The assembly is brought up to temperature causing the solder to melt and flow together. It is then cooled to resolidify and to form a continuous solder fillet. This method is

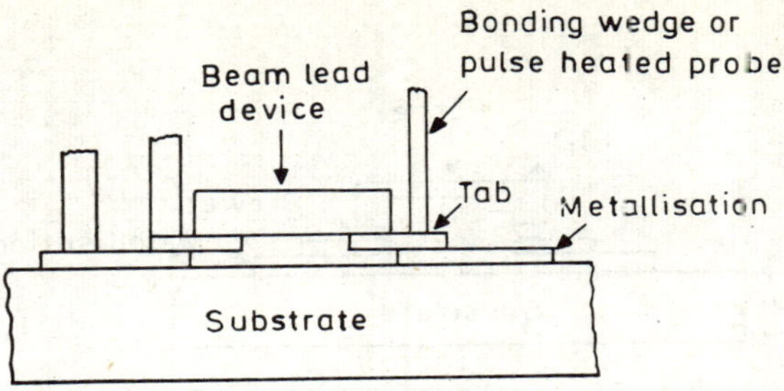

Fig. 11.12 Mounting of beam lead devices

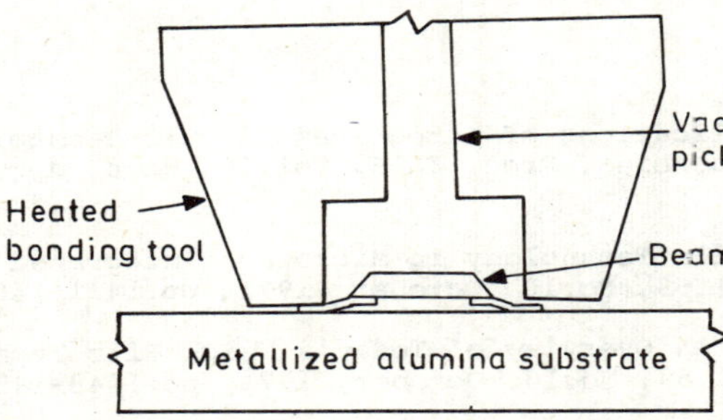

Fig. 11.13 Mounting of beam lead devices by 'wobble' tool

shown in Fig. 11.14.

It is possible to use localised heating and melt one side of the LID at a time. This can be done with a soldering iron but extra care must be taken and proper equipment used. This includes controlled temperature irons, extra fine points, good optics and a good positioning fixture. It is also possible to obtain a good solder joint by prewetting either the substrate or the LID. In general it is easier to get an adequate amount of solder on the substrate than on the feet of the LID.

If more than one LID device is to be attached to the substrate, a template can be made and the parts prealigned all of them reflow-soldered in one batch operation. This can be a considerable cost saver, and can eliminate the heating, destroying or unsoldering of one part while trying to put on another.

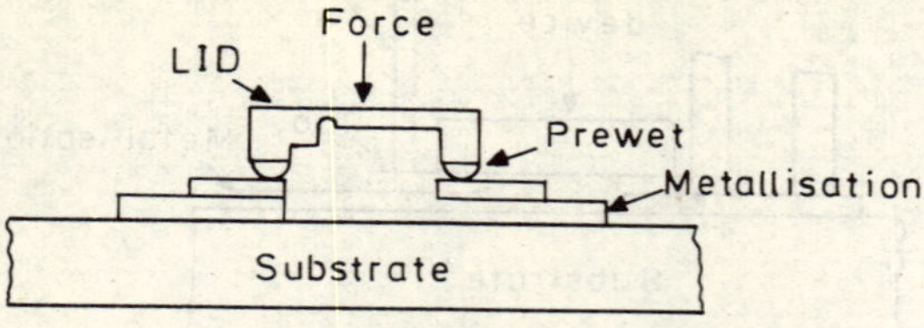

Fig. 11.14 Mounting of LIDs

References

1. Sobol, H., Applications of Integrated Circuit Technology to Microwave Frequencies, Proc. IEEE, Vol.59, No.8, August, 1971, pp. 1200-1211.

2. Caulton, M., Film Technology in Microwave Integrated Circuits, Proc. IEEE, Vol.59, No.10, October, 1971, pp.1481-1489.

3. Ressing, T.C., An Overview of Today's Thick Film Technology, Proc. IEEE, Vol.59, No.10, October, 1971, pp.1448-1454.

4. Thun, R.E., Thick Films or Thin Films, IEEE Spectrum, October, 1969, pp. 73-79.

5. Holland, L., Vacuum Deposition of Thin Films, London, England: Chapman and Hall (Proc. IEEE, October, 1971, Vol.59, No.10, p. 1402).

6. Gregor, L.V., Thin-Film Processes for Microelectronic Applications, Proc. IEEE, Vol. 59, No.10, October, 1971, pp.1390-1403.

7. Sobol, H., and A.E. Hilling, Applications of Integrated Circuit Technology to Microwave Frequencies, Proc. 1971, European Microwave Conf. Vol.II, pp. C 1/S:1-9.

8. Gonzales, A.J., and E.M. Pilofsky, Applications of Scanning Electron Microscopy to Thin Film Studies on Semiconductor Devices, Proc. IEEE, Vol. 59, No.10, October, 1971, pp.1429-1433.

9. Keister, F.Z., An Evaluation of Materials and Processes for Integrated Microwave Circuits, IEEE Transactions on Electron Devices, Vol.ED-15, No.7, July, 1968, pp.531-537.

10. Digiondomenico, O.J., and T.M. Foster, Hybrids for microwave gear get boost from thick film technology, Electronics, August 16, 1973.

11. Semiconductor devices for microwave integrated circuits, AEI Semiconductor Ltd., Lincoln, England, Issue 2, 4450-50/CCT Aug 1971.

12. Diode chips and carriers for hybrid circuits, Hewlett Packard Palo Alto, Calif., Technical Data Nov 1970.

13. Tarowsky, N., How to assemble hybrid microwave IC's, Microwaves, Vol. 8, No.8, Aug 1969 pp 52-62.

14. Aitchison, C.S., et al., Lumped-circuit elements at microwave frequencies, IEEE Trans. Microwave Theory Tech., Vol MTT-19, Dec 1971, pp 928-937.

15. Gelsing, R.J., and K.V. Steensel, A multilayer interconnection system with Au beam leads, Int. Conf. on microelectronics June 1969, IEE Conf. Publ. No. 54, pp 17-18.

16. Naylon, R., A comparison of the thermal impedance of flip-chip and beam lead assemblies, Int. Conf. on Microelectronics, June 1969, IEE Conf. Publ. No. 54, pp 21-22.

17. Bagrowski, J., S.G. Konsowski and G.D. Spencer, Interconnection of monolithic IC's through the use of advanced materials and techniques, IEEE Trans. Parts, Materials and Packaging, Vol. PMP-2, No.4, Dec. 1966, pp 90-98.

18. Beadles, R.L., Integrated Silicon device technology, Vol.XIV, Interconnections and Encapsulation, Research Triangle Institute Tech. Docummentary Report ASD-TDR-63-316, Vol. XIV, May 1967.

19. Christensen, H., Electrical contact with thermo-compression bonds, Bell Lab. Records, Vol.36, 1958, pp. 127-130.

20. Peterson, J.M., H.L. McKaig and C.F. DePrisco, Ultrasonic welding in electronic devices, IRE Convention record, Vol. 10(6), 1962 pp 3 - 12.

21. Adams, J.R. and H.B. Bonham, Analysis and development of thermocompression bond schedule for beam lead devices, IEEE Trans. Parts, Hybrids and Packaging, Vol.PHP-8, No.3, Sept. 1972, pp 22-26.

12

MONOLITHIC MICs

by

M.M. Hasan* and S.K. Mullick**

* Sections 12.1 to 12.3

** Section 12.4

Chapter Twelve

12.1 INTRODUCTION

Like low-frequency integrated circuits (ICs) microwave integrated circuits (MIC) can be also classified into two groups as:

1. Hybrid Microwave Integrated Circuits

2. Monolithic Microwave Integrated Circuits

At microwave frequencies, greater effort has been put on hybrid ICs than on monolithic ones because of difficult technological problems[1,2,3]. But looking at the growth of MICs and their applications in many systems, it appears that monolithic MICs may get more attention in the future. In fact for higher microwave frequencies and millimeter wave systems, monolithic MICs have been recommended. Encouraging results have been reported[4,5] for monolithic millimeterwave integrated circuits using semi-insulating (SI) GaAs.

Monolithic MIC techniques usually involve a very high resistivity or SI semiconductor substrate to provide resistive isolation and low losses. Active devices are fabricated locally on good quality semiconductor material at appropriate places and passive components are made on SI substrate using microstrips.

12.2 MICROSTRIPS ON SEMICONDUCTOR SUBSTRATES

Substrates for microwave integrated circuits (MIC) should have low loss at microwave frequencies, high dielectric constant and good adherence quality in relation to deposited metals. In view of the development of MICs, microstrip lines on semiconductor substrates are being extensively studied[8,9]. Silicon of resistivity greater than 2000 Ohm-cm and SI-GaAs of resistivity greater than 10^7 Ohm-cm have been used and microstrip lines and simple monolithic microwave integrated circuits have been reported [4,5,6]. The fabrication of microstrip lines involve metal evaporation, electroplating and photoresist techniques for defining the geometry. Aluminium and gold are commonly used as the conductor material.

The theoretical analysis of the microstrip transmission line assumes TEM mode and is based on Wheeler's theory[7] discussed in Chap.3. Hyltin[8] and Beneking[9] have described in detail the various aspects of microstrip transmission on semiconductor substrates. Welch and Pratt[10] have reported about the losses in such transmission systems.

Figure 12.1(a) shows the variation of characteristic impedance with W/d where W is the line width and d is the thickness of the substrate. Figure 12.1(b) shows variation of λ_o/λ_g versus W/d. These results are based on quasi-TEM mode analysis discussed in Chapter 3.

In an MIC, generally an impedance of 20 ohm to 100 ohm is used in various parts of a circuit. These impedance values can be readily obtained on a semiconductor substrate with a thickness of 10 mil or lower. For an X-band monolithic TR switch[5] on

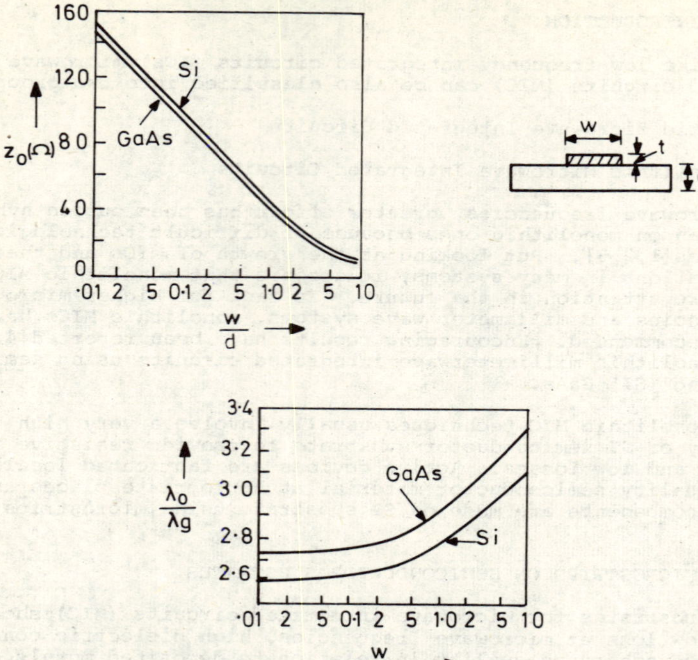

Fig. 12.1 (a) Impedance of microstrip line on high resistivity semiconductors. ρ(Si) = 1000 ohm-cm, GaAs is semi-insulating and t=5μ.(Ref.[9])

(b) Relative wavelength of microstrip line on high resistivity semiconductors (Ref.[9]).

high resistivity silicon, a 6 mil wide line was used on a 10 mil thick substrate to achieve a characteristic impedance of 50 ohms. However, it is difficult to maintain high resistivity of Si during the processing of monolithic MIC. And for lower resistivity, the line losses increase considerably as shown in Fig.12.2. It is seen that with Si of resistivity 5000 ohm-cm the dielectric loss of Si and alumina are camparable at a frequency 10 GHz or more[1]. This figure also indicates that monolithic MIC, where high resistivity Si or GaAs are used as substrate, are more suited for higher microwave frequencies and millimeteric waves[2,5]. Figure 12.3 shows the line loss versus Si resistivity and Fig.12.4 shows the conductor losses as a function of conductor thickness.

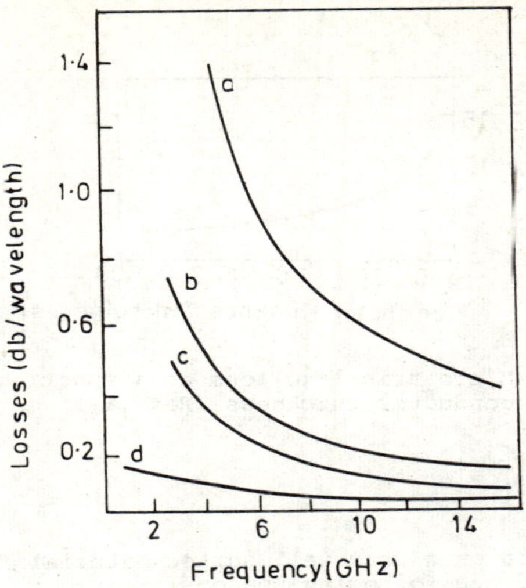

Fig. 12.2 Microstrip line losses as a function of frequency (Ref.[1])
(a) Dielectric loss in silicon substrate for ρ=1000 ohm-cm and d=10 mil
(b) Dielectric and conductor loss in silicon substrate for ρ=5000 ohm-cm and d=10 mil
(c) Dielectric loss in silicon substrate for ρ=5000 ohm-cm and d=10 mil
(d) Dielectric and conductor loss in alumina substrate for d=20 mil
(Reprinted from Electronics, June 10, 1968; Copyright McGraw-Hill Inc., 1968)

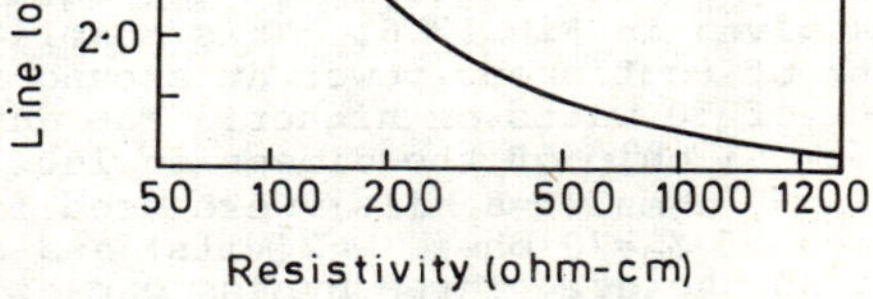

Fig. 12.3 Microstrip line loss as a function of silicon resistivity (Ref.[8])

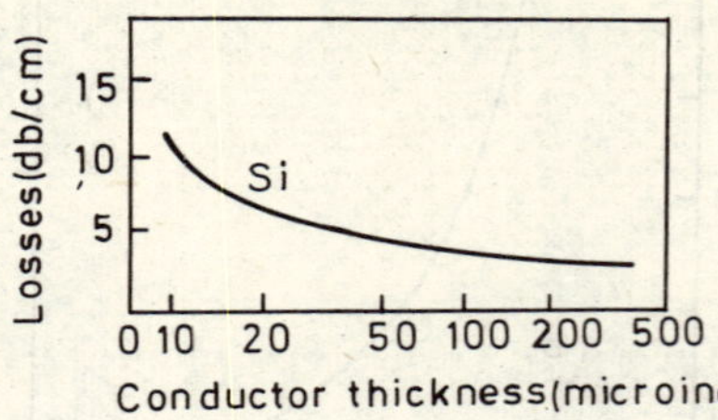

Fig. 12.4 Microstrip line loss as a function of conductor thickness (Ref.[8])

SI-GaAs seems to be a very well suited material for monolithic MICs because of the following reasons.

(i) The entire range of microwave semiconductor devices can be fabricated in GaAs.

(ii) The SI-GaAs of resistivity 10^8 ohm-cm is readily available. High resistivity substrates give better isolation and lower losses.

(iii) GaAs has a high dielectric constant (ε_r=12) and thin wafers (thickness 10 mils or less) can be used. This would cause considerable saving in the real estate for passive components.

12.3 SOME MONOLITHIC MIC CONFIGURATIONS

Several monolithic circuit modules are reported to have been successfully fabricated. In initial investigation the substrate chosen was high resistivity silicon but later work mostly uses SI-GaAs.

The details of the Si monolithic technology are described by Ertel[6]. An all Si TR-switch was made using surface oriented PIN diodes and microstrip lines (Fig.12.5). Its equivalent lumped element circuit is given in Fig.12.6. This monolithic TR switch can handle 1 watt of continuous power at around 9 GHz and it can sustain peak power of 50 watts or higher. The entire circuit was made on 1500 ohm-cm Si chip of the dimension 100x1000x10 mils. Microstrip lines of Z_o=50 ohms (W=6 mils) were used for rf transmission and other lines of Z_o=70 ohms (W=2 mils) and quarter wavelength long were used as chokes. Three bypass capacitors were provided for rf path to ground. Surface oriented diodes were made by diffusion in a controlled geometry. The diodes were used in series. The insertion loss, δ, of a series diode switch in a

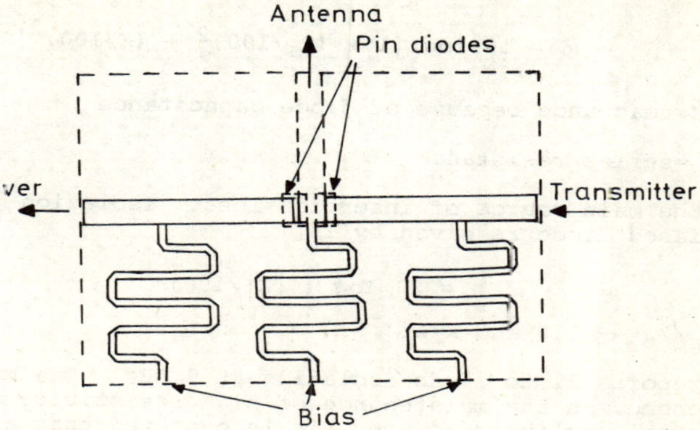

Fig. 12.5 Monolithic 9-GHz transmit-receive switch (Ref.[6])

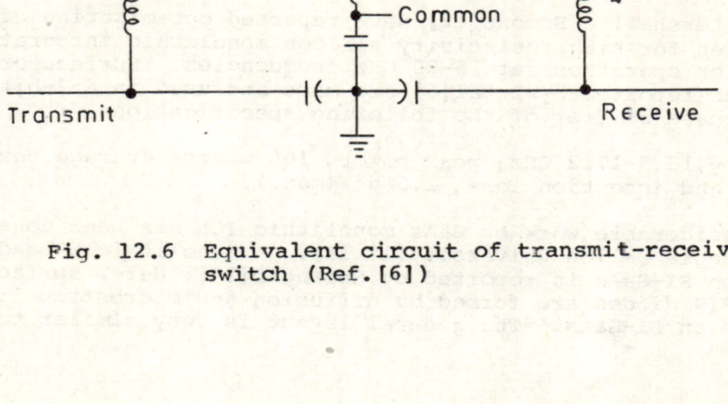

Fig. 12.6 Equivalent circuit of transmit-receive switch (Ref.[6])

50-Ω transmission line is given by[11]

$$\delta = 10 \log [1 + (R_s/100)^2 + (X/100)^2]$$

where X=reactance because of diode capacitance

and R_s=series resistance.

R_s is the main source of insertion loss. Isolation, η , for reverse biased diode is given by[11]

$$\eta \approx 20 \log (|X|/100)$$

when $R_s/Z_o << 1.0$ and $X/Z_o \geqslant 6.0$

Ertel reports 31 to 27 db isolation at 9 GHz. The major problem experienced was the maintenance of high resistivity of Si during the thermal cycling. However, it is expected that a better quality TR switch can be fabricated now because of the availability of Si in much higher resistivity range (10,000 ohm-cm and higher)

The monolithic MIC modules on GaAs developed at Texas Instruments have been reported in many papers[4,5]. It has also been emphasized that monolithic technology is more attractive for mm-wave systems. Mehal and Wacker[4] have reported a few circuits with details of the selective epitaxial growth techniques. Monolithic Schottky-barrier diode (Fig.12.7) and a 94 GHz mixer circuit were successfully fabricated. To fabricate the Schottky Barrier diodes, two separate epitaxial deposition steps were used. A heavily doped n^+ pocket, 10 microns thick and 1x2 mils in area is followed on top by a thin n-type layer of ≈ 0.5 μ. The n^+ deposition is typically doped to 2×10^{18} cm^{-3} ,while the n layer is 10^{17} cm^{-3}. Mixer circuit using these diodes is shown in Fig.12.8. It consists of two RF inputs to the diodes in a branch line 3-db coupler circuit and, an IF output terminal along with two quarter wavelength open circuited stubs providing an RF short to ground. The total chip size is 30x30 mils with a 4 mil thick substrate. A single ended 94 GHz mixer circuit is described by Mao et al.[6] which has used the fabrication technique identical to that described by Mehal & Wacker[4]. Mao et al. give more detail in their paper about the microstrip lines and monolithic circuits for mm wave integrated circuits.

Battershall & Emmons[12] have reported optimization of diode structures for high resistivity silicon monolithic integrated circuits for operations at 15-20 GHz frequencies. Surface oriented diodes as reported by Ertel[6] are made and used in a 3-bit monolithic phase shifter of the following specifications:

Frequency, 15.5-17.2 GHz; peak power, 100 watts; average power, 5 watts and insertion loss, 1.5 db (max.).

Considerable work on GaAs monolithic ICs has been done at the Carnegie Mellon University[13 14]. A monolithic X-band TR switch on SI-GaAs is reported by Eppinger[13]. Here, surface oriented PIN diodes are formed by diffusion and microstrip lines are made on SI-GaAs. The general layout is very similar to that

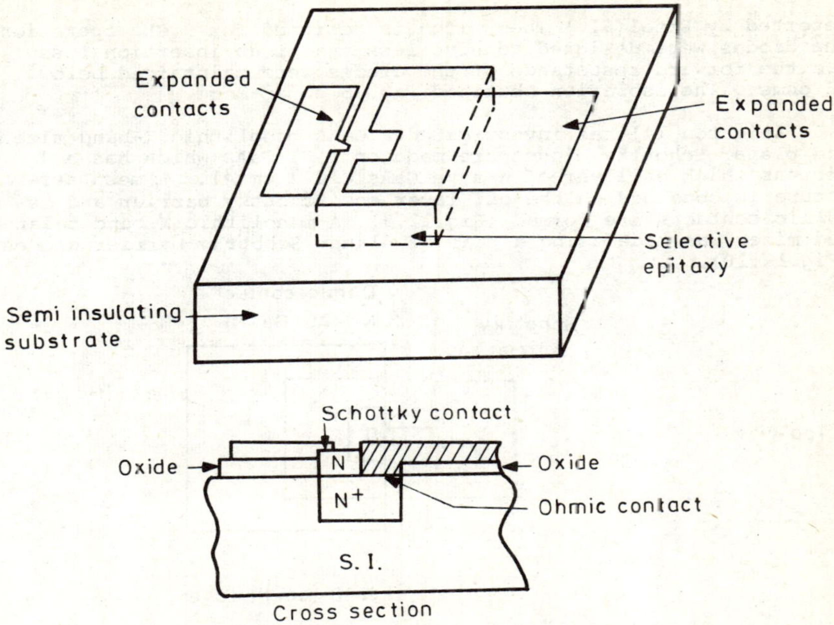

Fig. 12.7 Design structure of planar Schottky barrier diode (Ref.[4])

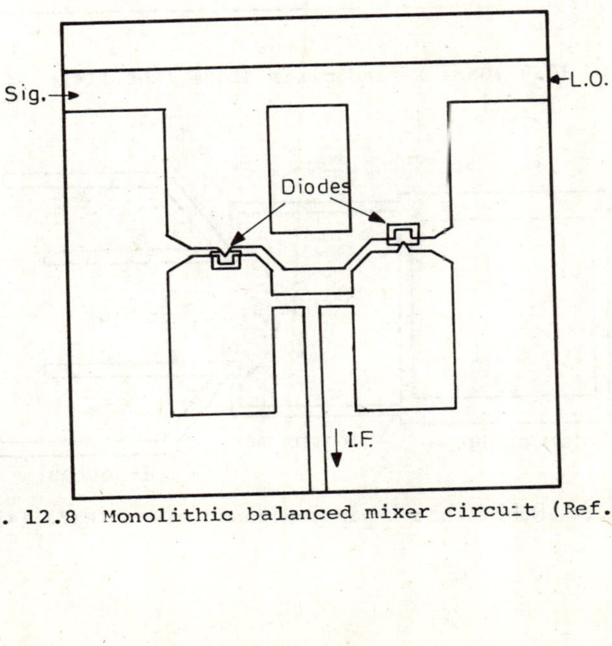

Fig. 12.8 Monolithic balanced mixer circuit (Ref.[4])

reported by Ertel[6]. The switch is designed for 9 GHz operation. The diodes were designed to give less than 1 db insertion loss i.e.the forward resistance of the diodes were maintained below 12 ohms. The isolation obtained was 25 db or more.

Leighton[14] has investigated a GaAs monolithic X-band mixer. The planar Schottky diodes are made on a SI-GaAs which has a 4 microns thick epilayer of n-type GaAs (10^{17} cm^{-3}). A mesa structure is obtained on the epi layer and Schottky barrier and metallic contacts are formed (Fig.12.9). A monolithic X-band balanced mixer was made using a pair of planar Schottky barrier diodes (Fig.12.10).

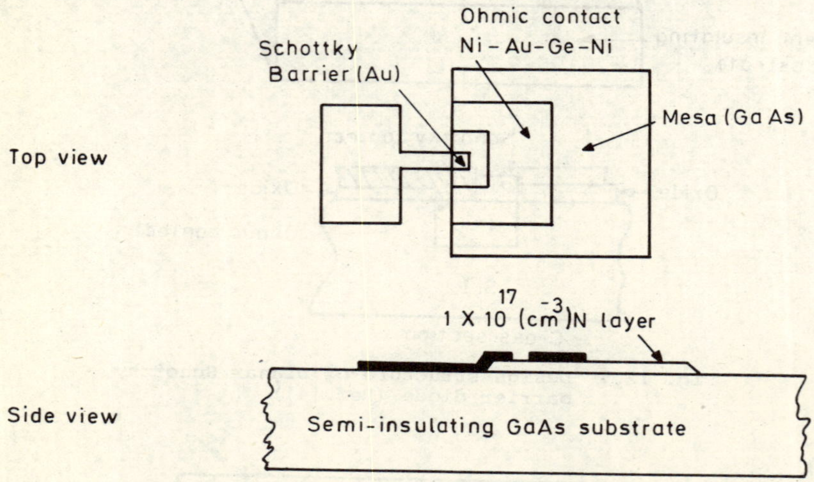

Fig. 12.9 GaAs X-band mixer diode (Ref.[14])

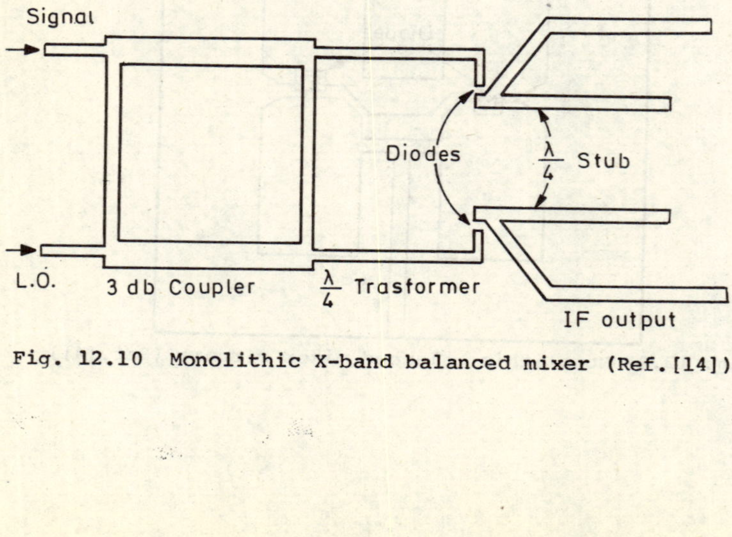

Fig. 12.10 Monolithic X-band balanced mixer (Ref.[14])

Another millimeter-wave balanced mixer monolithic circuit is described by Antell[15].

12.4 MICROSTRIP ON SILICON-SILICON OXIDE SUBSTRATES

12.4.1 Use of $Si\text{-}SiO_2$ substrate

Microstrip lines on semiconductor substrates have already been discussed in section 12.2. The necessity of employing a high resistivity substrate in order to keep the transmission losses low, has been pointed out. However, high resistivity substrates are not the best starting point for fabrication of most of the active devices. This incompatibility necessarily leads to more involved processing. An alternative transmission structure suitable for monolothic ICs that does not necessarily need high resistivity substrates consists of microstrip type transmission line on semiconductor-SiO_2 substrate. This structure has already been studied as a model for the interconnection metallization in monolithic ICs[16]. This study, based on a lumped circuit model has greater relevance in the design of very high speed digital machines using very fast semiconductor switching devices in LSI environment. The aspect which is of greater concern to us is the signal propagation characteristic of these lines as used in constructing useful devices for MICs. A detailed analysis of propagation characteristics of microstrip lines on $Si\text{-}SiO_2$ substrate based on the parallel plate waveguide model has been reported recently[17]. This study brings out the existence of three limiting forms of fundamental mode of propagation namely,(i) dielectric quasi-TEM mode, (ii) Skin effect mode and (iii) slow wave mode. The appearance of one or the other mode depends on the substrate resistivity and the operating frequency. For the line widths, resistivity of substrates and frequencies usually employed in MIC work, the slow wave mode appears to be of most interest. In the following we shall briefly review the propagation characteristics of the microstrip lines on $SI\text{-}SiO_2$ substrates with emphasis on the slow wave mode of propagation.

12.4.2 Lumped circuit model[16]

A lumped circuit model for a microstrip line of Fig.12.11 is shown in Fig. 12.12. This model contains five elements; R represents the sum of series resistances in the metal strip and that in the return path, C_1 represents the capacitance of a capacitor formed by the metal strip and a virtual ground plane very close to the top of semiconductor substrate, L represents the inductance of the line, and G and C_2 are determined by the thickness and resistivity of the substrate. This lumped circuit model has been justified on physical and experimental grounds.

From an argument based on the relaxation time of the surface charge at the boundary of the two media, one may conclude that the parameter C_1 may be found by assuming a virtual ground plane at the $Si\text{-}SiO_2$ boundary. Since the line width W is in general much wider than b_1, the following approximate equation may be used to calculate C_1:

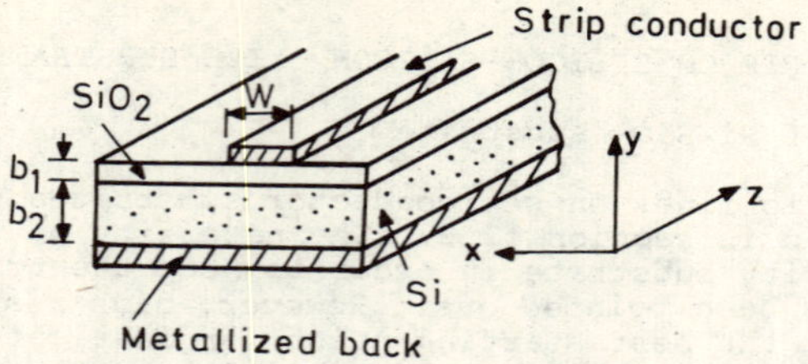

Fig. 12.11 Microstrip line on $Si-SiO_2$ system

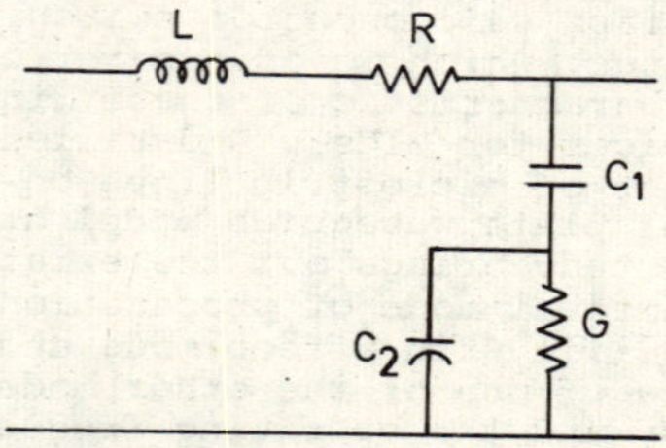

Fig. 12.12 Lumped equivalent circuit for microstrip in Fig. 12.11

$$C_1 = \varepsilon_1 W\{1+2b_1[1+\ln(\pi W/b_1)]/\pi W\}/b_1$$

Where ε_1 is dielectric constant of SiO_2. This C_1 calculation has been justified by experimental measurements with a Boonton bridge on a 1-mil line with 5000 A° SiO_2 layer. The measured C_1 is 21 pf/cm, while the calculated value is 19.2 pf/cm with $\varepsilon_1=4\varepsilon_o$.

The parameter C_2 which is determined by the thickness of the semiconductor substrate, will be much smaller than C_1 and for the frequency spectrum of interest ωC_2 will be much smaller than G.

When the current flows into the metal conductor in the longitudinal direction, one may consider, for the purpose of calculating the inductance, that there is a virtual ground plane for the current return path located somewhere in the substrate. This is a difficult question to answer in general. However, if there is a metal ground plane on the back of silicon substrate, then this metal plane instead of the bulk silicon will serve as the ground plane and one may find the inductance L from the following equation

$$L = Z_0 / v$$

where Z_o is the characteristic impedance of the transmission line made up by the metal conductor and the ground plane and v is the propagation velocity. Since both Z_o and v are proportionally effected by the dielectric constant of the media, we can consider both Z_o and v as if they were in free space. For a 1 mil wide line on top of an 8 mil thick substrate with a metal ground plane, the calculated value of 8.5 nH/cm compares well with the experimental measurement of 8.6 nH/cm.

The series resistance R should include series resistance of the metal line and the return path and some extra resistance due to skin effect. For the current return path either in the bulk silicon substrate or in the metal ground plane its thickness may be considered fixed while its width is a function of frequency. This spreading effect may be taken into account by introducing a spreading factor which is equal to the ratio of effective return path to the line width. At high frequencies the current in the ground plane will not spread far beyond the line width W, and the series resistance will be high.

The shunt conductance G may be determined from the conductivity, the thickness and the width of the silicon substrate path. Similar to the current return path this width is also a function of frequency.

The propagation constant γ of this structure is given by

$$\gamma^2 = (R + j\omega L) \{ [j\omega C_1 (j\omega C_2 + G)] / [G + j\omega (C_1 + C_2)] \}$$

This equation in turn yields the following approximations for the attenuation and phase constants for different frequency ranges.

1) Low Frequency Range : $\omega << R/L$

$$\alpha \approx \sqrt{(\omega C_1 R/2)} \; [\, 1 + \omega (L/R - C_1/G)/2 + \omega^2 C_1 (C_2/G + L/R)/4G \,]$$

and

$$\beta \approx \sqrt{(\omega C_1 R/2)} \; [\, 1 - \omega (L/R - C_1/G)/2 + \omega^2 C_1 (C_2/G + L/R)/4G \,]$$

2) Mid-Frequency Range : $R/L << \omega << G/C_1$

$$\alpha \approx \sqrt{LC_1} \; [\, R/L + \omega^2 C_1 (\, 1 + RC_2/4GL \,)/G]/2$$

and

$$\beta \simeq \omega \sqrt{LC_1} \; [\; 1 - RC_2/4LG + \omega^2 C_1 C_2/4G^2 \;]$$

3) High Frequency Range : $\omega >> G/C_2$

$$\alpha \simeq \sqrt{LC_2} \; [\; R/L + G/C_2 + G^2/4\omega C_1 C_2 \;]/2$$

and

$$\beta \simeq \omega\sqrt{LC_2} \; [\; 1 + G/4\omega^2 C_2 \; (\; G/C_2 - R/L \;)]$$

The validity of the above analysis is borne out by experimental measurements on such lines. Another line of inquiry that strengthens these claims is the study of admittance of these lines. This is verified by time domain reflectometric measurements

12.4.3 Parallel plate waveguide model[17,18]

A study of propagation along microstrip on Si-SiO_2 substrate, considered as a parallel plate waveguide, was first reported by Guckel et al.[2]. The parallel plate waveguide model is depicted in Fig.12.13. This model assumes the following :(i) The width of the line W is very much greater than the spacing b ($=b_1+b_2$) between the top conductor and the ground plane. (ii) The ground plane and the line metals are either lossless or have losses which are very small compared to losses in the substrate.

This study, based on the transverse resonance analysis of the parallel plate waveguide mode, established the existence of a slow surface wave mode of propagation. It was also shown that the phase velocity of this mode depends strongly on the substrate resistivity and hence such a structure could be potentially used as a variable delay line.

Later, fairly detailed theoretical and experimental investigations have been carried out by Hasegawa et al.[18,19,20]. Their analysis establishes the existence of three limiting forms

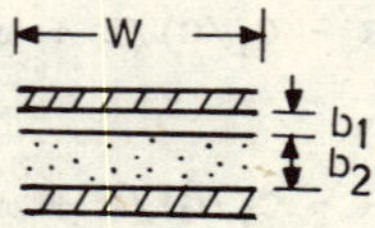

Fig. 12.13 Parallel plate waveguide model for microstrip in Fig. 12.11

of the fundamental TM mode supported by such a structure. These are:(i) Dielectric quasi TEM mode (ii) Skin effect mode and (iii) Slow-wave mode. These limiting forms arise out of different behaviour of the substrate when its resistivity and the frequency are varied.

For the first mode, the product of frequency and resistivity of Si substrate is large enough and provide a small dielectric loss and the substrate essentialy behaves as a dielectric and the line can be regarded as a microstrip line consisting of Si and SiO_2. The fundamental mode therefore is close to a TEM mode as long as λ is much greater than b. The equivalent lumped circuit for this mode is shown in Fig.12.14. Here L ,the inductance per unit length, is given by

$$L = \mu_o b/W$$

C_1 is the capacitance contributed by SiO_2 layer and is given by

$$C_1 = \varepsilon_o \varepsilon_{SiO_2} W/b_1$$

C_2 is the capacitance formed by the silicon layer and is given by

$$C_2 = \varepsilon_o \varepsilon_{Si} W/b_2$$

G_s, the shunt conductance arising because of the finite conductivity σ_2 of the Si-layer, is given by

$$G_s = \sigma_2 W/b_2$$

When the product of frequency and substrate conductivity is large enough to yield a small depth of penetration into silicon the substrate behaves like a lossy conductor and the line may be considered as a microstrip line on an imperfect ground plane made of silicon. The line becomes highly dispersive and we have the skin effect mode. The equivalent circuit for this mode is shown in Fig.12.15. Here L_1 is the inductance calculated by assuming

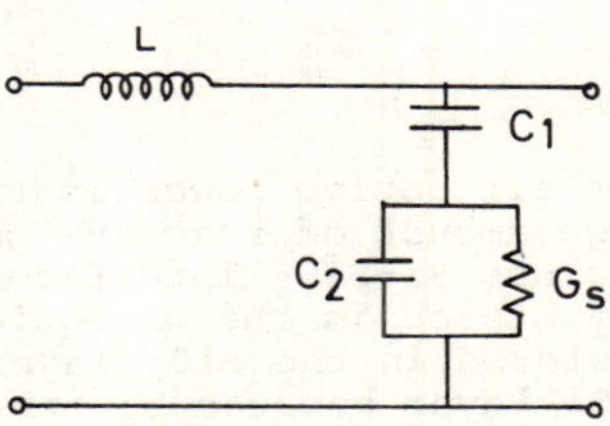

Fig. 12.14 Equivalent circuit for Quasi-TEM mode

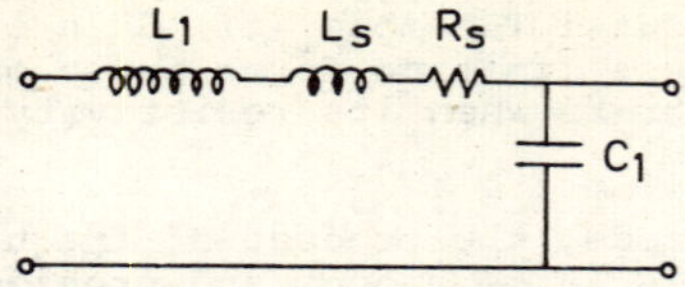

Fig. 12.15 Equivalent circuit for skin effect mode

the ground plane to be at the Si-SiO2 interface. Therefore ,

$$L_1 = \mu_o\ b_1/W$$

L_s and R_s arise due to penetration of fields in silicon because of the skin effect. This gives,

$$L_s = \mu_o\delta/2W \quad \text{and } R_s = 2\ \pi f L_s$$

where δ is the depth of penetration in silicon layer given by

$$\delta = \sqrt{2/(\sigma_2\omega\mu_o)}$$

C_1 is the capacitance contributed by the SiO_2 layer as before.

The slow wave mode arises when the frequency is not high and the resistivity is moderate. In contrast to the 'dielectric' and 'metallic' behaviour of the substrate as in the first two modes this mode corresponds to the semiconductor behaviour of the substrate. This mode is of particular significance because it appears within the resistivity and frequency ranges encountered in the monolithic MIC technology. Moreover, this mode experiences considerable smaller attenuation than what is expected of propagation in a homogeneous parallel plate waveguide filled with silicon of that prticular resistivity. In fact, the choice of the resistivity can be made optimum for minimum attenuation for this mode. This value of resistivity is given by

$$\rho_c = \sqrt{(\mu_o b_1 b_2/3\varepsilon_o\varepsilon_{sio_2})}$$

In this mode , almost all active power is transmitted through the SiO_2 layer but a large amount of reactive power is exchanged between the two layers across $Si\text{-}SiO_2$ interface. The average amount of magnetic energy stored in the waveguide is equal to that of electric energy stored in the SiO_2 layer, but the former is stored mainly in the Si layer because $b_2 >> b_1$. The sum of these energies is transmitted in the Z-direction. The propagation velocity slows down owing to the energy transfer across the interface. The equivalent circuit corresponding to this mode is shown in Fig.12.16. L, C_1 and G_s have been defined earlier. $R_s(f)$ represents the loss caused by the longitudinal components of the conduction current in the silicon substrate and can not be

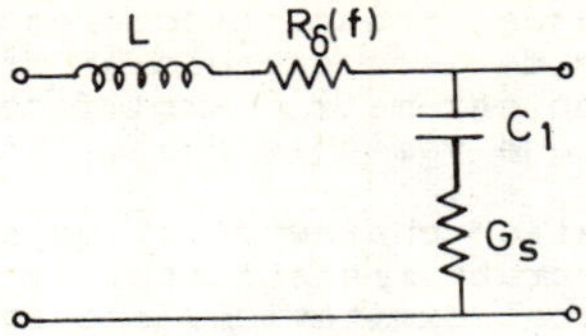

Fig. 12.16 Equivalent circuit for slow wave mode

evaluated by the quasi-static approach used in this analysis. It has been shown experimentally that $R_s(f)$ varies as the square of frequency[20].

Comparing with Fig.12.12, we note that the equivalent circuit for the slow wave mode is identical to the lumped circuit model discussed in section 12.4.2 except for absence of C_2. It may be recalled that C_2 is the capacitance of the Si layer and may be ignored depending on values of G_s and the frequency of operation. The spreading factor, which was used for thin lines originally studied for the lumped model[16], may be taken to be unity in the present case of wide lines.

12.4.4 Applications to MICs

Of the three limiting forms of the fundamental mode of propagation discussed above, the slow wave mode appears to be most promising for monolithic microwave integrated circuits because it exists for the frequency-resistivity combinations normally employed for such work. Already a number of interesting devices using the microstrip structure discussed above have been proposed. Two closely spaced parallel lines on such a structure can be used as a directional coupler, which can have output at the near end or the far end, a feature which is not possible by a conventional directional coupler with a homogeneous medium[21]. If one wishes one can divide the flow of power into the far end and the near end of the coupled line without output at the signal line. Also, it is possible to have this configuration serve as a three way power divider.

Another possible application of the slow wave mode is as a new type of slow wave circuit for solid state travelling-wave amplifier[19].

When the SiO_2 insulator thickness becomes comparable to the depletion width at the Si-SiO_2 interface, the capacitance C_1 then becomes voltage variable MIS capacitance, and the phase velocity on the structure becomes voltage (or field) dependent. Another structure that provides voltage variable propagation is a distributed Schottky barrier varactor[22]. Here the microstrip line is formed directly on the semiconductor, and the depletion layer obtained by reverse biasing, takes the place of the insulator and the depletion layer in the MIS microstrip. Because of the voltage variable capacitance, these structures can be used to form distributed parametric devices capable of frequency conversion, harmonic generation and parametric amplification. Other obvious possibilities are voltage variable delay lines and phase shifters.

Since the propagation characteristics are also affected by the semiconductor substrate resistivity, by inducing a change in resistivity by optical excitation, interesting distributed opto-electronic devices become a possibility.

References

1. Moroney, W.J., Microwave ICs, Electronics, Vol.41, June 10, 1968, pp.100-108.

2. Sobol, H., Applications of integrated circuit technology to microwave frequencies, Proc. IEEE, Vol.59, Aug.1971, pp.1200-1211.

3. Caulton, M., and H. Sobol, Microwave integrated circuit technology - a survey, IEEE Jr. of Solid State Circuits, Vol.SC-5, Dec. 1970, pp. 292-303.

4. Mehal, E.W. and R.W. Wacker, GaAs integrated microwave circuits, IEEE Trans.Electron Devices, Vol.ED-15, July 1968, pp. 513-516.

5. Mao, S. et al., Millimeter-wave integrated circuits, IEEE Trans. Electron Devices, Vol.ED-15, July 1968, pp. 517-523.

6. Ertel, A., Monolithic IC techniques produce first all silicon X-band switch, Electronics, January 23, 1967, Vol. 76-81.

7. Wheeler, H.A., Transmission-line properties of parallel strips separated by a dielectric sheet, IEEE Trans. Microwave Theory Tech. Vol.MTT-13, March 1965, pp. 172-185.

8. Hyltin, T.M., Microstrip transmission on semiconductor dielectrics, IEEE Trans. Microwave Theory Tech., Vol.MTT-13, Nov. 1965, pp. 777-781.

9. Beneking, H., Microwave Circuitry on semiconductor, Proc. 1971 European Microwave Conference, Paper No. C2/1.

10. Welch, J.D., and H.J. Pratt, Losses in microstrip transmission systems for integrated microwave circuits, NEREM Record, 1966, pp. 100-101.

11. Garver, R.V., Theory of TEM diode switching, IRE Trans. Microwave Theory Tech., Vol.MTT-2, May 1961, pp. 224.

12. Battershall, B.W. and S.P. Emmons, Optimization of diode structures for monolithic integrated microwave circuits, IEEE Trans. Electron Devices, Vol.ED-15, July 1968, pp.507-512.

13. Eppinger, T.M., Integrated X-band microstrip transmit-receive switch on semi-insulating gallium arsenide, Ph.D. Thesis, Carnegie-Mellon University Pittsburgh, Pa., Feb. 1971.

14. Leighton Jr., W.H., Monolithic X-band microstrip line mixers on semi-insulating gallium arsenide, Ph.D. Thesis, Carnegie-Mellon University, Pittsburgh, Pa. May 1970.

15. Antell, G.R., Monolithic mixers for millimeter wavelengths, Proc. 1971 European Microwave Conference, Paper No. C2/3.

16. Ho, I.T. and S.K. Mullick, Analysis of transmission lines on integrated-circuit chips, IEEE J. Solid State Circuits, Vol.SC-2, Dec. 1967, pp. 201-208.

17. Guckel, H., et al., A parallel plate waveguide approach to microminiaturized, planar transmission lines for integrated circuits, IEEE Trans. Microwave Theory Tech., Vol.MTT-15, Aug. 1967, pp. 468-476.

18. Hasegawa, H., et al., Slow wave propagation along a microstrip line on Si-SiO_2 system, Proc. IEEE Vol.59, Feb.1971, pp. 297-299.

19. Yanai, H. et al., Slow-wave propagation along the transmission line on Si-SiO_2 system, presented at the 8th Int.Conf. on Microwave and Optical Generation and Amplification (MOGA 70), Amsterdam, Sept. 1970.

20. Hasegawa, H. et al., Properties of microstrip line on Si-SiO_2 system, IEEE Trans. Microwave Theory Tech., Vol.MTT-19, Nov. 1971, pp.869-881.

21. Ho, I.T. and S.K. Mullick, Integrated Circuit directional coupler, U.S. Patent No.3 500 255, March 1970.

22. Hughes, G.W. and R.M. White, MIS and Schottky barrier microstrip devices, Proc. IEEE, Vol.60, 1972, pp. 1460-61.

13

TECHNOLOGY OF MONOLITHIC MICs

by

M.M. Hasan* and Amarjit Singh**

* Sections 13.1 to 13.4
** Sections 13.5 and 13.6

Chapter Thirteen

13.1 PROCESSES INVOLVED IN THE FABRICATION OF MONOLITHIC MICs

Processes involved in the fabrication of monolithic MICs are essentially identical to the processes required for making discrete microwave semiconductor devices[1, 2, 3]. These are epitaxial growth, growth of oxide layer, diffusion, metallization photolithography etc. However, one very important consideration in MICs is the starting semiconductor substrate material. The substrate has to be a material of high resistivity in order to give good isolation and low losses. In the case of microstrip circuits on semiconductor - SiO_2 substrate, the losses are dependent more on the quality of the SiO_2 layer and the substrate is not required to have extremely high resistivity. The two materials which are commonly used as substrates are high resistivity silicon (Si), and semi-insulating gallium arsenide (SI-GaAs). MICs such as a PIN diode TR switch (see Sec. 12.3) can be readily fabricated on a high resistivity semiconductor substrate by simple diffusion and metallization. However, circuits involving other devices would require selective epitaxial growth of a semiconductor layer of desirable doping level. Various steps involved in selective epitaxial growth are shown in Fig. 13.1. Selective epitaxial growth, whenever required, becomes the most critical step and successful device fabrication is dependent on the quality of the epitaxially grown pockets. An alternative approach[5] is to use good quality Si or GaAs chips and grow on them SI-GaAs, in single crystal or polycrystalline form, so that good quality Si or GaAs islands are embedded in the grown SI-GaAs. This arrangement is shown in Fig. 13.2.

The details of the subsequent steps in fabrication of monolithic MICs depend on the starting semiconductor material. In the case of high resistivity Si (resistivity greater than 5000 Ω - cm), the subsequent steps are identical to those in Si integrated circuits. In the case of SI-GaAs (resistivity greater than 10^8 Ω - cm), after the selective epitaxy, the subsequent processes are identical to those for technology of GaAs devices. However, photolithography and metallisation are common to both the Si and GaAs MICs. In the case of Si MICs, SiO_2 can be grown thermally. The method of pyrolitic deposition of SiO_2 and Si_3N_4 can be used for either Si or GaAs. In general for Si and GaAs MICs, the process steps are similar but the techniques are different.

13.2 EPITAXIAL GROWTH OF SEMICONDUCTOR LAYER

In the fabrication of many devices (like varactor diode, IMPATT diode, transistor etc.) an epitaxial layer of semiconductor is required on the substrate of the same material. The conductivity type (n or p) of the epi-layer may be the same or different from the substrate material. However, the resistivity is generally designed to be different from that of the substrate. Thus the epilayer may be n on n^+ or p on p^+ or any other combination. Sometimes a heavily doped epilayer is formed for making ohmic contact.

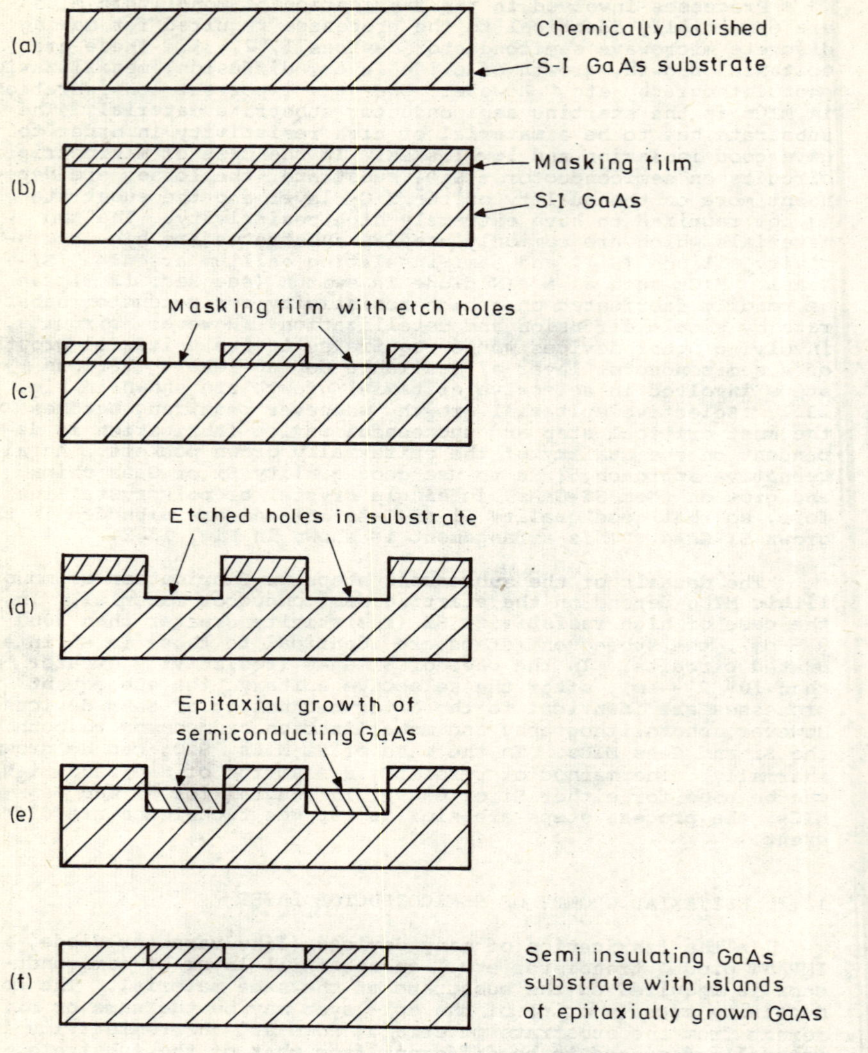

Fig. 13.1 Main process steps in selective epitaxial growth

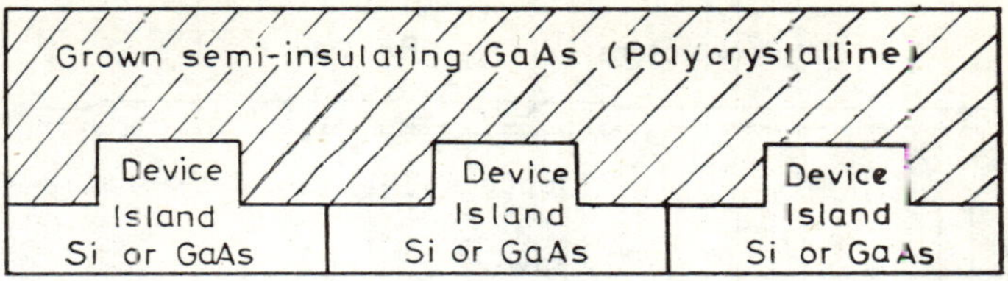

Fig. 13.2 Alternative approach for monolithic GaAs MICs

13.2.1 Epitaxial process for silicon

In this process Si wafer kept on a graphite susceptor, is heated to a high temperature (1000 - 1200°C) in a quartz tube using radio frequency induction heating. Silicon atoms are deposited on the substrate as a result of chemical decomposition of silicon compound supplied inside the quartz tube. They get arranged as an extension of the existing crystal structure of the substrate. The commonly used chemical processes are the following [6, 7, 8]:

(1) Hydrogen reduction of halide ($SiCl_4$)

In this system a mixture of silicon tetrachloride ($SiCl_4$) and hydrogen gas is passed through the quartz tube containing hot silicon wafer. Silicon atoms are liberated as a result of the reaction:

$$SiCl_4 + 2H_2 \xrightleftharpoons{1200°C} Si + 4HCl$$

A typical set up used for this method is shown in Fig. 13.3.

(2) Pyrolysis of silane (SiH_4)

Thermal decomposition of silane at high temperature gives off silicon and hydrogen as:

$$SiH_4 \xrightarrow{1000°C} Si + 2\,H_2$$

During the process of epitaxial growth dopant gas can be also supplied in proper quantity so that the resulting layer of Si is either p-type or n-type of the required resistivity. Phosphine (PH_3) or arsine (AsH_3) gas can be used for obtaining a n-type epitaxial layer of Si, and Diborane (B_2H_6) for a p-type layer.

13.2.2 Epitaxial process for GaAs

Growth of an epitaxial film of GaAs is more difficult than silicon layer because there are no suitable gallium compounds that are gaseous at room temperature and atmospheric pressure.

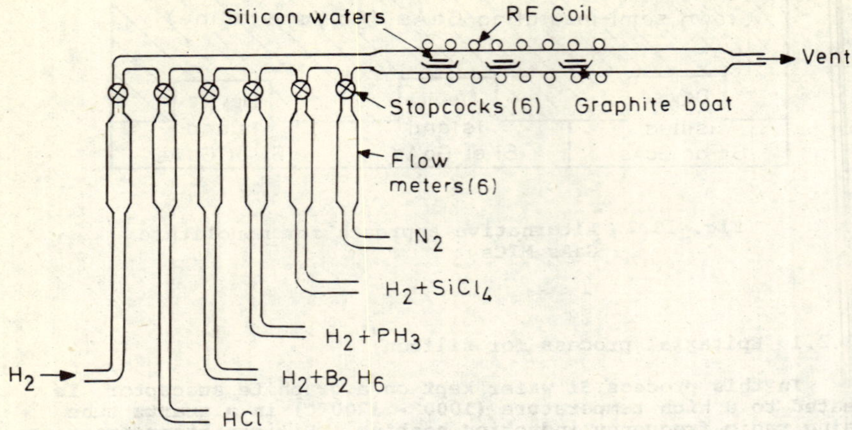

Fig. 13.3 Epitaxial growth system for Si

Both the As and Ga gaseous compounds must be supplied to the reactant system under closely controlled conditions. In this case the technique is called vapour phase epitaxy. The commonly used deposition techniques based on gaseous compounds are as follows [9]:

In one method, a gaseous gallium compound is formed and mixed with an arsenic compound. Gaseous gallium trichloride ($GaCl_3$) is formed by passing HCl or chlorine over gallium in the reactor system. The chlorine can be obtained from the decomposition of $AsCl_3$. A three-zone furnace is used as shown in Fig. 13.4. The three zones are gallium zone (775°C), reaction zone (850°C) and deposition zone (750°C). The As and gaseous Ga compounds are mixed in the reaction chamber and passed over the GaAs substrate. Deposition of epilayer occurs when the GaAs substrate is at 750 - 800°C. This technique gives high-purity films (donor impurities about $10^{15}/cm^3$, mobility about 6000 to 7000 cm^2 V-Sec) on semi-insulating or heavily doped GaAs substrates. Hydrogen sulfide and selenide are used as the donor source materials and dimethyl zinc or pure zinc as the acceptor source.

The second method uses a two zone furnace and a condensed source of GaAs in liquid gallium as shown in Fig. 13.5. The source mixture is presynthesized before deposition by exposing liquid Ga to $AsCl_3$. If the As vapour pressure exceeds the decomposition pressure of the GaAs - in - gallium solution at the operating temperature (about 850°), a GaAs skin forms on the gallium and serves as the source of GaAs. After this skin is formed, the substrate is exposed to the source reaction products at about

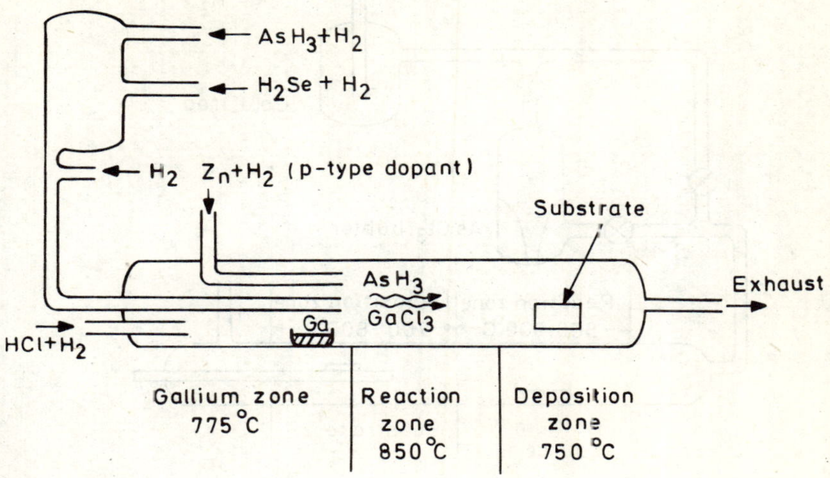

Fig. 13.4 First method of vapour phase epitaxy for GaAs

750°C. This method is known to have produced better quality films of GaAs than the previously described method[11]. The film can be doped using the sources mentioned earlier.

These methods are generally known as vapour - phase deposition techniques. The liquid - phase deposition techniques can also be used. The details are available in reference[10]. This system uses pure Ga alongwith polycrystalline GaAs pieces and high purity hydrogen gas. The reactor furnace has arrangement for tilting so that the GaAs substrate is covered with the saturated solution. Liquid phase epitaxy has advantages over the vapour phase where highly doped epitaxial films are needed.

13.3 GROWTH OF DIELECTRIC LAYERS

Dielectric layers are used in MIC technology for the following purposes:

(a) for protection and surface passivation of the device.

(b) as a mask against diffusion of impurities or against chemical etching.

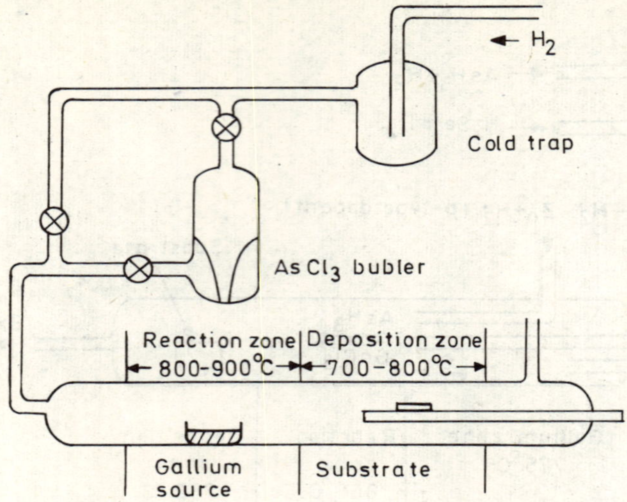

Fig. 13.5 Second method of vapour phase epitaxy for GaAs

(c) as a source of impurity for diffusion when a doped laye: is deposited on the semiconductor.

(d) as dielectric in lumped capacitors.

(e) for microstrips on low resistivity substrates as discussed in Sec. 12.4.

The commonly used dielectric layers are silicon dioxide and silicon nitride. While fabricating silicon devices, SiO_2 layers are readily formed by thermal oxidation of silicon but the same is not true with germanium and gallium arsenide devices. Therefore chemical deposition techniques are used for planar Ge and GaAs devices.

Techniques used for silicon dioxide layers are listed below [6, 7, 8]:

(1) Thermal oxidation of silicon using (i) Dry oxygen, (ii) Wet oxygen and/or (iii) Steam.

(2) Chemical deposition by (i) Pyrolysis of alkoxysilane or (ii) Thermal oxidation of silane.

(3) Reactive sputtering.

(4) Electrochemical oxidation of silicon(anodic oxidation).

13.3.1 Thermal oxidation of silicon

Silicon substrate is kept in a furnace at high temperature in the range of 1000-1200°C and oxygen gas is passed either in dry state or through a water bubbler. In the wet oxygen method, water in the bubbler may be either at room temperature or between 90 - 95°C. The water temperature may be raised to the boiling point and steam may be passed over the silicon surface. These processes are primarily controlled by the diffusion of oxidizing species (oxygen or steam) to the silicon-oxide interface and the reaction is described by the parabolic law: $x^2 = Bt$, where x is the oxide layer thickness, t is the time and B is the rate constant. The dry oxygen process is the slowest but gives a better quality of oxide layer. Steam oxidation is the fastest process. Chemical reactions are as follows:

$$Si + O_2 \longrightarrow SiO_2$$

$$Si + 2\,H_2O \longrightarrow SiO_2 + 2H_2$$

Table 13.1 gives properties and growth rates of thermally grown oxide layer.

Table 13.1 Properties and growth rates of thermally grown oxide layer[8]

Oxidation type	Temp. °C	Density gm/cm^3	Rate constant $(micron)^2/min$
Dry O_2	1000	2.27	1.48×10^{-4}
	1200	2.15	7.64×10^{-4}
Wet O_2	1000	2.18	38.5×10^{-4}
	1200	2.21	117.5×10^{-4}
Steam	1000	2.08	54.5×10^{-4}
	1200	2.08	159.5×10^{-4}

13.3.2 Chemical deposition of SiO_2 [6, 12]:

Chemical deposition techniques may be used to grow SiO_2 layers at reduced temperatures. A typical reaction is the pyrolytic decomposition of an alkoxy-silane at low temperatures (500 - 700°C). One commonly used organic compound is tetraethylorthosilicate (TEOS). The following reaction takes place:

$$Si\,(OC_2\,H_5)_4 \longrightarrow SiO_2 + SiO + CO_2 + \text{gaseous organic radicals}$$

At 600°C, a clear amorphous oxide layer can be deposited at rates of 250A°/min.

A second method of chemical deposition of oxide layer is by thermal oxidation of silane (SiH_4) in oxygen as per the following reaction:

$$SiH_4 + 2\,O_2 \rightleftharpoons SiO_2 + 2H_2O$$

This reaction is preferable to that using TEOS since no gaseous organic radical or carbon is formed and the temperature for reaction is lower. Silicon substrate is kept at 300 - 500°C. Another important reason of preference of silane is that silicon nitride can be also obtained by substituting amonia for oxygen, as indicated in the next section. The rate of oxide growth of the silane oxidation process at 400°C is about 800A°/min.

13.3.3 Techniques for silicon nitride[13]

Silicon nitride has certain advantages over silicon dioxide although its processing is more difficult. These advantages are: (i) Si_3N_4 is structurally more compact are more impervious and, therefore, has better masking properties against the diffusion of the common impurities of donors and acceptors. (ii) Migration of alkali and other metal ions which causes instability in MOS FET is less in Si_3N_4. (iii) Growth temperature of nitride is between 600 and 1000 °C.

Typical properties of silicon oxide and silicon nitride are compared in Table 13.2. For Ge and GaAs devices two alternatives available are use of Si_3N_4 or deposited layer of SiO_2.

Table 13.2 Properties of SiO_2 and Si_3N_4[13]

Properties	Si_3N_4	SiO_2
Resistivity at 25°C(ohm-cm)	10^{15}	$10^{14} - 10^{16}$
Dielectric strength (V/cm)	10^7	$(5-7) \times 10^6$
Dielectric constant	7.5 - 9.0	3.8 - 3.9
Index of refraction	2.05	1.452 - 1.466
Density (gm/cm^3)	3.1	2.15 - 2.25
Energy gap (eV)	5.1	8

Amorphous films of Si_3N_4 are prepared using the following methods:

1. Chemical deposition, or
2. Reactive Sputtering.

Chemical deposition is preferred because a common nitride/oxide system can be made. The deposition process is usually carried out in a flow system by passing reactants over heated substrate surfaces where the chemical reaction takes place yielding an adherent dense deposit. Anyone of the following systems could be used: (i) Silane and ammonia, (ii) Silane and nitrogen, (iii) Silicon tetrachloride and ammonia, or (iv) Silicon tetrachloride and a mixture of N_2 & H_2. Chemical reactions in these systems are:

i) $3\ SiH_4 + 4\ NH_3 \longrightarrow Si_3N_4 + 12\ H_2$

ii) $3\ SiH_4 + 2\ H_2 \longrightarrow Si_3N_4 + 6\ H_2$

iii) $3\ SiCl_4 + 4\ NH_3 \longrightarrow Si_3N_4 + 12\ HCl$

iv) $3\ SiCl_4 + 2\ N_2 + 6\ H_2 \longrightarrow Si_3N_4 + 12\ HCl$

Evaluation of dielectric layer

The dielectric layer should have the following properties. It should be thick enough to act as an effective mask against diffusion. The layer to be used as surface protection must not contain appreciable concentration of mobile ions. When used as an insulator it must have a high breakdown strength and a high resistivity.

Thickness of dielectric layer can be estimated by using colour charts. Colour charts for SiO_2 and Si_3N_4 are available in the literature. Another method is to make a step in the dielectric layer and count the fringes in a metallograph using monochromatic light.

Other properties of the dielectric layer can be obtained from metal-insulator-semiconductor (MIS) structure. A very common measurement is capacitance versus voltage plot at low and high frequencies.

13.4 DIFFUSION

13.4.1 Introduction [6,7,8,14,15,16]

Solid-state diffusion of donor and acceptor impurities is the most important technology for the formation of p-n junctions in semiconductor material. Boron as acceptor and phosphorus, antimony and arsenic as donors are the commonly used impurities for silicon devices. Zinc as acceptor and tin as donor are used for diffusion in GaAs.

Diffusion coefficient of an impurity is expressed as

$$D = D_o \exp\{-E_{ac}/kT\} \tag{13.1}$$

where D = diffusion coefficient at T°K (cm^2/sec)

D_o = diffusion coefficient when $T = \infty$

E_{ac} = activation energy of the diffusing impurity (eV)

T = temperature of diffusion (°K), and

k = Boltzmann constant.

Diffusion coefficients of impurities are very sensitive functions of temperature.

Diffusion of impurities is carried out at elevated temperatures (800 - 1250°C) under controlled atmosphere in furnaces where temperatures are controlled within ± 0.5°C. A typical diffusion system for silicon is shown in Fig. 13.6. Precise control of temperature and time is essential to obtain a predictable junction depth in a doped semiconductor.

Solid, liquid and gaseous sources are used for diffusion of impurities, though liquid and gaseous sources are preferred because of better control and reproducible results. A list of source materials for Si are given in Table 13.3.

Table 13.3 Source materials for diffusion in Si

Source	Boron	Phosphorus	Arsenic
Solid source	B_2O_3, BN_3	P_2O_5, P_3N_5	As_2O_3
Liquid source	BBr_3	$POCl_3$	$AsCl_3$
Gas source	B_2H_6	PH_3	AsH_3

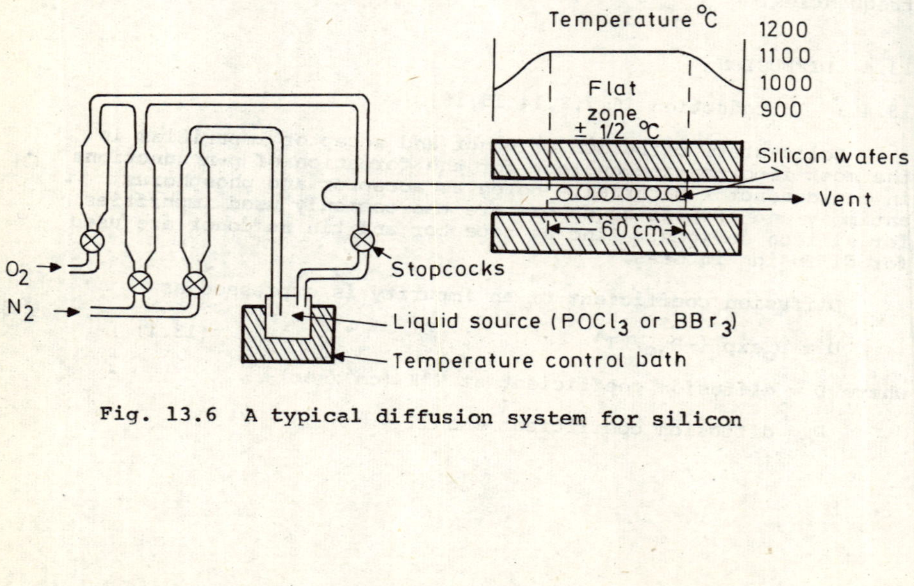

Fig. 13.6 A typical diffusion system for silicon

A semiconductor technologist is seriously concerned about the impurity profile inside the semiconductor material after diffusion. The impurity profile obtained at the end of the first diffusion is going to be disturbed during the subsequent thermal cycling which is unavoidable while fabricating a device. Furthermore, even if the furnace temperature is very precisely controlled, the diffusion coefficient, D, cannot be taken as constant as demonstrated by several investigators[17, 18]. When the concentration of the diffusing impurity is greater than the intrinsic carrier concentration at the temperature of diffusion, then enhanced diffusion takes place. The diffusion coefficient becomes a function of the impurity concentration and D may increase up to several times its normal value at that temperature. In the case of shallow diffusion of phosphorus, it has been observed that D may increase by an order of magnitude depending upon the concentration of diffusing phosphorus inside the substrate.

Besides this, there are (i) effects of lattice strain on diffusion as observed in the case of "emitter dip" when phosphorus is diffused in a boron diffused Si base layer[18], (ii) effects of out-diffusion of impurities, and (iii) effects of re-distribution of impurities during thermal oxidation. If all these effects are taken into account simultaneously, the problem of analytical determination of the impurity profile becomes fairly complicated. However, it is important to obtain precise impurity distribution in the semiconductor material for process control and accurate characterisation of devices.

13.4.2 Impurity distribution

A simple mathematical model of the diffusion phenomenon in semiconductors is given by Fick's first and second laws. In a one dimensional case these can be written as,

$$F = -D \frac{\partial C}{\partial x} \quad \text{and} \tag{13.2}$$

$$\frac{\partial C}{\partial t} = \frac{\partial}{\partial x} \left(D \frac{\partial C}{\partial x}\right) \tag{13.3}$$

where F = particle flux (atoms/cm^2/sec)

D = diffusion coefficient (cm^2/sec)

C = impurity concentration (atoms/cm^3)

x = distance from the surface (cm), and

t = time (sec).

Here diffusion is assumed to take place through a unit cross-sectional area (1 cm^2). When the diffusion coefficient is constant and equal to D, then

$$\frac{\partial C}{\partial t} = D \frac{\partial^2}{\partial x^2}(C) \tag{13.4}$$

For the situation where there is a constant source (or infinite source) and the medium is semi-infinite, the impurity concentration is

$$C(x,t) = C_o \{1-(2/\sqrt{\pi})\int_o^{x/2\sqrt{Dt}} \exp(-m^2)\,dm\}$$

$$= C_o \operatorname{erfc}(x/2\sqrt{Dt}) \quad (13.5)$$

where C_o is surface concentration (atoms/cm^3).

If instead of an infinite source, there exist Q atoms of impurity per unit area on the surface of the semiconductor and this semiconductor is heated in an atmosphere which prevents evaporation of the impurity atoms, then the impurity profile as a result of diffusion will be a Gaussian distribution i.e.,

$$C(x,t) = C_s(t) \exp(-x^2/4Dt) \quad (13.6)$$

$$C_s(t) = Q/\sqrt{\pi Dt} \text{ (= surface concentration)} \quad (13.7)$$

The gradient of the Gaussian impurity profile at x denoted by 'a' is given by

$$a = (-x/2Dt)\,C(x,t) \quad (13.8)$$

13.4.3 P-N junction formation

When acceptor impurity is diffused into a donor-doped semiconductor substrate (or vice-versa), the net impurity distribution is given as:

$$C(x,t) = C_o \operatorname{erfc}(x/2\sqrt{Dt}) - C_B \quad (13.9)$$

where C_B is substrate impurity concentration (atoms/cm^3). A p-n junction is formed at $x = x_j$ where $C(x,t) = 0$. x_j is given by

$$x_j = 2\sqrt{Dt}\,\operatorname{erfc}^{-1}(C_B/C_o) \quad (13.10)$$

13.4.4 Two-step diffusion process [6,14,15]

As is evident from the solid solubility curve of Trumbore, the complementary error function impurity distribution, resulting from diffusion with infinite source, does not give enough flexibility in the value of the surface concentration; whereas the surface concentration in the Gaussian distribution is a function of diffusion time and temperature and can be varied widely. The device engineer requires a technique which gives more flexibility in controlling the surface concentration so that base width and base impurity distribution can be controlled at will. Therefore a two-step diffusion process is often used in semiconductor device fabrication technology. In the first step, impurities are diffused into the semiconductor with infinite source to a small depth. This process is called deposition step. The resulting impurity profile is the complementary error function eqn.(13.5). The total number of impurity atoms Q penetrated in a 1-cm^2 section of the semiconductor is given by:

$$Q = \int_0^\infty C_o \, \mathrm{erfc}(x/2\sqrt{D_1 t_1}) \, dx = (2/\sqrt{\pi}) \, C_o \sqrt{D_1 t_1} \quad (13.11)$$

where D_1, t_1 correspond to the diffusion coefficient and time, respectively, for the deposition step. Note that by increasing the time only the penetration depth increases but not the surface concentration.

Once the impurities have been introduced into the semiconductor, they are then diffused deeper and this step is called drive-in diffusion. The impurity profile as a result of drive-in diffusion step is Gaussian and is given by

$$C(x,t_1,t_2) = (Q/\sqrt{\pi D_2 t_2}) \exp(-x^2/4D_2 t_2) \quad (13.12)$$

where D_2, t_2 correspond to the diffusion coefficient and time of drive-in diffusion. The temperature of deposition and drive-in are not the same, therefore D_1 and D_2 have different values.

The final surface concentration is given by:

$$C(o,t_1,t_2) = C_s = (2/\pi) \sqrt{(D_1 t_1/D_2 t_2)} \, C_o \quad (13.13)$$

Equation (13.12) has been obtained by assuming that Q is confined in a thin layer at the surface of the semiconductor. Therefore departure from eqn.(13.12) may be expected, if the condition $\sqrt{D_2 t_2} \gg \sqrt{D_1 t_1}$ is not satisfied.

Kennedy and Murley [19] have given a rigorous mathematical analysis of the two-step diffusion process. According to them the final impurity distribution is:

$$C(x,t_1,t_2) = C_o (2/\pi) \int_{\sqrt{B}}^\infty \exp(-m^2) \, \mathrm{erf}(am) \, dm \quad (13.14)$$

where a is $\sqrt{(D_1 t_1/D_2 t_2)}$ and B is $x/(2\sqrt{D_1 t_1 + D_2 t_2})$.

The surface concentration after the second diffusion is given by:

$$C_s = C_o \, (2/\pi) \tan^{-1} (\sqrt{D_1 t_1/D_2 t_2}) \quad (13.15)$$

Equation(13.14) gives reasonable results and it is commonly used for predicting impurity profile in the case of two-step diffusion. The integral has been evaluated by Smith for various values of a's and B's. The figures are reproduced in Runyan[11] and Ghandhi[8].

Considerable departure from complementary error function profile has been observed in the case of phosphorus and boron diffusion in silicon [17,18]. There is a flattening of the impurity distribution profile near the surface. Such diffusion has been characterized by concentration dependent diffusion coefficient and the diffusion equation is:

$$\frac{\partial C}{\partial x} = \frac{\partial}{\partial x} \{D_{eff} (C) \frac{\partial C}{\partial x}\} \tag{13.16}$$

where D_{eff} is effective diffusion coefficient which incorporates enhancement factors H and P, and is given as, $D_{eff} = D\,H(C)\,P(C)$ D is normal diffusion coefficient independent of concentration, H(C) is enhancement factor due to field-aiding effect, and is given as,

$$H(C) = 1 + 1/\sqrt{1+4\ (n_i/C)^2} \tag{13.17}$$

P(C) is enhancement factor due to increase in vacancy concentration, and is given as[20],

$$P(C) = 1 + (C/C_R)^2 \tag{13.18}$$

n_i is intrinsic concentration at the temperature of diffusion,and C_R is a characteristic concentration dependent on the type of impurity[16].

To obtain accurately a profile of the diffused impurity, eqn.(13.16) has to be solved numerically using a digital computer[21]. Such a profile is required in order to get accurate estimation of junction depth, which in turn makes possible a better characterization of diffused devices.

Sheet resistance of diffused layer is measured using four-point probe and the junction depth is measured by angle lapping and staining technique.

13.5 ION IMPLANTATION

For purposes of impurity doping the diffusion process has become the work-horse of planar technology. There has been a steady growth in understanding of the processes giving good control over the impurity profile. However, there is a basic limitation arising out of the fact that diffusion is essentially a process dependent on impurity gradients.

New avenues have been opened up by the use of beams of ions impinging on semiconductor materials, and acting as dopants. This technique offers a qualitatively different degree of control over the impurity profile, and other related parameters.

In this section the basic principles, main characteristics and some applications of this technique have been discussed.

13.5.1 Basic principles and main features

The process of ion-implantation consists basically of the following steps:

a. Generation of a beam of ions of the desired element.

b. Acceleration of the beam to the required energies (usually of the order of 20 to 300 Kev).

c. Exposure of the target material to the ion beam at a given orientation and for a given period.

d. Annealing of the substrate to counteract the lattice damage caused by the ion beam.

The main features of this technique are the following[22,23]:

a. Doping is possible at low temperatures. Even the subsequent annealing can be done at temperatures which are considerably lower than those normally used for diffusion. The process is therefore applicable to materials which cannot withstand high temperatures or those in which the diffusion process is so slow as to require extremely high temperatures.

b. Whereas for alloying or diffusion the doping concentration depends upon a number of parameters, here a very close control over doping is possible by controlling the beam current and the duration of the exposure. For a doping concentration of the order of 10^{15} per cm^3, an exposure of the order of 20 seconds is required with ion beam current density of the order of 10 micro amperes per cm^2.

c. Whereas in the case of diffusion the usual doping profile is a monotonically decreasing function of distance from the surface, in the case of ion implantation the profile can have a peak at a controllable depth below the surface. This depth depends upon the choice of crystal orientation and the energy to which the ions have been accelerated.

d. There is very little lateral spread of the impurity ions.

e. Ions can be implanted through a layer of oxide or an extremely thin layer of conductor.

f. The junction depth is much more uniform since local variations or impurities have less effect than in the case of diffusion.

13.5.2 Range versus energy relationship

The penetration depth of ions is dependent on the initial kinetic energy and the nature and orientation of the target material. In the case of amorphous materials the doping profile is a Gaussian curve. This is also the case for a crystallographic direction with small indices.

In a crystalline target when one looks along a direction with small indices, the nuclei are seen to be arranged in line. The interactions of an ion beam proceeding along such a direction, with the nuclei in the lattice are then minimised.

If the beam makes an angle with a crystallographic direction of the above kind, but the angle is less than a critical angle of the order of 3 to 5 degrees; the ions are channelled by the repulsive force of the net charge as the ions penetrate deeper into the electronic shells. The ions are however progressively scattered out of the channel. The doping profile is thus a combination of a peak at a certain depth below the surface, superimposed upon a background provided by the de-channelled ions, as shown in Fig. 13.7 [22].

As the ion beam penetrates the target, there are two kinds of energy loss, namely (a) that due to interaction with electrons, and (b) that due to interaction with nuclei. For electrons the stopping power is proportional to the velocity or the square root of the kinetic energy. For nuclei the stopping power depends mainly upon the masses of the ions. For high energies the stopping power of the electrons is dominant. In this case the range is approximately proportional to the square root of the kinetic

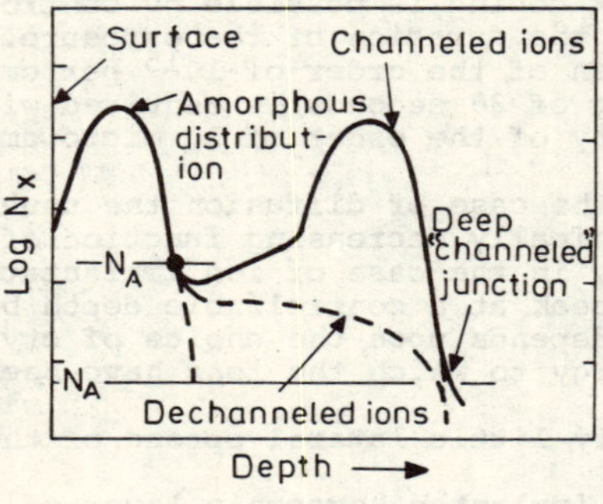

Fig. 13.7 Doping profile obtained by ion implantation (From [22])

energy. Most of the lattice damage however takes place at the low energies. Typical curves showing the depth of the peak as the function of energy are shown in Fig. 13.8 [22].

13.5.3 Lattice damage and annealing

In order that the implanted impurities may function in the desired manner in the semiconductor devices, it is necessary that they may be at electrically active sites (usually substitutional sites in the lattice), and also that the mobility and life time of the majority carriers may be adequate. It therefore, becomes necessary to anneal the material making it possible for impurity ions to migrate to substitutional sites and for the lattice damage to be minimised.

The percentage of impurity ions which will reach active sites is critically dependent upon the annealing temperature and the initial conditions of the substrate, especially the extent to which it was amorphous. Experiments with silicon have given the following general results [25].

a. If the silicon is in an amorphous state before ion-implantation, then annealing for 30 minutes at 650°C after ion-implantation, recrystalises the layer and gives a high degree of electrical activity (between 50 and 100 per cent).

b. If the silicon is not driven amorphous before annealing, less electrical activity (of the order of 10 per cent) may result. The annealing in this case may have to be performed at temperatures in the 800-900°C range.

c. In case the silicon is made amorphous to a certain depth and the ions penetrate deeper than this value, an anneal at 650°C as mentioned above, results in a transition region where electrical properties of the majority carriers will be dominated by the presence of traps. This can be avoided by annealing at temperatures of the order of 800 - 900°C.

13.5.4 Applications

The special features of ion implantation which have been discussed above are utilised in fabricating devices which are beyond the reach of the diffusion process. Double drift region avalanche diodes are one category of microwave semiconductor devices where ion implantation has been used with notable succes. These devices are discussed in Sec. 9.4. Apart from this, the technique is also being used in fabrication of MOS devices for accurate control of threshold voltage and in bipolar transistors for giving extremely narrow base width of the order of 0.25 microns.

13.6 ELECTRON BEAM TECHNOLOGY FOR PATTERN DELINEATION

In the case of photolithography, diffraction effects limit the resolution to dimensions of the order of a wavelength of

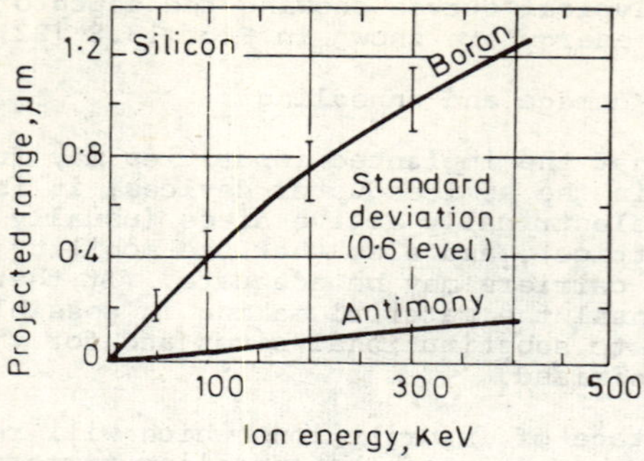

Fig. 13.8 Location of the peak as a function of ion energy (From [22])

light. By using electron beams and taking advantage of the much smaller wavelength associated with them, a considerably higher degree of resolution is possible. The spot size could thus be less than 500A°. Further the electron beam can be deflected and intensity modulated according to a programme. Because of these reasons, electron beam delineation offers a promising approach towards higher degree of miniaturisation or larger scale integration.

The simplest method of forming a pattern in a sequential way would be to use a raster scan as in television. However, the bandwidth required is given by

$$B_W = L^2 F/2d_s{}^2$$

where L is side of square raster, F is frame rate, and d_s is spot diameter. Thus if L is 1 cm, d_s is 2 μm and F is 30, the value of B_W obtained is 375 MHz.

In order to reduce the bandwidth, the scanning may be done only area-by-area and controlled by a digital computer. In this, after scanning a given area the electron beam moves to another one under command from the computer. Thus the bandwidth can be reduced to a value of the order of 1 MHz. In the total system, apart from the digital computer one requires an input/output buffer register, and a digital to analogue converter, for coupling the computer to the deflection system. The electron beam equipment is essentially similar to a scanning electron microscope [26].

The elements of time which have to be taken into account in order to determine the total time required for exposing a wafer are the following:

1. Time required for the beam to deposit the requisite amount of charge on the electronresist layer.

2. Time required in going from one element to the next.

3. Time delay from computer to the digital-to-analogue converter.

4. Time for deflection circuit to settle to the new position.

It has been estimated that to cover 5 x 10^5 element, using an electron beam with a current density of the order of 1 amp. per cm the total time required would be of the order of 30 seconds.

Apart from a sequential exposure using a scanning electron microscope, another technique which is sometimes employed is that of parallel exposure. An electronic image is formed on the substrate using as an object a specially prepared mask which acts as a photo cathode. De-magnification may be introduced if desired. The mask itself is prepared by an electron bean which can have a minimum line width of the order of 0.25 microns.

The electron spot size depends upon spherical aberration, chromatic aberration, space charge expansion and transverse thermal velocities. Taking all of these into account a spot size of less than 100A° could be obtained. However, other limitations to the ultimate resolution are created by the phenomenon of backscatter in the electron resist layer. This increases the spot size by an order of magnitude. Other points to be taken into account are distortions in deflection.

The fine resolution possibilities of an electron beam pattern delineation have been exploited for junction field effect trasistors in gallium arsenide and for transducers for acoustic wave amplifiers.

References

1. Caulton, M., Film Technology in Microwave Integrated Circuits . Proc. IEEE, Vol. 59, Oct. 1971, pp 1481-1489.

2. Caulton, M. and H. Sobel, Microwave Integrated Circuit Technoloty - a survey, IEEE J. Solid-State Circuits, Vol. SC-5, Dec. 1970, pp 292-302.

3. Sobol, H., Technology and design of hybrid microwave integrated circuits , Solid State Technology, Vol. 13, Feb.1970, pp. 49-57.

4. Mehal, E.M. and R.W. Wacker, GaAs integrated microwave circuits , IEEE Trans. Electron Devices, Vol. ED-15, July 1968, pp 513-516, and other papers in this issue.

5. Jordan, A.G., Private Communication.

6. Burger, R.M. and R.P. Donovan (Eds.), Fundamental of Silicon Integrated Device Technology, Vol. I, Prentice-Hall, Inc., New Jersey, 1967.

7. Warner, R.M., Jr. and J.N. Fordem Walt (Eds-Motorola Inc.), Integrated Circuits - Design Principles and Fabrication, McGraw Hill Book Co., New York, 1965.

8. Ghandhi, S.K., Theory and Practice of Microelectronics, John Wiley & Sons, New York, 1968, Chs. 4,5,6.

9. Tietjin, J.J. and J.A. Arnick, The Preparation and Properties of Vapor-Deposited Epitaxial $GaAs_{1-x}P_x$ using Arsine and Phosphine , Jr. of Electrochem. Soc., Vol. 113, 1966, p.724.

10. Kang, C.S. et al., Preparation and properties of high purity epitaxial GaAs grown from Ga solution , App. Phys. Letters Vol. 11, No. 5, Sept. 1967, pp. 171-173.

11. Lawley, K.L., Film-making - a delicate job performed under pressure, Electronics, November 13, 1967 pp 114-115.

12. Jordan, E.L., A Diffusion Mask for Germanium , J. Electrochem. Soc., Vol. 108, 1961, p. 478.

13. Doo, V.Y., D.R. Nicholas, and G.A. Sivey, Preparation and Properties of Pyrolytic Silicon Nitride , J. Electrochem. Sec., Vol. 113, Dec. 1966, pp.1247-1281.

14. Runyan, W.R., Silicon Semiconductor Technology Ch. 7, McGraw-Hill, 1965.

15. Grove, A.S., Physics and Technology of Semiconductor Devices , John Wiley (1967), Ch. 2,3.

16. Boltaks, B.I., Diffusion in Semiconductors, Academic Press (1963) Chs. 4,7.

17. Tsai, J.C.C., Shallow phosphorus diffusion profiles in silicon , Proc. IEEE, Vol. 57, pp.1499-1506, Sept. 1969.

18. Nicholas, K.H., Studies of anomalous diffusion of impurities in silicon Solid-State Electronics, Vol. 9, pp.35-47, January 1966.

19. Kennedy, D.P. and P.C. Murley, Impurity atom distribution from a two-step diffusion process , Proc. IEEE, Vol. 52, pp. 620-621, May, 1964.

20. Thai, N.D., Concentration dependant diffusion of boron and phosphorus in silicon , Jr. of App. Phys. Vol. 41, pp.2859-2866, 1970.

21. Hasan, M.M. and S. Venkateswaran, Diffusion in Semiconductors, Electronics Letters, Vol. 6, pp. 366-368, June 11, 1970.

22. Brewer, G.R., The application of electron/ion beam technology to microelectronics , IEEE Spectrum, Vol.8, No.1, January 1971, pp 23-37.

23. Dill, H.G., R.M. Finnila, A.M. Leupp, T.N. Toombs, The Impact of Ion Implantation on Silicon Device and Circuit Technology , Solid State Technology, December 1972, pp 27-35.

24. Gibbons, J.F., Ion Implantation in Semiconductors - Part I Range Distribution Theory and Experiments , Proc. IEEE, Vol. 56, March 1968, pp. 295-319.

25. Gibbons, J.F., Ion Implantation in Semiconductors - Part II Damage Production and Annealing, Proc. IEEE, Vol.60, No.9, September 1972, pp. 1062-1096.

26. Ozdemir, F.S., E.D. Wolf and C.R. Budkey, Computer-Controlled Scanning Electron Microscope System for High-Resolution Microelectronic Pattern Fabrication , IEEE Transactions on Electron Devices, Vol. ED-19, No.5, May 1972, pp.624-628.

14

SOME SYSTEM APPLICATIONS OF MICs

by

K.C. Gupta* and Amarjit Singh**

* Sections 14.1 and 14.2
** Section 14.2

Chapter Fourteen

14.1 MICs IN PHASED ARRAY RADARS

14.1.1 Phased array radar

A phased array radar uses a phased array antenna in place of mechanically steered antennas in conventional radar systems. A phased array antenna system comprises of a large number of identical antenna elements arranged spatially in a two dimensional array. The phase of the r.f. power fed to each of elements can be adjusted electrically to yield the desired phase shifts among the radiations from the various elements. This phase relationship determines the direction of the radiated beam. Thus by incorporating a phase shifter with each element of the array, the radiated beam can be scanned electrically by adjusting the settings of the phase shifters.

The number of antenna elements in a phased array depends upon the beam width requirement. An approximate relation is

$$\text{Number of elements} = 10000/[\text{Beamwidth (in degrees)}]^2$$

Thus if a beam width of 1° is required the number of antenna elements needed is 10,000. Since each element requires a phase shifter, the number of phase shifters required for a typical radar system is of the order of ten thousand.

Another significant feature of phased array radar is that it provides an option to incorporate solid state microwave sources for transmitter power. A low power source can be connected to each element of the array and power summation obtained in space. This will require typically ten thousand small power microwave sources in place of a single high power microwave tube as employed in most of conventional radar systems, and in some phased array radars. In the later case, power dividers are used to feed power to individual antenna elements.

14.1.2 Concept of T-R module

The source of r.f. power and the phase shifter required for each element can be placed in a module situated near the antenna elements. This module situated near the antenna elements has to perform some other functions also. These include a provision for receiving the signals returned from targets, arrangement for switching the antenna between the transmit and receive functions and a method of interfacing the beam steering networks with a computer. These functions may be accomplished by a variety of means using amplifiers, frequency converters and control methods. A usable grouping is based on the availability of power amplifiers and low noise amplifiers at the frequency of operation and is called a 'Transmit-Receiver Module' (or T-R module).

T-R module is one of best examples of circuit functions suitable for fabrication in MIC form. The following factors are responsible for this:

a) As mentioned above, identical T-R modules are required in substantially large numbers.

b) Maximum spacing between array elements in a phased array system is limited to a fraction of a wavelength to avoid beam degradation. Consequently T-R modules are required to fit into a relatively small physical space.

c) Many applications of phased arrays involve sophisticated and complex electronic systems in which high reliability is demanded.

14.1.3 Configurations for T-R module

Design of a Transmit-Receive Module depends upon the choice of r.f. source (or transmitter chain) and the front end of the receiver circuit. As discussed earlier, this choice depends on the frequency range.

Low frequency module

At the low microwave frequencies, transistor power amplifiers and low noise amplifiers are available, and a simple module configuration may be used. Such a module is shown in Fig.14.1.

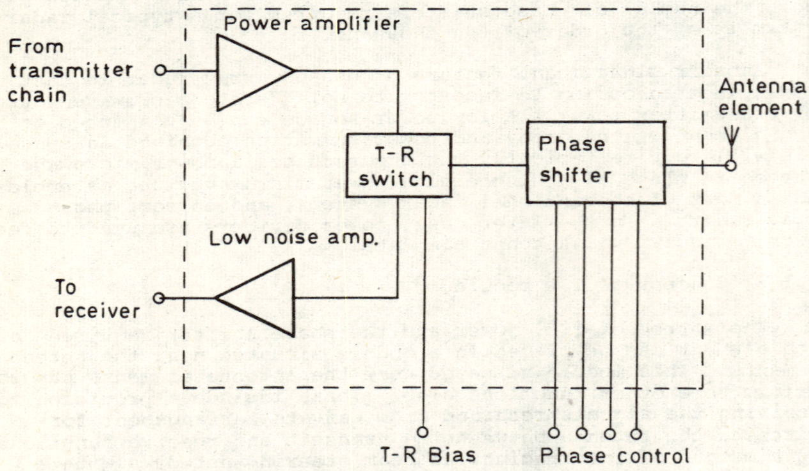

Fig. 14.1 Simplest form of T-R module

In this set up the phase shifter is placed next to the antenna. The T-R switch is a single pole double throw electronic switch. Switching is normally obtained by changing the d.c. bias on PIN diodes. A significant disadvantage of the configuration shown in Fig. 14.1 is that the insertion loss of the phase shifter attenuates the power output of the power amplifier and adds directly to the noise figure of the receiver. Also the phase shifter should have a power handling capacity greater than the output of the power amplifier. These difficulties may be avoided by using the module configuration shown in Fig. 14.2.

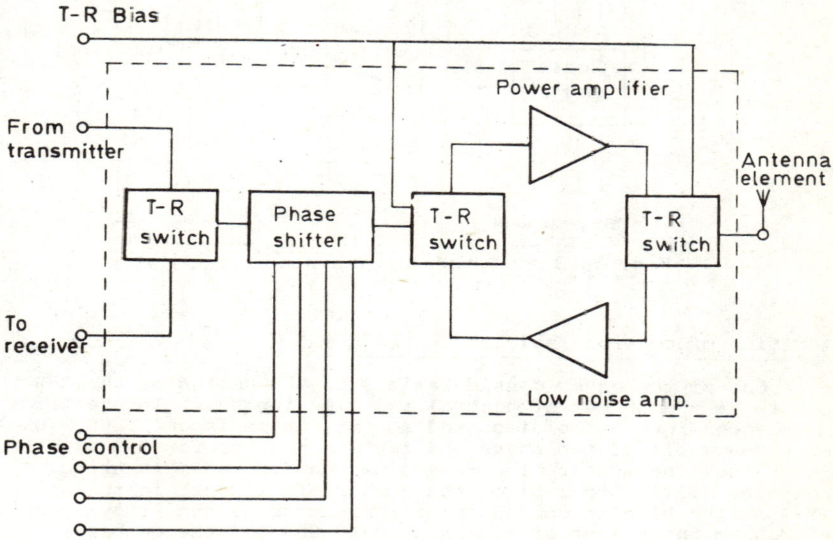

Fig. 14.2 T-R module with a low power phase shifter

In the circuit configuration shown in Fig. 14.2, the power amplifier and the low noise amplifier are nearest to the antenna element (of course a T-R switch is required immediately next to the antenna element to separate the transmitter channel from the receiver channel). In this arrangement, the insertion loss specification of the phase shifter may be relaxed, allowing a lower cost version. The number of T-R switches needed for this set up is three. This can be reduced to one if two separate phase shifters are used as shown in Fig.14.3. In this case, since the signal transmission through the phase shifter is only in one direction, non-reciprocal phase shifters may also be used. In the two configurations discussed earlier it is necessary to use reciprocal phase shifters. However, the diode phase shifters which are commonly used for T-R modules are invariably reciprocal. Another advantage of the module configuration shown in Fig. 14.3 is that the insertion losss associated with the two T-R switches in the receiver and the transmitter chains is avoided. However, the replacement of one phase shifter and two T-R switches by two phase shifters should be examined in an economic optimisation of the system.

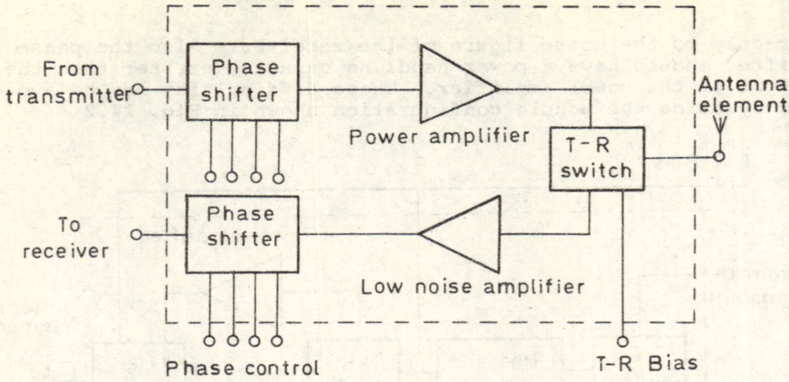

Fig. 14.3 T-R module with non-reciprocal phase shifters

Addition of control logic to T-R module

One of the other considerations in the design of the T-R module is the number of control and bias inputs. In the three module configurations discussed so far, there is one input for each phase bit of the phase shifter and another for T-R switching. Thus the number of wires to be run from beam steering computer will be four times the number of elements in the array. (T-R switch bias is common for all elements of the array). For a 1° beam the number of elements is 10,000 and the number of wires is about 40,000.

The number of connecting wires running from the beam steering computer to the array can be reduced by addition of control logic to the module. This is generally considered essential for a practical system. A typical arrangement that could be used is shown in Fig.14.4. Here two inputs regarding phase information are fed to the control logic. These are a digital word (or pulse train) input that is common to each column and another word which is common to each row. These two inputs are added to arrive at the desired phase shift. This arrangement reduces the number of wires from the beam steering computer from some 40,000 wires in a 100 x 100 array to about 200 wires, one for each row and each column. In addition, there are two wires for reset and T-R signals common to all the phase shifters. When a non-planar configuration or an optical feed is used, there is a third input to each module for spherical correction. However, since this correction does not depend on the beam position, the information can be prefed and stored in the control logic hardware corresponding to each phase shifter.

Another degree of complexity in the T-R module could be required for a particular radar system. In order to achieve low side lobe levels on the receive patterns, amplitude tapering of the signal from each antenna element is used. This is achieved by placing an adjustable attenuator following each low noise

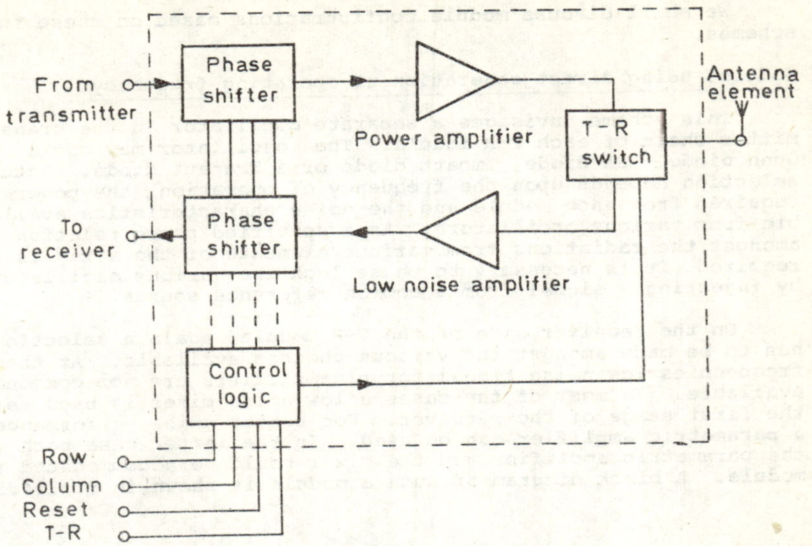

Fig. 14.4 T-R module with control logic included

amplifier. Electrically the adjustable attenuators could also be placed next to the phase shifter in Fig.14.4. This allows one to have different tapers, one for the transmit and another for the receive patterns. One could also incorporate an arrangement to vary the power output of the power amplifier in order to achieve various antenna patterns on transmit.

The control logic could also incorporate a scheme for automatic testing of the module.

High frequency module

Transistor power amplifiers are at present not available for frequencies above 4 GHz or so. As a result the module configurations described above cannot be used in C-band or X-band. There are two distinct schemes for generation of r.f. power at these frequencies.

1) Direct generation at the frequency of operation by using Gunn Oscillators, LSA Oscillators, Impatt diode oscillators or Trapatt diode oscillators.

2) Generation of larger power at lower frequency (by transistor or trapatt oscillators) and frequency multiplication using varactors or step recovery diodes.

We shall discuss module configurations based on these two schemes.

Modules using direct generation at operating frequency

This scheme envisages a separate oscillator in the transmitter chain of each T-R module. The oscillator may use a Gunn diode, LSA diode, Impatt diode or a Trapatt diode. Actual selection depends upon the frequency of operation, the power required from each module and the noise characteristics available from various oscillators. As a specified phase relation amongst the radiations from various elements of the array is required, it is necessary to phase lock the various oscillators by injecting a signal from a common reference source.

On the receiver side of the T-R module, again a selection has to be made amongst the various choices available. At these frequencies low noise transistor preamplifiers are not commonly available. In most of the cases a low noise mixer is used as the first stage of the receiver. For better noise performance a parametric amplifier can be used. In the latter case both the parametric amplifier and the mixer could be accommodated in module. A block diagram of such a module is shown in Fig.14.5.

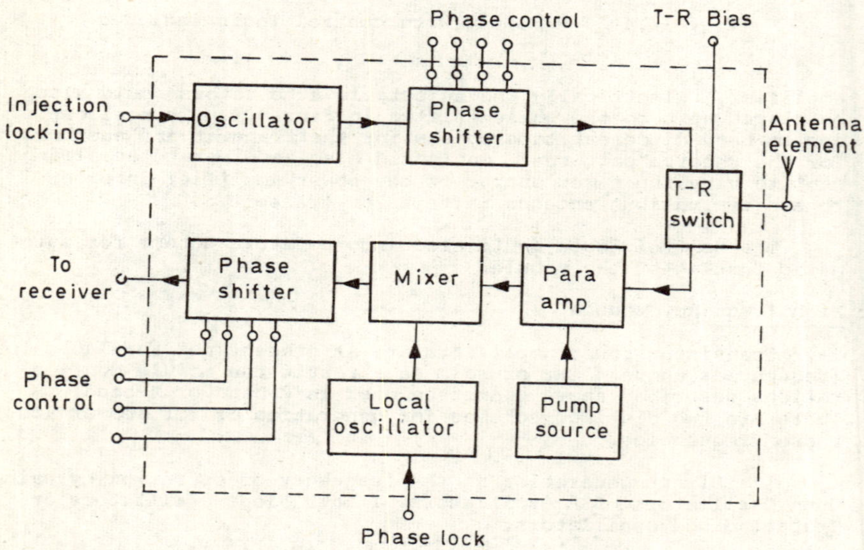

Fig. 14.5 High frequency T-R module with parametric amplifier

In the case where the parametric amplifier is not required the module configuration can be simplified as shown in Fig.14.6.

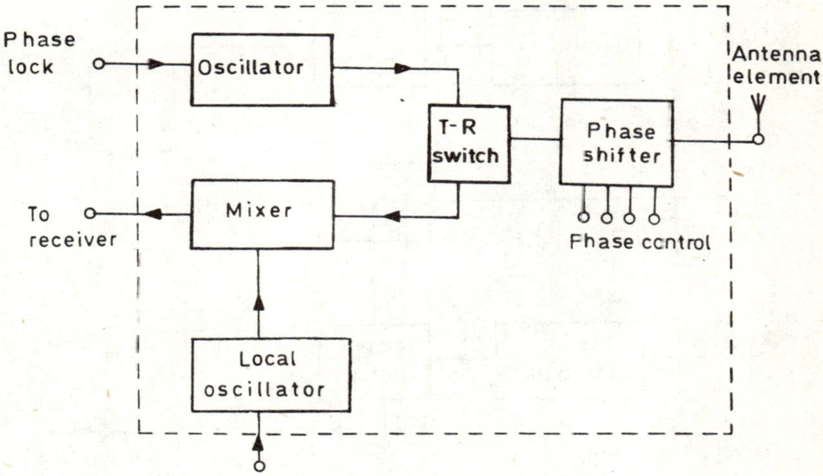

Fig. 14.6 A simple configuration for high frequency T-R module

When the mixer circuit is included in the module, the summation for received signal is done at IF. In such cases, the phase shift for the received signal could also be obtained by including the phase shifter in the local oscillator chain as shown in Fig.14.7. It is also necessary to phase lock the local oscillators in various modules. This is true for configurations shown in Figs. 14.5 and 14.6 (in all the cases when summation is done at IF).

Module using frequency multipliers

The design of these modules is similar to modules using oscillators (discussed above) except for the fact that the phase locked oscillator is replaced by a power amplifier at a lower frequency and a frequency multiplier. A simple arrangement of this type is shown in Fig.14.8.

As in the earlier cases, the mixer also can be incorporated in the module itself. The number of r.f. connections can be reduced by generating the local oscillator signal within the module itself. In most radar applications, the transmitter output and the local oscillator are not required at the same time, so these signals can be duplexed. A module using this approach is described later on.

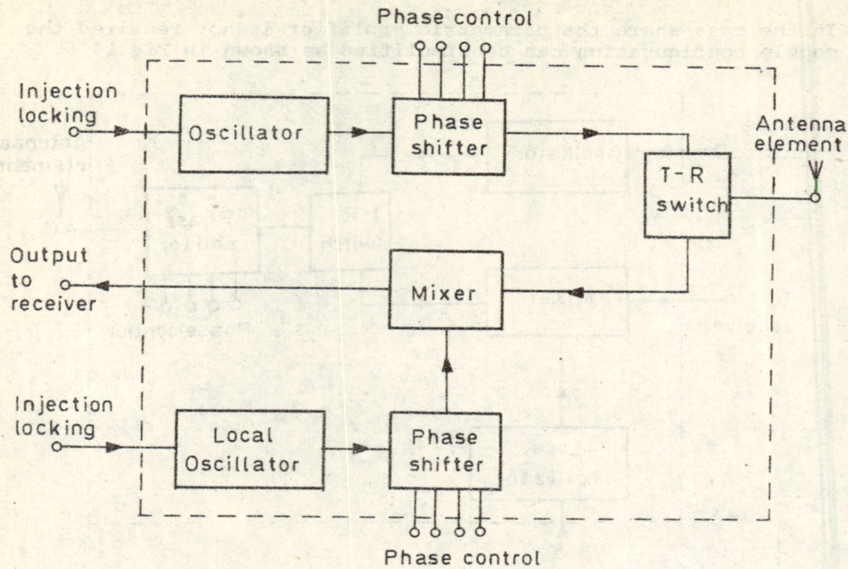

Fig. 14.7 T-R module with a phase shifter in local oscillator chain

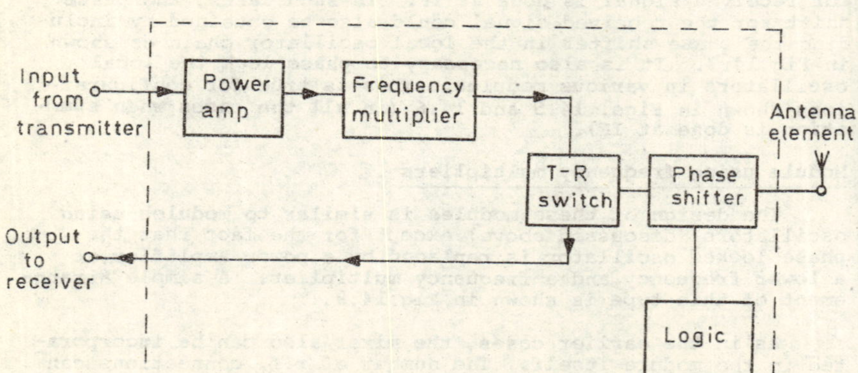

Fig. 14.8 T-R module using a frequency multiplier

14.1.4 T-R modules reported in literature

T-R module at X-Band

Block diagram of a T-R module operating at X-band [1,2] is shown in Fig.14.9. The module uses a multiplier each in transmitter and local oscillator chain. Phase shifters at S band are used before the multipliers. During the transmit interval, a 2.250 GHz signal is fed to the module. This signal is amplified and passed through a phase shift network and fed into the power amplifier. The output of the power amplifier goes to the X4 frequency multiplier. Multiplier output is passed through the T-R switch to the antenna.

The local oscillator signal is generated in a similar manner. During the receive interval of the cycle, the drive frequency is at 2.125 GHz and it is also passed through the power amplifier, phase shifter and the frequency multiplier. The return signal coming into the antenna is routed by the T-R

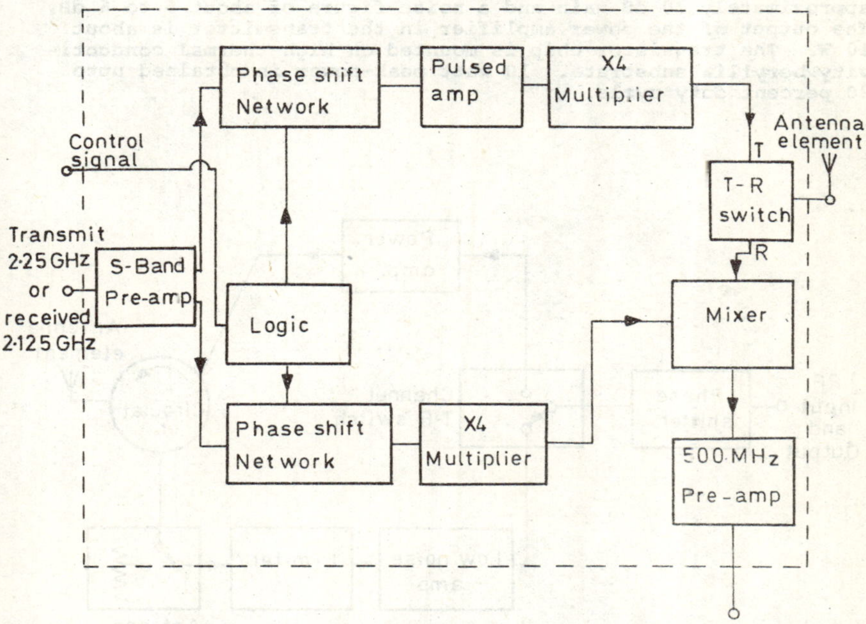

Fig. 14.9 An X-band T-R module (Ref.[1])

switch into the balanced mixer. The output at 500 MHz IF frequency is amplified.

Two phase shifters are used in this module. This is due to the fact that the phase shift on received signal is applied through local oscillator. For a given positive phase shift applied in the transmit chain, there must be in fact, a negative phase shift in the local oscillator chain (regardless of whether the IF frequency is above or below the incoming signal).

In this module, the r.f. power generated, i.e. output of X4 frequency multipliers, is 1 watt. Power radiated is 0.9 watts. Output of the local oscillator chain is 2mW.

L-band transmit-receive module

An L-band transmit receive module is described in [1]. A simplified block diagram of the module is shown in Fig. 14.10. Low noise amplifier is a two stage integrated transistor amplifier designed for optimum noise figure performance. This amplifier is fabricated on an alumina substrate, with microstrip transmission lines for impedance matching. The amplifier has approximately 20 dB gain and a noise figure of about 5 to 6 dB. The output of the power amplifier in the transmitter is about 10 W. The transistor chip is mounted on high-thermal conductivity beryllia substrate. 10 watt peak-power is obtained upto 20 percent duty cycle.

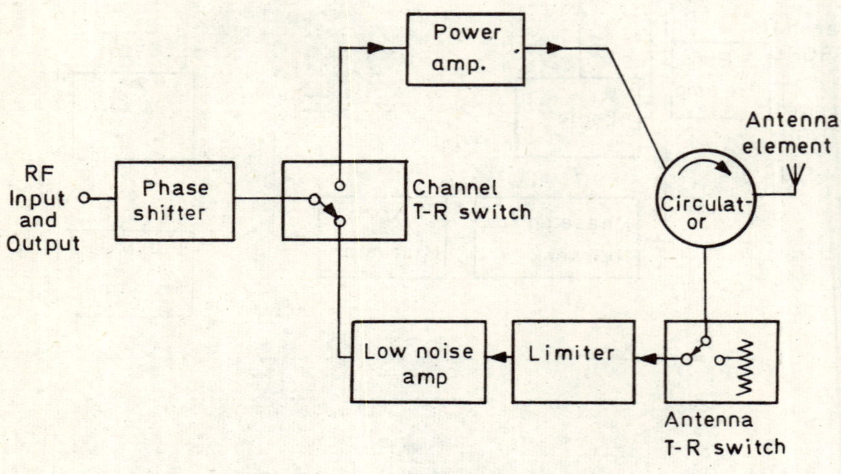

Fig. 14.10 An L-band T-R module (Ref.[1])

S-Band module

An S-band CW power module for phased arrays has been described [3,4]. This is transmitter part of the T-R module and converts 100 mW at 1.5 GHz to 15 W CW power at 3 GHz. The block diagram of the power module is shown in Fig.14.11.

The L-band input signal is first fed to a preamplifier that is followed by a driver amplifier using transistor TA 7005. Three stage amplifier is used for driver. The power output from the driver is then equally divided into two channels, each containing a power amplifier and a varactor doubler. Three TA 7305 transistors mounted in microstrip carriers are directly paralleled to form the power amplifier.

Direct interstage matching is employed between the preamplifier and driver and between the power amplifier and doubler sections. A quadrature hybrid is used for the power divider to simplify the interstage matching problem between the driver and the two power amplifiers. This configuration presents a constant load impedance to the driver amplifier independent of the magnitude of the input mismatch at the power amplifiers, provided that mismatches are equal for both amplifiers. Because of the phase shift in the hybrid, the reflected signals arrive at the load resistance of the divider and are dissipated there.

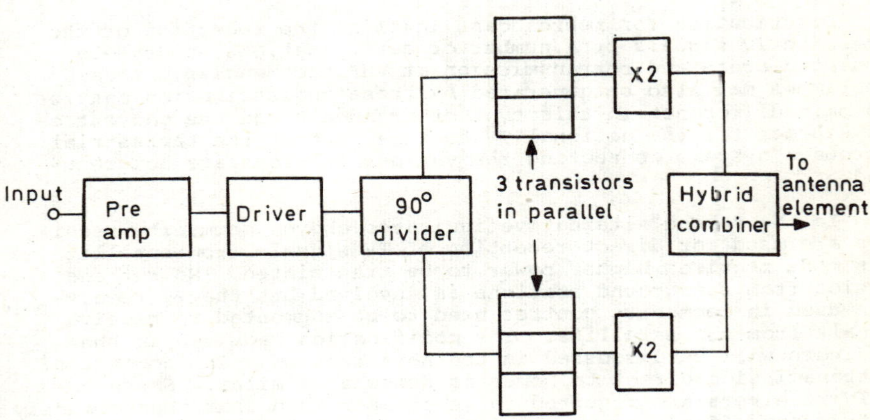

Fig. 14.11 An S-band T-R module-transmitter part (Ref.[3])

At input port, the two reflected signal are 180° out of phase. Thus the input VSWR of the divider is quite small and essentially independent of the input VSWR of the power amplifiers. The phase shift of the divider is doubled in the multiplier, which causes the power outputs from the two chains to be out of phase by 180 degrees. Therefore, a hybrid ring is used to combine the two chains.

14.2 MICs IN SATELLITE TELEVISION SYSTEMS

14.2.1 Satellite television systems

In a satellite television system, a transponder on the satellite is used to relay the TV programmes to TV receivers or the redistribution centres located over large areas, such as, the whole of the Indian subcontinent. The programmes may originate at one or several points and are transmitted to the satellite by an upward microwave link by the main earth station. Three different modes may be used for down link transmission. These are,

1) Distribution for rebroadcast

2) Direct to community centres

3) Direct to homes

Distribution for rebroadcast involves the reception of the satellite TV signals by a number of earth stations in densely populated areas and retransmission at VHF frequencies. Local programmes may also be generated by these redistribution centres. The main difference in this type of broadcast and the conventional schemes for the national TV hookups is that the terrestrial microwave systems connecting the various stations are not required.

In the second alternative indicated above, community receivers are used for direct reception of TV signals from satellite. This mode requires higher power to be transmitted. No retransmission from the ground stations is involved but the TV receivers used in community centres need to be augmented to receive signals from the satellite. The modification required in the receivers will be discussed in the next section. The concept of TV transmission direct to homes is somewhat similar. Since home TV receivers are required to be cheaper than what the community centres can afford, power required to be radiated from the satellite is still higher.

A comparison of the three possible methods for the downward transmission [6] is given in Table 14.1. The choice between these three alternatives depends upon the population density in the particular area. It is also possible to use a combination of any two or all the three alternatives. In such a case the system is named as a hybrid system.

India has plans for using ATS-F satellite to be launched by National Aeronautics and Space Administration (NASA U.S.A.) for

TABLE 14.1 A comparison of three methods for downward transmission

		Distribution for rebroadcast	Direct to community centres	Direct to homes
I Satellite	Weight	1000 lbs	1600 lbs	2000 lbs
	Frequency	3.7-4.2 GHz	770-890 MHz	770-890 MHz
	ERP	37 dbW	53 dbW	70 dbW
	Modulation	FM	FM	VSM
	Antenna	5.7X2.2 ft	30 ft dia	30 ft dia
	Cost	Rs 80M	Rs 104M	Rs 160M
II Receiver	Antenna	30 ft dia	10 ft dia	5.2 ft dia
	Noise Figure	1.9 db	6 db	5 db
	Converter	FM to AM	FM to AM	None
	Cost	Rs 800,000	Rs 1200	Rs 800

an experimental instructional television (ITV) programme all over the country[7]. This programme known as Satellite Instructional Television Experiment (SITE) will be conducted by Indian Space Research Organisation (ISRO) jointly with NASA. The experiment will use ATS-F satellite for a period of one year to broadcast 4 to 6 hours of instructional TV programmes every day to about 5000 villages located in various parts of the country. ATS-F is the first synchronous satellite with capability of direct broadcast of TV programmes to modified TV sets in 800-900 MHz band. The satellite has on board a large 30 ft. deployable antenna coupled with a high power transmitter (80 watts) and an accurate 3-axis altitude control system ($\pm$ 0.1°) to make direct reception possible. SITE programme is scheduled to begin in 1975. The satellite will broadcast one video channel accompanied simultaneously by two audio channels.

It is proposed that SITE be followed by a national satellite TV system which has come to be known as INSAT (Indian National Satellite) [8]. Three detailed system studies of INSAT have been carried out by ISRO in collaboration with General Electric Company USA (1969), in collaboration with Hughes Aircraft Corporation USA (1969), and then in collaboration with Massachusetts Institute of Technology USA (1970). These studies envisage the launching of a single, high power synchronous satellite for television broadcasting and long distance telecommunications. However, the frequency band for TV broadcast will have to be decided in accordance with the recommendations made at the

World Administrative Radio Conference (WARC) of July 1971. The frequencies allotted for satellite TV broadcast are (i) 620 to 790 MHz (ERP less than 49.5 dbW), (ii) 2500 MHz to 2690 MHz (ERP up to 62 dbW)and (iii) 11.7 to 12.2 GHz (no restriction on ERP). Because of the low transmitter power allowed for (i) and advanced technology required for (iii),it is expected that the frequency range of 2.50 to 2.69 GHz would be used for satellite TV broadcast in INSAT.

The INSAT programme will require a large number of TV receivers capable of receiving direct TV broadcast from satellite. If we assume that about half of the total number of villages are covered by direct reception and that each village has two TV sets on the average, the total number of such receivers will exceed 500,000.

14.2.2 Augmented TV receivers for reception from satellite

TV broadcast from satellites differs from conventional terrestrial TV broadcast in the following respects.

1) Frequencies used for satellite TV are much higher i.e.either in the upper region of U.H.F. band (800-960 MHz) or in the microwave region. As pointed out in the previous section the transmission from ATS-F is planned at 850 MHz. The subsequent TV broadcast from a national satellite may be in the microwave range.

2) In order to conserve on radiated power, TV transmission from satellite uses frequency modulation for video signals.

3) Signal received from direct TV broadcast are much weaker.

For these reasons, normal TV receivers require modification at the front end if they are to be used for direct reception of satellite TV. For this purpose a suitably designed 'front-end' may be added in front of conventional TV receivers. A typical block diagram of such a front-end is shown in Fig.14.12. Various other alternatives involving minor modifications are, of course, possible. The subsystem shown within the dotted block constitutes the microwave part of the front-end.

Availability of low noise microwave transistors allows an r.f. amplifier to be easily incorporated for signal frequencies upto around 4 GHz. However a design using a low noise balanced mixer with Schottky barrier diodes may not need an r.f. amplifier and would lead to lower cost. Usually the frequency stability considerations require a local oscillator driven by a crystal controlled oscillator. A reference frequency of about 80 MHz is used as in this frequency range inexpensive crystal-cutting techniques have already been developed (for colour TV).

There have been several studies [9,10,11] on the design and development of front-ends for satellite TV receivers. As with other consumer products, the most important consideration in the design of the front-end is cost, especially the cost of microwave circuitry and to a lesser extent that of handling FM TV signals. In most of the other microwave equipments (for defense

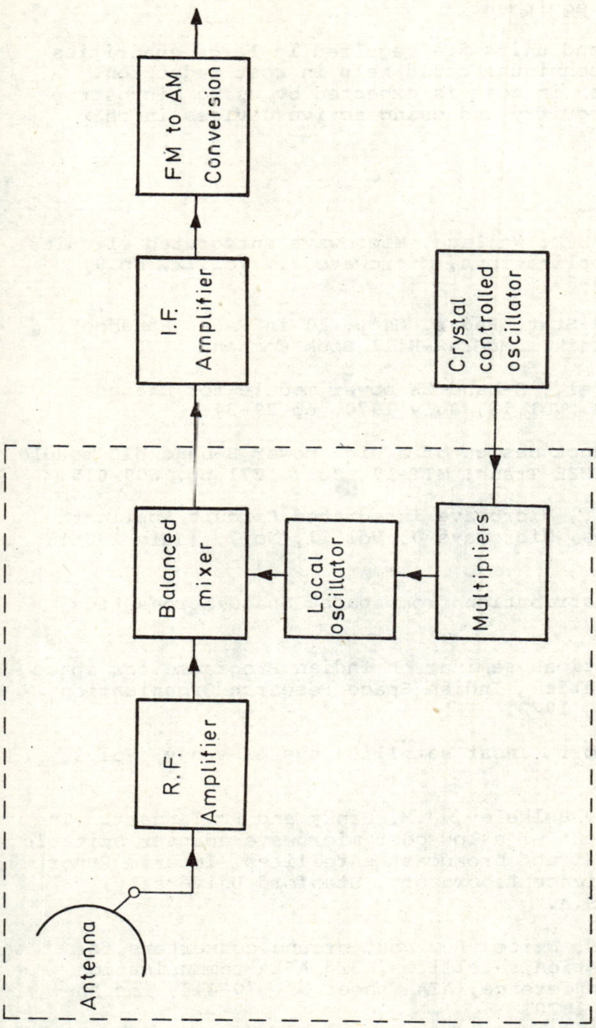

Fig. 14.12 Block diagram of the front-end converter for reception of TV signals from satellites.

and commercial applications) emphasis is on performance, ruggedness and reliability. Also, as the number of units normally required is relatively small, cost reduction for large production does not become the major consideration. Thus the design of the 'front-end' calls for a design philosophy which is not common to other microwave equipment.

Since the front-end units are required in large quantities (500,000 or so) MIC techniques could help in cost reduction. A substantial reduction in cost is expected by using microstrip design for passive circuitry and using active devices in chip form.

References

1. Vincent, B.T., and M.E. Wallace, Microwave integrated circuits for Phased array applications, Microwave J., Vol.12, No.9, (Sept. 1969) pp. 53-59.
2. Hyltin, T.M., Solid-State Radar, Chap. 30 in Radar Handbook, Edited by M.I. Skolnik. McGraw-Hill Book Company N.Y.
3. Belohoubek, E., et al., S-band CW power module for phased arrays, Microwave J.,Vol.13, July 1970, pp.29-34.
4. Vaccaro, F.E., Product design of a high power S-band MIC module for phased arrays,IEEE Trans. MTT-19, July 1971,pp. 609-615.
5. Arnold, R.D., et al., Microwave Integrated Circuit Applications to Radar Systems, Microwave J.,Vol.11, No.7, (July 1968) pp. 45-54.
6. Lowenhar, H., TV distribution from space, Space/Aeronautics, March 1969, pp. 48-57.
7. Report on communications seminar on Indian Programme for space research and application, Indian Space Research Organisation, Ahmedabad 7-12 Aug., 1972.
8. INSAT Technical Report, Insat satellite system study, Vol.2, March 1971.
9. Lusignan, B.B., P.Z. Bulkeley, J.M. Janky and R. Taggart. The design and development of a low cost microwave adapter suitable for reception from direct broadcast satellites, Interim Report July 1969, Radio Science Laboratory, Stanford University, Stanford Calif., U.S.A.
10. Hesler, J.P. and O.S. White, Low cost ground converters for high power communication satellites, 3rd AIAA communication satellite systems conference, AIAA paper No. 70-440, Los Angeles Calif., April 1970.
11. Lusignan, B.B., Low cost ETV Satellite Receiver, 3rd AIAA communication Satellite Systems Conf., AIAA paper No. 70-439, Los Angeles Calif., April, 1970.

15

BIBLIOGRAPHY ON MICs

by

G.T. Mirchandani, B.G. Sharma and R.N. Agnihotri

Chapter Fifteen

1967

1. Alderham, J., Electronic applications for glass and ceramic, Elecnc Prod (Aug 1967) 38-50.

2. Aslaksen, E.W., Integrated microwave power distribution network, IEEE Int Electron Dev Meeting (Oct 1967) 116 and IEEE Tr Electron Dev (Sept 1968) 15 (9) 679-682.

3. Blight, R.E., Microstrip hybrid couplers and their integration into balanced mixers at X and K-bands, IEEE Int Microwave Symp (1967) 136-138.

4. Botka, A.T., Microwave integrated circuits, IEEE Integrated Circuits Seminar Proc (1967) 3-12.

5. Botka, A.T., and Moroney, W.J., High isolation, fast switching integrated microscopic RF switches, Int Solid St Ccts Conf. Dig Tech Pap (1967) 56-57.

6. Brand F.A. and Gelnovatch V., Microwave Solid State devices and circuits, Microwave J (Jul 1967) 10 (8) 22, 25, 82-109.

7. Brenner, H.E. Micromin for microwaves. Special report. Microwaves (Oct 1967) 6 (10) 21-34.

8. Brinton, J.B. Miniature component family uses thin film circuitry. Microwaves (Mar 1967) 6 (3) 140-141.

9. Brinton, J.B. Problems and progress of MICs. Microwaves (Mar 1967) 6 (3) 55-57.

10. Cernaglia, N.P. et al., Beam-lead Schottky barrier diodes for low noise integrated microwave mixers. IEEE Int Electron Dev Meeting (1967) 114.

11. Collard, J.R. et al., Development of L-and C-band Solid State sources. Final rep. 16 Mar 66-30 Jun 67 on phase I. U.S. Govt R & D Rep. Ad-822441 (Oct 1967) 57.

12. Comeforo, J.E., Properties of ceramics for electronic applications. Elecnc Engr (Apr 1967) 26 (4) 82-87.

13. Cooke, H.F. and Policky, G.J., Computer design and optimization of UHF and microwave circuits, IEEE Int Solid St Ccts Conf Dig (1967) 106-107.

14. Dalley, J.E., Microwave hybrid integrated circuits, IEEE Int Conv Rec (1967) 15 (6) 30-36.

15. Daly D.A. et al., Lumped elements in MICs. IEEE Tr Microwave Theor Tech (Dec 1967) 15 (12) 713-721, and IEEE Int Microwave Symp Dig (1967) 136-141.

16. Das, R., Ferroelectric phase shifters., IEEE Int Microwave Symp (1967) 185-187.

17. Ertel ,A. ,Monolithic circuit techniques produce first Au-Silicon X-band switch.Electronics (23 Jan 1967) 40 (2) 67-81.

18. Gelnovatch, V.G.,Microwave application,Microelecncs Comes of Age.Symp Summ.IEEE (1967) MES-12 ,2p & IEEE National Conf (1967) Paper MES-12 ,1-3.

19. Gilden , M. ,Miniature integrated hybrid harmonic generators for microwave frequencies. Int Solid St Ccts Conf Dig Tech Pap (1967) 62-63.

20. Granberry, D.S. and Sterling, C.E.,Packaging an X-band MIC. IEEE Int Electron Dev Meeting (Oct 1967) 118.

21. Guckel , H. et al., Parallel-plate waveguide approach to microminiaturized ,planar transmission lines for integrated circuits. IEEE Tr Microwave Theor Tech (Aug 1967) 15 (8) 468-476.

22. Hair, H.A.,Planar ferrite devices for MICs. IEEE Int Conv Rec (1967) 15 (Pt.6)37-38.

23. Hershenov,B., All garnet-substrate microstrip circulators for integrated circuits. IEEE Int Microwave Symp (1967) 142-144 and Proc IEEE (May 1967) 55 (5) 696-697.

24 Hershenov ,B., Microstrip junction circulators for microwave integrated circuits.IEEE Int Microwave Symp (1967) and IEEE Tr Microwave Theor Tech (Dec 1967) 15 (12) 748-750.

25. Hooper ,R.C.and Wheeler ,C.A. ,Fabrication of monolithic MICs Materials and processes considerations. AIME Tech Conf on Preparation and Properties of Electronic Materials, New York (1967).

26. Hoover ,J.C. et al., Recent advances in MICs combining semiconductor and ferrimagnetic elements. IEEE Int Conv Rec (1967) 15 (Pt.6) 39-40.

27. Horton , J.B., Thin film X-band varactor quadrupler.IEEE Int Microwave Symp (1967) 145-146 and IEEE Tr.Microwave Theor Tech (Dec 1967) 15 (12) 752-754.

28. Hyltin, T.M., Microwave Integrated electronics .IEEE Int Electron Dev Meeting (Oct 1967).

29. Kaup ,H. ,Characteristics of microstrip transmission lines. IEEE Tr Elecnc Comp (Apr 1967) 16 (2) 185-193.

30. Kilpatrick,T.H.,Micromin for microwaves here and now.Microwaves (Oct 1967) 6 (10) 22-34.

31. Kyle,R.H. and Midford T.A., Integration of an impatt device. Proc IEEE (Dec 1967) 55 (12) 2190-2191.

32. LaGrange ,J.B., Integrated circuits for radar application. IEEE Microelecncs Symp (June 1967) E 2-1 to E 2-4.

33. Leeds,M.P.,Microwave IC's come of age. Electronics (30 Oct 1967) 40 (22) 107-113.

34. Little, W.A. ,et al.,Hybrid integrated circuit digital phase shifters. Int Solid St Ccts Conf Dig Tech Pap (1967) 58-59.

35. Mallery,P. Thermal pulse bonding of beam leads. Proc Elecnc Comp Conf (1967) 276-282.

36. Mao,S. et al., Integrated microwave receiver. Interim Tech Rep No. 4. Contract No. AF 33(615)-5102 (Aug 1967).

37. Mao,S. et al., Integrated microwave transit time oscillators. IEEE Int Electron Dev Meeting (Oct 1967) 116.

38. Matthei,W.G.,Devices for hybrid microwave integrated circuits. SWIEEE Conf (Apr 1967).

39. Matthei,W.G.,Low noise microwave receivers. IEEE Nat Conv Rec (1967) Pt 7, 7-12.

40. Mayo,R.F.,et al., Microstrip circuitry for integrated transistor amplifiers. IEEE Int Microwave Symp (1967) 132-135.

41. Mehal,E.M.,Over horizon-GaAs IC's. Electronics (13 Nov 1967) 40 (23) 134-136.

42. Microwave Association. Microwave Integrated circuits. Progress Rep. Microwave Assoc (Jul 1967).

43. Moroney,W.J.and Botka,A.T.,Integrated microwave switching. IEE Conf on Integrated Circuits (1967) 15-28.

44. Nergaard,L.S.,Potential of integrated electronics for microwave systems. Int Solid St Ccts Conf Dig Tech Pap (1967) 40-41.

45. Okean, H.C.,Microwave amplifiers employing integrated tunnel diode devices. IEEE Tr Microwave Theor Tech (Nov 1967) 15 (11) 613-622.

46. Okean, H.C.,Properties of a TEM transmission line used in microwave integrated circuit applications. IEEE Tr Microwave Theor Tech (May 1967) 15 (5) 327-328.

47. Portnoy,W.M.,and Hyltin,T.M.,X-band balanced mixer integrated circuit. IEEE Int Solid St Ccts Conf Dig (1967) 38-39.

48. Romme,G.T. and Hair,H.A.,Thin ferrites for integrated microwave devices. 4th,5th Quarterly Tech Report U.S.Govt.R&D Rep.Contract No. AF 33(615)-3332 (1967) by AF Lab.,Res & Tech.Div,AF Syst.Commd.

49. Roome,G.T. et al., Planar 4-port switchable circulator for MICs.Int Solid St Ccts Conf Dig Tech Pap (1967) 54-55.

50. Roome,G.T. et al., Thin ferrites phase shifters for integrated microwave devices. J App Phy (1 Mar 1967) 38 (3) 1411-1412.

51. Saunders,T.E. and Stark,P.D.,Integrated 4 GHz balanced transistor amplifier. IEEE J Solid St Circuits (Mar 1967) 2 (1) 4-10.

52. Schwarzmann,A.,Microstrip plus equations adds up to fast designs. Electronics (2 Dec 1967) 40 (25) 109-112.

53. Sclater,N.,Civilian markets beckon microwave ICs. Elecnc Des (6 Dec 1967) 25 (23) 34.

54. Small, T.W.,Microwave microelectronics. IEEE Symp on Microelecncs Applications (1967) 15-18.

55. Sobol,H.,Extending IC technology to microwave equipment. Electronics (20 Mar 1967) 40 (6) 112-124.

56. Sobol,H.,Microwave modules using integrated circuits. SWIEEE Conf (Apr 1967).

57. Sobol,H. and Hughes,J.J., Measurement of the permitivity of insulating films at microwave frequencies. IEEE Tr Microwave Theor Tech (Jun 1967) 15 (6) 377-378.

58. Stinehelfer,H.E.,Computer analysis of microwave integrated switches. IEEE Int Microwave Symp (1967) 147-149.

59. Stover,H.L. et al., Microwave integrated circulators. IEEE Int Electron Dev Meeting (Oct 1967)116.

60. Tatsuguchi,I.and Aslaksen,E.W.,Integrated 4 GHz balanced mixer assembly. Int Solid St Ccts Conf Dig Tech Pap (1967) 60-61 and IEEE J Solid St Ccts (Mar 1968) 3 (1) 21-26.

61. Temme,D.H.,Microwave control circuits. IEEE Nat Conv Rec (1967) Pt.5,22-30.

62. Texas Instruments Inc.,Study of Solid State integrated Microwave circuits (Scientific Report).U.S.Govt R & D Report No. N69-17852 (1967) 190p.

63. Varian Associates,LEL Div.USA.,Miniature components family uses thin-film circuitry. Microwaves (Mar 1967) 6 (2) 140-142.

64. Vendelin,G.D.,High dielectric substrates for microwave hybrid integrated circuitry. IEEE Int Microwave Symp Dig (1967) 125-128 and IEEE Tr Microwave Theor Tech (Dec 1967) 15 (12) 750-752.

65. Vincent,B.T.,Integration techniques for microwave circuitry Proc.Ist Cornell Bien.Conf on Engng Appl of Electron Phenomena (1967) 176-205.

66. Wacker,R.W. and Mao,S.,Surface oriented GaAs Gunn oscillators and Schottky barrier diodes. IEEE Int Electron Dev Meeting (1967) 114.

67. Webster,R.R.,Integrated Microwave oscillators,amplifiers,

switches and converters. Int Solid St Ccts Conf Dig Tech Pap (1967) 38-39.

68. Webster,R.R.,Microwave integrated circuits (Abstract).IEEE Int Conv Rec (1967) 15 (Pt.6) 55.

69. Wolters,K.C. and Clar,P.L.,Microstrip transmission lines on high dielectric constant substrates for hybrid MICs.IEEE Int Microwave Symp (1967) 129-131.

70. Yamashita,E.and Mittra,R.,Radiation impedance approach to the analysis of a thin film inductor in a microwave integrated circuit. IEEE J Solid St Ccts (Jun 1967) 2 (2) 47-49.

71. Zuleeg,R.,Silicon-sapphire transistors point way to MICs. Electronics (20 Mar 1967) 40 (6) 106-108.

1968

1. Allen,J.L. and Taft,D.R.,Ferrite elements for hybrid MIC systems.IEEE J Solid St Ccts (Jun 1968) 3 (2) 67-73 and IEEE Tr Microwave Theor Tech (Jul 1968) 16 (7) 405-411.

2. Andresciani,V. and Fiordiponti.,Characteristics of linear integrated circuits and microwave transistors with particular reference to the "S" parameter (In Italian). Adv in Elecncs Proc Elecncs Cong 15 (1968) 225-243.

3. Arnold,R.D. et al.,MIC applications to radar systems. Microwave J (Jul 1968) 11 (7) 45-52.

4. Aslaksen,E.W.,Integrated microwave power distribution network.IEEE Tr Electron Dev (Sep 1968) 15 (9) 680.

5. Atwood,W.and Stinehelfer,H.E.,Multistub filter for microstripline. IEEE Tr Microwave Theor Tech (Jul 1968) 16 (7) 477-480 and IEEE J Solid St Ccts (Jun 1968) 3 (2) 139-142.

6. Austin,J.E.,Development of integrated microwave components for ground-based phased arrays. Rep Submitted to Lincoln Labs,Lexington Mass.(Jan 1968).

7. Axler,M.F.,Integrated microwave switches, a matter of time. Electrotechnology (USA) (Sep 1968) 82 (3) 45-51.

8. Battershall, B.W.and Emmons,S.P.,Optimization of diode structures for monolithic MICs.IEEE J Solid St Ccts (Jun 1968) 3 (3) 107-112 and IEEE Tr Microwave Theor Tech (Jul 1968) 16 (7) 445-450.

9. Belohoubek,E.F. et al.,Solid state microwave surveillance radiosonde sources. Final rep 1 Jun 67-30 Sept 68. U.S.Govt. R & D Rep AD-845171 (Dec 1968) 56p.

10. Blau,R.,Microwave integrated circuit avalanche diode oscillator. IEEE Tr Microwave Theor Tech (Mar 1968) 16 (3) 196-197.

11. Blight,R.E. and Howell,C.M.,C-Through X-band microstrip coupler for use in broadband integrated circuits. Int Solid St Ccts Conf Dig Tech Pap (1168) 126-127.

12. Bonfeld,M.D. et al.,Integrated microwave FM discriminator IEEE Int Microwave Symp (1968) 139-146.

13. Botka,A.T. et al.,Integrated X-band radar receiver front end. Microwave J (Jul 1968) 11 (7) 65-71.

14. Botka,A.T. et al.,MIC front end for low power radars. Int Solid St Ccts Conf Dig Tech Pap (1968) 50-51.

15. Brand,F.,MICs-who needs them .Microwave J (Jul 1968) 11 (7) 24-26.

16. Brand F. et al.,Straight talk on microwave micromin, Microwaves (Jan 1968) 7 (1) 43-56.

17. Brehm,G.E.and Mao,S.,Varactor tuned integrated Gunn oscillator. IEEE J Solid St Ccts (Sep 1968) 3 (3) 217-220 and IEEE Int Solid St Ccts Conf (1968) 78-79.

18. Brenner,H.E.,Use a computer to design suspended-substrate ICs. Microwaves (Sep 1968) 7 (9) 38-45.

19. Bryant,T.G.and Weiss,J.A.,Normal mode impedances of a coupled pair of microstrip transmission lines. IEEE Int Microwave Symp Dig (1968) 117-122.

20. Bryant,T.G.and Weiss,J.A.,Parameters of microstrip transmission lines and of coupled pairs of microstrip lines. IEEE Tr Microwave Theor Tech (Dec 1968) 16 (12) 1021-1027.

21. Bura,P.et al.,Design considerations for an integrated 1.8 GHz parametric amplifier. IEEE Tr Microwave Theor Tech (Jul 1968) 16 (7) 424-428 and IEEE J Solid St Ccts (Jun 1968) 3 (2) 86-90.

22. Caroli,I.,Microwave components in integrated form for radio links (in Italian). Adv in Elecncs-Proc Elecncs Cong 15 (1968) 359-367.

23. Caulton,M.,Thin film microwave components. NEREM Rec (1968) 136-137.

24. Caulton,M.et al.,Hybrid integrated lumped-element microwave amplifiers. IEEE Tr Electron Dev (Jul 1968) 15 (7) 3 399-406 and IEEE Tr Microwave Theor Tech (Jul 1968) 16 (7) 397-404 and IEEE J Solid St Ccts (Jun 1968) 3 (2) 59-66.

25. Caulton,M. et al.,UHF film integrated circuits. Ist Quart. Rep. U.S.Govt R & D Rep. USECOM Contract DAAB07-68-C-0296-1 (Nov 1968).

26. Chase,I.L.and Gelnovatch,V.G.,Synthesis of distributed wide band transistor amplifiers by computer techniques for application to MICs.USAEC Tech Rep No. ECOM-3067 (Dec 1968).

27. Chinchillo,A.R. et al.,Investigation of various metallizations on alumina for hybrid MICs. GOMAC conf (Oct 1968).

28. Clar,P.L.,Microwave integrated circuit technology, Electrotechnology (Oct 1968) 82 (4) 37-42.

29. Cohn,S.B.,Microwave band pass filters containing high-Q dielectric resonators. IEEE Tr Microwave Theor Tech (Apr 1968) 16 (4) 218-227.

30. Cohn,S.B.,Slot-line an alternative transmission medium for integrated circuits. IEEE Int Microwave Symp (1968) 104-109.

31. Cooke,R.E. et al.,Gunn effect oscillator modules using microstrip circuits. IEEE Int Electron Dev Meeting (Oct 1968).

32. Cooper,H.W., et al.,Microwave integrated Impatt radiator. IEEE Int Microwave Symp Dig (1968) 72-76.

33. Cristal, E.G. et al.,Design of microwave filter networks. U.S.Govt R & D Rep AD-827402 (Jan 1968).

34. Cristal, E.G. et al.,Straight talk on microwave micromin; panel discussion. Microwaves (Jan 1968) 7 (1) 43-56.

35. Cuilwik,A.W.,Synthesis of microstrip transmission networks utilizing semiconductor dielectrics,Ph.D. Thesis,New York Univ.N.Y. School of Engng & Sci (1968) 167p.

36. Downing,O.J.and Hoyle,L.J.,Microwave integration.U.S.Govt R&D Rep N69-36754 (23 Apr 1968) 8p.

37. Dunn,V.E.and Domenico,A.J.,Recent Advances in microstrip circulators. IEEE Int Microwave Symp Dig (1968) 248-254 and IEEE Tr Microwave Theor Tech (Dec 1968) 16 (12) 1060-1061.

38. Emery,F.E.,Monolithic microwave integrated circuitry.NEREM Rec (1968) 138-139.

39. Emery,F.E.and Noel,P.L.,Recent experimental work on silicon microstrip microwave transmission lines. IEEE Tr Microwave Theor Tech (Jul 1968) 16 (7) 483-484.

40. Emmons,S.P. et al.,Monolithic Ku-band diode phase shifter. IEEE Int Electron Dev Meeting (Oct 1968).

41. Foxell,C.A.P.,Semiconductor devices for microwave integrated circuits. Abs.only.IEE Colloq on MIC (1968) 1p.

42. Gelnovatch,V.G. and Burke,T.F.,Computer aided design of wide band integrated microwave transistor amplifiers on high dielectric substrates. IEEE Tr Microwave Theor Tech (Jul 1968) 16 (7) 429-439 and IEEE J Solid St Ccts (Jun 1968) 3 (2) 91-101.

43. Gilden,M.and Blight,R.,30 GHz ADO integrated into an X-band parametric amplifier. NEREM Rec (1968) 98-99.

44. Goodman,P.C.,Wideband stripline matched power divider. IEEE Int Microwave Symp Dig (1968) 16-20.

45. Gourse,S.J.and David,R.F.,How to use lumped constants in microwave circuit design. Microwaves (Mar 1968) 7 (3) 47-51.

46. Greene,C.K.,Microstrip non-reciprocal tunable YIG filter. IEEE Tr Microwave Theor Tech (Jul 1968) 16 (7) 484-486.

47. Greenwald, C. et al., Printed thick film MICs. NEREM Rec (1968) 140-141.

48. Gundlach,R.,Ferrites attraction is magnetic and growing.Electronics (14 Oct 1968) 41 (21) 104-118.

49. Gunshinan,B.F., How to set up a micromin Lab.Microwaves (Dec 1968) 7 (12) 36-50.

50. Harrison,G.R. and Allen,J.L.,Integrated circuit technology applicable to functional microwave modules. IEEE Int Conv Rec (1968).

51. Hartwig,C.P. et al.,Frequency dependent behaviour of microstrip. IEEE Int Microwave Symp Dig (1968) 110-116.

52. Hartwig,C.P. et al.,Microstrip technology.Proc Nat Elecncs Conf Chicago (1968) 9-11,314-317.

53. Herskowitz,G.H.ed.,Computer aided integrated circuit design. McGraw-Hill (1968) Chapter II.

54. Huang,H.C.and Mackenzie,L.A.,Gunn diode operated in the hybrid mode. Proc IEEE (May 1968) 56 (5) 1232-1233.

55. Hyltin,T.M.,Microwave integrated electronics for radar and communication systems. Microwave J (Feb 1968) 11 (2) 51-55.

56. Institution of Electrical Engineers,London.,Colloquium on microwave integrated circuits and on microwave solid state sources. IEE Conf Publication, (1968) 7 papers.

57. Johnson K.M.,High performance integrated microwave circuit frequency quadrupler. IEEE Tr Microwave Theor Tech (Jul 1968) 16 (7) 420-424,IEEE J Solid St Ccts (Jun 1968) 3 (2) 82-86 and Int Solid St Ccts Conf Dig Tech Pap (1968) 84-85.

58. Johnson,K.M.,X-band integrated circuit mixer with reactivity terminated image. IEEE Int Microwave Symp Dig (1968) 131-138 and IEEE Tr Microwave Theor Tech (Jul 1968) 16 (7) 388-397 and IEEE J Solid St Ccts (Jun 1968) 3 (2) 50-59 and IEEE J Electron Dev (Jul 1968) 15 (7) 450-459.

59. Jones,R.R. et al.,Magnetically tunable microstrip filter structures. IEEE Microwave Symp Dig (1968) 282-290.

60. Joy,J.F. et al.,Microminiature S-band integrated receiver. IEEE Int Electron Dev Meeting (Oct 1968).

61. **Karp,A.et al., Circuit properties of microwave dielectric** resonators. IEEE Tr Microwave Theor Tech (Oct 1968) 16 (10) 818-828.

62. Katoh,H. and Ayaki,K.,Strip-line circuits for microwave IC and capacitor loss. Natnl Conv Rec of IECE (Japan) (1968) 466.

63. Keister,F.Z.,Evaluation of materials and processes for integrated microwave circuits. IEEE Tr Microwave Theor Tech (Jul 1968) 16 (7) 469-475 and IEEE J Solid St Ccts (Jun 19 **1968) 3 (2) 131-137.**

64. Kelley D. et al.,Microstrip filters and couplers. IEEE Tr Microwave Theor Tech (Aug 1968) 16 (8) 560-562.

65. **Knerr,R.H.,New type of waveguide-to-stripline transition. IEEE Tr Microwave Theor Tech (Mar 1968) 16 (3) 192-194.**

66. Knight,S.P. et al.,Thin film lumped-element integrated S-band power amplifiers. Int Solid St Ccts Conf Dig Tech Pap (1968) 86-87.

67. Kramer,A.G. et al.,Integrated X-band CW radar front end.Microwave J (Jul 1968) 11 (7) 55-62.

68. Kramer,A.G. et al.,Integrated X-band transceiver for radar applications. Int Solid St Ccts Conf Dig Tech Pap (1968) 56-57.

69. **Lamar,A.J.and Taft,D.R.,Ferrite elements for hybrid microw-** ave integrated systems. IEEE Tr Electron Dev (Jul 1968) 15 (7) 467-473.

70. Leede M.B.Microwave ICs come of age.Electronics (Oct 1968) **40 107-113.**

71. Lim,Y.C.and Moore,R.A.Properties of alternately charged **coplanar parallel strips by conformal mappings. IEEE Tr El-** ectron Dev (Mar 1968) 15 (3) 173-180.

72. MacDonald,W.J. and Wheeler,C.A.,Microwave hybrid circuit. Fabrication-processing considerations. Int Soc for Hybrid Microelecncs-Hybrid Microelecncs Symp (1968) 23-29.

73. **Macleay,R.Q.,PNP^{+} variable coupling,directional coupler and** reflective phase shifter. Int Solid St Ccts Conf Dig Tech Pap (1968) 124-125.

74. Mao,S. et al.,Millimeter-wave integrated circuits. IEEE J Solid St Ccts (Jun 1968) 3 (2) 117-123 and IEEE Tr Microwave Theor Tech (Jul 1968) 16 (8) 455-461.

75. **March,S.,Wideband stripline hybrid ring.IEEE Tr Microwave** Theor Tech (Jun 1968) 16 (6) 361.

76. Masse D.,Broadband microstrip junction circulators-Proc IEEE (Mar 1968) 56 (3) 352-353.

77. Masse,D. et al.,Microstrip versus balanced stripline.Proc IEEE (Aug 1968) 56 (8) 1359-1360.

78. Matthei,W.G.and Crowe W.,New MIC modules.IEEE Wescon Tech Pap (1968) 2/1-1 to 2/1-6.

79. **Matthews,E.W.and Rector,R.M.,Hybrid techniques at microwave** frequencies.Int Soc for Hybrid Microelecncs-Hybrid Microelecncs Symp (1968) 439-449.

80. **Mehal,E.W.and Wacker,R.W.,GaAs integrated microwave circuits.** IEEE J Solid St Ccts (Jun 1968) 3 (2) 113-116 and IEEE Tr Microwave Theor Tech (Jul 1968) 16 (7) 451-454.

81. Miller,A. et al.,Sapphire based integrated microwave components and circuits. IEEE-Elecncs Comp Conf-Proc.(May 1968) 367-373.

82. **Moroney,W.J.,Impact of microwave integrated circuits on microwave device and circuit development.** IEE Colloquium on Microwave Integrated Ccts (1968) 1P.Abs.only given.

83. **Moroney,W.J.,Microwave ICs Pt.1 New problem but newer solution Electronics (10 Jun 1968) 100-108.**

84. Moroney,W.J.Microwave integrated hybrid circuit design and fabrication. IEEE Elecncs Comp Conf-Proc (May 1968) 356-360.

85. Morris,J.E.,3-dB directional coupler for microwave integrated circuits. IEEE Tr Microwave Theor Tech (Jul 1968) 16 (7) 476-477 and IEEE J Solid St Ccts (Jun 1968) 3 (2) 138-139.

86. **Napoli,L.S.and Hughes,J.J.,Low noise integrated S-band receiver. Microwave J (Jul 1968) 11 (7) 38-42.**

87. Napoli L.S. et al.,Low noise integrated X-band receiver.Int Solid St Ccts Conf Dig Tech Pap (1968) 48-49 and Microwave J (Jul 1968) 11 (7) 37-42.

88. Nelson,W.W.,Hybrid integration techniques for multi-octave, **multiple-throw microwave switches,IEEE Int Solid St Ccts Conf** Dig (1968) 12-13.

89. Norton,L.E. et al.,High power 1 GHz coplanar Gunn-effect oscillators. IEEE Tr Microwave Theor Tech (Jul 1968) 16 (7) 480-481.

90. Noyce,R.N.,Making integrated electronics technology work. IEEE Spectrum.(May 1968) 5 (5) 63-66.

91. **Okean,H.C.and Weingart,H.,S-band integrated parametric ampli**fier having both flat gain and linear phase response. IEEE Int Microwave Symp Dig (1968) 155-163 and IEEE Tr Microwave Theor Tech (Dec 1968) 16 (12) 1057-1059.

92. **Opp,F.L.and Hoffman,W.F.,Design of digital loaded-line phase-**shift networks for microwave thin film applications. IEEE Tr microwave Theor Tech (jul 1968) 16 (7) 462-468 and IEEE J Solid St Ccts (Jun 1967) 3 (2) 124-130.

93. Pipin,J.E.,Microwave ferrite devices.Microwave J (Apr 1968) 29-45.

94. Portnoy,W.M. and Leedy,H.M.,Monolithic single ended X-band mixer circuit. IEEE J Solid St. Ccts (Mar 1968) 3 (1) 31-32.

95. Presser,A. et al., Hybrid integrated L-band transmitter. IEEE Tr Microwave Theor Tech (Jul 1968) 16 (7) 482-483.

96. Pucel,R.A. et al.,Losses in microstrip. IEEE Tr Microwave Theor Tech (Jun 1968) 16 (6) 342-350 and (Dec 1968) 16 (12) 1064.

97. Riezeman M.J.,Parley analyses 3 paths to world of MICs. Elecnc Des (7 Nov 1968) 16 (23) 25-30.

98. Robinson, G.H., Microwave microelectronic techniques applied to phase shifters and other control devices. Final technical report 29 May-29 Nov. 1968. U.S.Govt R & D Rep N69-28165 (Nov 1968) 40p.

99. Roome, G.T. and Hair,H.A.,Thin ferrite devices for microwave integrated circuits. IEEE Tr Microwave Theor Tech (Jul 1968) 16 (7) 411-420 and IEEE J Solid St Ccts (Jun 1968) 3 (2) 73-82.

100. Roome,G.T. et al.,Ferrite devices for microwave integrated systems. Int Solid St Ccts Conf Dig Tech Pap (1968) 52-53.

101. Schilling,W.,Microwave beams. Electronics (9 Dec 1968) 41 (25) 90-91.

102. Schilling,W.,Real world of micromin substrates. Pt.1.Microwaves (Dec 1968) 7 (12) 52-56.

103. Silvester,P.TEM wave properties of microstrip transmission lines. Proc IEE (Jan 1968) 115 (1) 43-48.

104. Sobol,H. and Carley,D.R.,Microwave integrated circuits. Nachrichtentechnische Fachberichte (1968) 35,385-387.

105. Solomon,A.H.,Integrated microwave components and subsystems using beam-lead semiconductor devices. IEEE WESCON Conv (Aug 1968) Pap 2/2.

106. Special Issue.,Microwave Integrated circuits.IEEE Microwave Theor Tech (Jul 1968) 16 (7) 382-485, and IEEE J Solid St Ccts (Jun 1968) 3 (2) 46-148.

107. Stewart, R.G. and Giuliano,M.N.,X-band integrated diode phase shifters. IEEE Int Microwave Symp Dig.(1968) 147-154.

108. Stinehelfer,H.E.,Accurate calculation of uniform microstrip transmission lines. IEEE J Solid St Ccts (Jun 1968) 3 (2) 101-106 and IEEE Tr Microwave Theor Tech (Jul 1968) 16 (7) 439-444.

109. Stinehelfer,H.E. et al.,Etched thick film microwave circuits NEREM Rec (1968) 134-135.

110. **Tatsuguchi,I.and Aslaksen,E.W.,Integrated 4 GHz balanced mixer assembly.** IEEE J Solid St Ccts (Mar 1968) 3 (1) 21-26.

111. **Troughton,P.,High Q factor resonators in microstrip. Elecn** cs Lett (29 Nov 1968) 4 (24) 520-522.

112. **Vendelin,G.D.,Ku band integrated receiver.Int Solid St Ccts** Conf Dig Tech Pap (1968) 54-55 and IEEE J Solid St Ccts (Sep 1968) 3 (3) 255-257.

113. Vincent,B.,Microwave integrated circuits. Texas Instrum Inc Seminar Rep.(1968).

114. **Vosburgh,M.C.,Application of solid state and microwave Integrated circuit technologies to phased array radars. Microwave J (Sep 1968) 11 (9) 41-46.**

115. **Wagner,R.J. et al.,X-band impatt microstrip power sources. IEEE J Solid St Ccts (Sep 1968) 3 (3) 221-225.**

116. Webster,W.M.,Integrated microwave modules-prospectus.IEEE J Solid St Ccts (Jun 1968) 3 (2) 46-50 and IEEE Tr Electron Dev (Jul 1968) 15 (7) 446-450 and IEEE Tr Microwave Theor **Tech (Jul 1968) 16 (7) 384-388.**

117. Wells,J.F. and McPhun,M.K.,Lumped components for microwave integrated circuits. Abs.only. IEE Colloq on MIC (1968) 1p.

118. Wolters,K.C. et al.,Analysis and experimental evaluation **of distributed overlay structures in MICs. Int Microwave** Symp Dig (1968) 123-130.

119. Yamashita,E.,Variational method for the analysis of microstrip-like transmission lines. IEEE Tr Microwave Theor Tech (Aug 1968) 16 (8) 529-535.

120. Yamashita,E. and Mittra,R.,Variational method for the analysis of microstrip lines. IEEE Tr Microwave Theor Tech (Apr 1968) 16 (4) 251-256 and (Aug 1968) 16 (8) 529-535.

121. **Yamashita,E.and Yamazaki,S.,Parallel strip line embedded in** or printed on a dielectric sheet. IEEE Tr Microwave Theor Tech (Nov 1968) 16 (11) 972-973.

1969

1. Adams,D.K. and Ho,R.Y.C.,Filtering,frequency multiplexing, **and other microwave applications with inverted-common-coll**ector-transistor circuits.IEEE Int Microwave Symp Dig (1969) 14-20.

2. Aitchison,C.S.,Microwave integrated circuits.(Abs.only)European Microwave Conf (1969) 48.

3. Allen,C.M. et al.,Integrated parametric amplifier module with self contained solid state pump source. IEEE Proc Microelecncs Symp (1969) p.d5.1-d5.2.

4. Bandy,G.C. et al.,MERA modules-how good in an array? Microwave (Aug 1969) 8 (8) 39-49.

5. Barrett,M.and Viens,J.,Fully integrated varactor tuned avalanche oscillator in Ku band. IEEE Proc Microelecncs Symp Microwave Microelecncs (1969) B6.1-6.3.

6. Bauer,R.J.,MICs for active aperture radar. IEEE Proc Microwave Microelecncs (1969) p d2.1-d 2.3.

7. Baughan,K.M. et al., Characteristics of a 100 mW X-band Gunn oscillator for long pulse operation in microstrip. European Microwave Conf (1969) 238-241.

8. Belohoubek,E.F.and Rosen,A.,Coupled microstrip line varactor doublers. IEEE Tr Microwave Theor Tech (May 1969) 17 (5) 286-288.

9. Belohoubek,E.F. et al.,Hybrid integrated 10-watt CW broad band power source at S-band. IEEE J Solid St Ccts (Dec 1969) 4 (6) 360-366 and IEEE Int Solid St Ccts Conf Dig (1969) 122-123.

10. Bertiger,B.R. et al.,Microwave lumped element networks. IEEE Microelectronics Symp Proc (1969) A5.1-6.

11. Blight,R.E.,3:1 bandwidth low noise mixer. IEEE Int Microwave Symp (1969) 21-25.

12. Braginsky,A.I. and Buck, D.C., Polycrystaline ferrite films for microwave applications. IEEE Tr Magnetics (Sep 1969) 5 (3).

13. Brand,F.A.,MICs-their economic impact.Microwave J (Sep 1969) 12 (9) 28-30.

14. Brymner,R. and Meland,E.,Thin film chromium resistors for microwave integrated circuits. IEEE Microelecncs Symp (1969) p.C4.1-C4.3.

15. Bunker,J.and Ferraton,Y.,MIC X-band B.I.T.E. assembly. IEEE Microelecncs Symp (1969) p.d3.1-d3.3.

16. Bunyan,A.L. et al.,Family of integrated twin-channel crystal video receivers for S,C, and X-bands. European Microwave Conf (1969) 552-557.

17. Bura,P. et al., Integrated parametric amplifiers in Impatt-diode pumping. IEEE Int Microwave Symp Dig (1969) 426-431.

18. Cabrera,E. and March,S., Ultrabroadband components in microstrip. IEEE Proc Microelectronic Symp (1969) p.a2.

19. Caulton,M.and Poole,W.E.,Designing lumped elements into microwave amplifiers. Electronics (14 Apr 1969) 42 (8) 100-110.

20. Caulton,M. et al.,Measurements on the properties of microwave integrated circuits.Int Microwave Symp Dig (1969) 38-44.

21. Cohn,S.B.,Slot line on a dielectric substrate. IEEE Tr Microwave Theor Tech (Oct 1969) 17 (10) 768-778.

22. Cooke,R.E.,Dispersion characteristics of microstrip transmission lines. European Microwave Symp (1969) p.2.

23. Craven,G. et al.,Integrated microwave systems employing evanescent mode waveguide components. European Microwave Conf (1969) 285-289.

24. Csurgay,A.I.,on the application of multivariable network functions to the analysis and design of microwave active circuits. European Microwave Conf Proc (1969) 62-65.

25. Cunningham,J.E.(ed.),Microwave integrated circuit techniques Texas Instrum Application Rep Bull CA-64 (Mar 1969).

26. Cunningham,J.E. and Cooke,H.F.,Merging technologies-microelectronics and microwaves.Elecncs Des News (Oct 1969) 61-68.

27. Curtis,W.C.and Honda,H.,Current applications of MICs (Abs only) NEREM Rec (1969) 229 and RCA Def Electron Prod (Microwave Tech) Tech Pap (1969) 13-15.

28. Dalley,J.E.,Strip line directional coupler utilizing a non-homogeneous dielectric medium. IEEE Tr Microwave Theor Tech (Sep 1969) 17 (9) 706-712.

29. Denison,E.and Spilling, J.F.,X-band integrated circuit marine radar beacon. European Microwave Conf (1969) 10-13.

30. Denlinger,E.J.,Radiation from microstrip resonators. IEEE Tr Microwave Theor Tech (Apr 1969) 17 (4) 235-236.

31. Di Piazza,G.C.,Microwave lumped element networks. IEEE Proc. Microelecnc Symp (1969) Pap A-5.

32. Disman,R.I. and Shewaga,W.J.,Next generation miniature microwave tuners. Microwave J (Jan 1969) 12 (1) 74,76.

33. Durnwirth,R.K.and Schilling,W., Design checklist for micromin substrates and circuits. Microwaves (Dec 1969) 8 (12) 54-57.

34. Flaherty,J.M.Microcircuit phased-array electronic counter measures system. Microwave J (Sep 1969) 12 (9) 69-74.

35. Flaherty,J.M.,MIC parametric amplifier. Product feature, Microwave J (Sep 1969) 12 (9) 84.

36. Foreman,R.S.,Integrated S-band parametric amplifier. IEEE Int Microwave Symp Dig (1969) 419-425.

37. Gerst,C.W. and Paciorek,L.J.,Stripline MICs, Microwave J (May 1969) 12 (5) 43-54.

38. Glance,B.,Microstrip IMPATT oscillator with high locking figure of merit.Proc IEEE (Nov 1969) 57 (11) 2052-2053.

39. Granberry,D.,Packaging techniques for MICs. Summary, IEEE Microelecncs Symp (1969) p.C5.1-C5.3.

40. Graziano,A.X.,Integrated Ku-band microwave discriminator using a high Q YIG reference cavity in microstrip. IEEE Proc Microelecncs Symp Microwave Microelecnes (1969) B7.1-B7.3.

41. Hanley,G.L.,Parallel line microstrip coupler design for tight coupling values. European Microwave Conf (1969) 307.

42. Harrison,G.R. and Aucremann,R.C.,Integrated microwave circuits for airborne communications arrays.IEEE Int Conv Dig (1969) 38-39.

43. Harrison,G.R. and Gsteiger,K.E.,MICs fabrication, design, performance,and application. Proc Nat Elecncs Conf 25 (1969) 170-175.

44. Hart,G.and Viola,R.,Design of L,C and S-band solid state array modules .European Microwave Conf Proc (1969) 443-446.

45. Hartwig,C.P.,Microstrip design considerations.Proc Microelecncs Symp-Microwave Microelecncs (1969) A1-1-2.

46. Hartwig,C.P. and Ready,D.W.,Ferrite films vs.ferrite substrates for MICs. IEEE Microelecncs Symp (1969) p.C1.

47. Hecht,S.et al.,Microstrip diplexer using bandstop filter. IEEE Proc Microelecncs Symp.Microwave Microelecncs (1969) A4.1-3.

48. Ho,R.Y.C.and Adams,D.K.,Have you tried active microwave filters .Microwaves (Jul 1969) 8 (7) 44-49.

49. Holst,P. and Lemke,M.,Ferrite substrates for microwave integrated systems.IEEE Tr Magnetics (Sep 1969) 5 (3) 478-480.

50. Hornsby,J.S.and Gopinath,A.,Numerical solutions of inhomogeneously filled guides with symmetrical microstrip line. European Microwave Conf (1969) 114-117.

51. Houston,T.W.et al., Computer aided design of MICs. IEEE WESCON (1969) pap 6/6 3p.

52. Hower,M.M.and Sodomsky,K.M.,Hybrid integrated transistor amplifier for high volume production. IEEE Int Microwave Symp Dig (1969) 439-444.

53. Huang,H.C.et al.,Gunn devices in MICs. IEEE Proc Microelecncs Symp (1969) p.b5.1-3.

54. Hughes,J.J. et al.,Novel technique for measuring Q factor of thin film lumped elements at microwave frequencies. Elecncs Lett (Oct 1969) 5 (21) 535.

55. Hunton,J.K.et al., Microwave integrated circuit balanced mixer with broad band width. IEEE Proc Microelecncs Symp (1969) p.a3.

56. Huntt,R.L. et al., Ferrimagnetic substrates for integrated microwave circuits. IEEE Tr Magnetics (Sep 1969) 5 (3) 482.

57. IEEE Proceedings of the 1969 Microelectronics symposium. Microwave Microelectronics. IEEE (1969) 22 papers.

58. Iveland,T.D.,Microstrip bandpass filters with dielectric resonators. European Microwave Conf (1969) 200-203.

59. Jackson,R.H.Fabrication and testing of large number of integrated microwave circuits. IEEE Int Conv Dig (1969) 36-37.

60. Johnson,A.K.and Zysman,G.I.,Coupled transmission-line networks in an inhomogeneous dielectric medium. IEEE Int Microwave Symp Dig (1969) 329-337.

61 Johnson,G.D.and Hyltin,T.M.,Application of microelectronic technology to radar and communication systems (abs.only). NEREM Rec (1969) 229.

62. Johnson,K.M.,Recent advances in MIC solid state signal sources. IEEE Proc Microelecnes Symp,Microwave Microelecncs (1969) B 3.1-3.

63. Jones,R.R.and Wicker,L.R.New ferrite microstrip devices. Microwaves (Jan 1969) 8 (1) 32-40.

64. Jones,S.,Frequency stable microstripline X-band Gunn oscillator. Proc IEEE (Mar 1969) 53 (3) 364-366.

65. Kamnissis,C.and Minton,R.,RF integrated amplifiers in high power UHF broad band structures. IEEE Proc Microelecncs Symp Microwave Microelecncs (1969) B2.1-2.

66. Katoh,H.and Ayaki,K.,Study of microwave IC capacitors. Papers of Tech Group on Commun Systm IECE Japan (May 1969) MW 69-9.

67. Kind,G.and Heeks,J.S.,Bulk-effect modules pave way for sophisticated uses. Electronics (3 Feb 1969) 42 (3) 94-96.

68. Knox,R.M.,Integrated microcircuits for UHF range. IEEE WESCON (1969) paper 3/5 13p.

69. Kruppa,W.,Design considerations for integrated microwave linear amplifiers. Ohio State Univ Ph D Thesis (1969) 272p.

70. Lange,J.,Interdigitated stripline quadrature hybrid. IEEE Int Microwave Symp (1969) 10-13 and IEEE Tr Microwave Theor Tech (Dec 1969) 17 (12) 1150-1151.

71. Lee,R.F.and Sodomsky,K.F.,Hybrid integrated L-band digital phase shifter. IEEE Int Microwave Symp Dig (1969) 26-30.

72. Lehrfeld,S.S.and Schilling,W.,Real world of micromin substrates. Part 2 to 5. Microwaves (Jan 1969) 8 (1) 44-47,(Mar 1969) 8 (3) 57-59, (Sep 1969) 8 (9) 36-38, (Dec 1969) 8 (12) 54-57.

73. Lemke,M. ,Microwave integrated devices and subsystems on ferrite substrates. European Microwave Conf (1969) 53-56.

74. Leppavouri,S. et al.,Investigation of MICs made by thin film techniques. State Inst of Tech Res Rep II (Finland) (1969) No. 25,3-18.

75. **Letron,Y.and Guidevaux,J.,Use of ferrites and garnets in microwave microelectronics.** (in French). Microelecncs Rel (Nov 1969) 8 (4) 319-324.

76. Liu,S.G.,Microstrip high power L-band avalanche diode oscillator. IEEE Int Microwave Symp Dig (1969) 261-265.

77. Maesel,M. ,Numerical solution to the characteristics of coupled microstrip. European Microwave Conf (1969) 118-121.

78. **Mannersalo,K.and Stubb,T.,Certain aspects of thin microwave** components. Electricity in Finland (1969) 42 (1) 23-27.

79. March,S. ,Microwave micromin bibliography. Microwaves (Dec 1969) 8 (12) 59-64 and (Jan 1970) 8 (1) 53-56,122-124.

80. March,S.and Cabrera,E. ,Ultrabroad band components in microstrip. IEEE Proc Microelecnc Symp (1969) Pap A-2.

81. Mariani,E.A. et al.,Slot line characteristics. IEEE Tr Microwave Theor Tech (Dec 1969) 17 (12) 1091-1096 & IEEE Int Microwave Symp Dig (1969) 99-105.

82. **Meier,P.J.and Okean,H.C. Integrated microwave amplifier-converter. Tech rep.Jan-Aug 1969.U.S.Govt R & D Rep AD-862509 (Nov 1969) 59p.**

83. Meland,E.and Brymer,F.,Integral thin film chromium resistors for MICs. IEEE Proc Microelecncs Symp (1969) Pap C4.

84. Mitchell,W.S.E.,Developments in microwave transistors and integrated circuits. Elecncs Engng (Dec 1969) 41 (502) 37-39.

85. **Mittra,R.and Itoh,T.,Theory of shielded microstrip lines.** European Microwave Conf Proc (1969) 14-17.

86. Miyakawa,T. et al.,Microwave integrated circuit doubler. Fujitsu Sci Tech J (Mar 1969) 5 (1) 89-109.

87. Moore,R.A. et al.,Miniature analog stripline X-band phase shifter. IEEE Proc Microelecncs Symp (1969) Pap C3.

88. Moyers,W.D.,Designing thin film hybrid microcircuits and how to get them built (Abs only). NEREM Rec (1969) Session 228.

89. **Muto,T.and Sakurai,S.,Thin film technology** (microwave) in Japanese. J Instn Elecncs Comm Engrs Japan (Nov 1969) 52 (11) 1367-1372.

90. Nelson,T. et al.,Miniature analog stripline X-band ferrite phase shifter. **IEEE Proc Microelecncs Symp.Microwave** Microelecncs (1969) C3.1-C3.3.

91. **Nergaard,L.S.,Potential of integrated electronics for microwave systems.** RCA Def Electron Prod (Microwave Tech)Tech Pap (1969) 13-15.

92. **Noel,P.L.and Emery,F.E.,Integrated Ku-band phase shifters.** IEEE Proc Microelecnc Symp.Microwave Microelecncs (1969) B4.1-4.

93. Okamura,S.,Ferrite elements (MIC)(in Japanese). J Electr Commun Engrs Japan (Nov 1969) 52 (11) 1377-1381.

94. Okamura,S.,Introduction (to microwave ICs)(in Japanese). J Instn Elecncs Comm Engrs Japan (Nov 1969) 52 (11) 1346-1349.

95. **Ota,T.et al.,Substrates. (MIC) in Japanese. J Instn Elecncs** Comm Engrs Japan (Nov 1969) 52 (11) 1381-1384.

96. Oxley,T.H. et al.,Design and performance of an integrated X-band receiver. European Microwave Conf (1969) 3-5.

97. **Paul,D.H.,Computer-aided-design of broadband microwave band pass filters. European Microwave Conf Proc (1969) 174-177.**

98. Penney,J.A.Ferrite devices-state of the art survey. European Microwave Conf (1969) 18.

99. **Pinac,J.C.and Skinner,B.S.Hybrid integrated circuit microwave** telemetry transmitter and command receiver. IEEE G-MTT Int Microwave Symp (1969) 450-458.

100. Poole,W.E.,UHF integrated power amplifiers. IEEE Int Solid St Ccts Conf Dig (1969) 126-127.

101. Robinson,G.H.,Microwave microelectronic techniques applied to phase shifters and other control devices. Final report 29 Jan 1968-29 Aug 1969. U.S.Govt R & D Rep N70-17302 (Nov 1969) 117p.

102. Robinson,G.H. and Allen,J.L.,Slot line application to miniature ferrite devices. IEEE Tr Microwave Theor Tech (Dec 1969) 17 (12) 1097-1101 & IEEE Int Microwave Symp Dig (1969) 106-109.

103. **Rodrigue,G.P.,Review of planar ferrite devices. IEEE Int Mi**crowave Symp Dig (1969) 79-98.

104. Rogers,C.B. and Cooke,R.E.,Microstrip circulators for MICs. (Abs only) European Microwave Conf (1969) 47.

105. Rosenberg,G.W. and Nordbotten,A.Experimental investigation of bulk semiconductor instabilities for microwave application. Final Sci.rep.1 Feb 1968-30 Jun 1969. U.S.Govt R & D Rep AD-865666 (Oct 1969) 24p.

106. **Ruth,S.D.,Hybrid-thick and thin. Elecnc Engr (Oct 1969) 28 (10) 60-65.**

107. Saunders,A.F. et al.,Integrated circuits at microwave frequencies. **Royal Radar Estab. Newsletter & Res Rev (1969)** No.8,5p.

108. **Schable,G.L.and Keen,R.S.,Aluminium metallization-advantages** and limitation for integrated circuit applications. Proc IEEE (Sep 1969) 57 (9) 1564-1570.

109. **Scherer,E.F. and Barnett,M.J., Broadband multistage avalanche** amplifier at X-band. IEEE Solid St Ccts Conf Dig (1969) 122-123.

110. **Schilling,R.B.,Lumped-element hybrid integrated microwave** systems. IEEE Annual Integrated Circuits Seminar 6 (1969) 5-11.

111. Schneider,M.V.,Dielectric loss in integrated microwave circuits. Bell Syst Tech J (Sep 1969) 48 (7) 2325-2332 and Proc IEEE (Jun 1969) 57 (6) 1206-1207.

112. **Schneider,M.V.,Microstrip lines for MICs.Bell Syst Tech J** (May-Jun 1969) 48 (5) 1421-1444.

113. Schneider,M.V. ,S-and X-band receiver designs pack all circuit functions on one substrate. Commun Des Dig (May 1969) 3 (5) 6-7.

114. Schneider,M.V. et al.,Microwave and millimeter wave hybrid integrated circuits for radio systems. Bell Syst Tech J (Jul-Aug 1969) 48 (7) 1703-1726.

115. Schwarzmann,A.,Approximate solutions for coupled pair of microstrip lines in MICs. Microwave J (May 1969) 12 (5) 79-82.

116. Smith,J.I.,Even-and odd-mode capacitance parameters for coupled lines in suspended substrates. IEEE Int Microwave Symp Dig (1969) 324-328.

117. **Stinehelfer,H.,Microstrip circuit designs, Microwave Assoc** Burlington Mass. Tech Rep AFAL-TR -69-10 (1969) p.57,63-65.

118. Strull,G.,Influence of innovation on microwave integration. IEEE Int Conv Dig (1969) 42-43.

119. **Takagi,T.,Application in electronics field (MIC) in Japane**se. J Elecncs Comm Engrs Japan (Nov 1969) 52 (11) 1449-1457.

120. **Tarowsky,N.,How to assemble hybrid MIC. Microwaves (Aug 1969)** 8 (8) 52-62.

121. Tauchi,S. and Ihochi,T.,Fabrication elements (Microwave integrated) in Japanese. J Instn Elecncs Comm Engrs Japan (Nov 1969) 52 (11) 1372-1377.

122. **Terry,L.E.and Wilson,R.W.,Metallization systems for silicon** integrated circuit. Proc IEEE (Sep 1969) 57 (9) 1580-1586.

123. Texas Instruments Inc.,Study of solid state integrated microwave circuits. Final report. US.Govt R & D Rep N69-36999 (Jan 1969) 41p.

124. Troughton,P.,Evaluation of alumina substrates for use in the microstrip MICs. European Microwave Conf (1969) 49-52.

125. Uenohara,M.,Introduction to MIC applications and future trends. J Instn Elecncs Comm Engrs Japan (Nov 1969) 52 (11) 1437-1441.

126. Vendelin,G. et al.,Comments on "microstrip vs. balanced stripline." Proc IEEE (Oct 1969) 57 (10) 1778-1780.

127. Vergnolle,C.and Rozec,X.,General properties of transmission phase shifters. European Microwave Conf (1969) 536-539.

128. Vincent,B.T.and Wallace,M.E.,MICs for phased-array applications. Microwave J (Sep 1969) 12 (9) 53-59.

129. Walker,G.H. et al.,Theory and practice of coupled microstrip systems. European Microwave Conf (1969) 57.

130. Wang,P.H.Y.,Five-octave microwave thin film amplifier. IEEE Int Solid St Ccts Conf Dig (1969) 128-129.

131. Watanabe,T.,Circuit elements (Microwave integrated) in Japanese. J Instn Elecncs Comm Engrs Japan (Nov 1969) 52 (11) 1354-1359.

132. Wen,C.P.,Coplanar wave guide : a surface strip transmission **line suitable for non-reciprocal gyromagnetic device appl**ications. IEEE Tr Microwave Theor Tech (Dec 1969) 17 (12) 1087-1090 & IEEE Int Microwave Symp (1969) 110-115.

133. Wen,C.P.,Integrated circuit metallized plastic symmetrical millimeter trough waveguide system with non-reciprocal elements. RCA Rev (Dec 1969) 30 (4) 724-728.

134. Wernlund,L.D.,Ferrite substrate materials for MICs. European Microwave Conf (1969) 58-61.

135. Wheeler,C.A. and Ayer,D.R.,Evolution of strip transmission line. Microwave J (May 1969) 12 (5) 31-40.

136. Whicker,L.R. et al.,Advances in planar ferrite devices . IEEE Proc Microelecncs Symp. Microwave Microelecncs (1969) C2.1-C2.4.

137. White,M.L.,Encapsulation of integrated circuits. Proc IEEE (Sep 1969) 57 (9) 1610-1615.

138. Yamashita,E.and Atsuki,K.,Design of transmission line dimensions for a given characteristic impedance. IEEE Tr Microwave Theor Tech (Aug 1969) 17 (8) 638-639.

139. Zaininger,R.H.and Wang,C.C.,Thin film dielectric materials for microelectronics.Proc IEEE (Sep 1969) 57 (9) 1564-1570.

140. Zysman, G.I. and Johnson, A.K., Coupled transmission line networks in an inhomogeneous dielectric medium. IEEE Tr Microwave Theor Tech (Oct 1969) 17 (10) 753-59.

1970

1. Aho, O. et al., Construction of a 7 GHz microwave integrated balanced mixer. Proc 4th Colloq on Microwave Communc (1970) Pt 4 MT 1/1-8.

2. Alley, G.D., Interdigitated capacitors for use in lumped element microwave integrated circuits. IEEE Int Microwave Symp Dig (1970) 7-13 and IEEE Tr Microwave Theor Tech (Dec 1970) 18 (12) 1028-1033.

3. Alley, G.D. et al., Thin film lumped constant microwave integrated filter structures. IEEE Int Microwave Symp Dig of Tech Pap (1970) 2-6.

4. Bady, I., Temperature stability of ferrite substrates in MICs. Int Magnetics Conf Dig (1970) 1p. and U.S. Govt R & D Rep AD-709597 (Jul 1970) 21p and IEEE Tr Magnetics (Sep 1970) 6 (3) p 674.

5. **Baily, W.L. and Kleppinger, R.E., Thin film lumped element** microwave integrated circuits. IEEE Proc Elecnc Comp Conf (1970) (20) 497-502.

6. Basawapatna, G., Microwave integrated circuits for avalanche diode applications. IEEE Int Conv Dig (1970) 312-313.

7. Belohoubek, E. et al., S-band CW power module for phased arrays. Microwave J (Jul 1970) 13 (7) 29-34.

8. Bunker, J., Manufacturing techniques for MICs. IEEE Int Conv Dig (1970) 318-319.

9. Bunker, J., Microwave integrated circuit fabrication, assembly, in-process, test and manufacturing techniques. Proc Int Conf Microelecncs, Paris (6-10 Apr 1970) 2p.

10. Caulton, M., Lumped element approach to microwave integrated circuits. Microwave J (May 1970) 13 (5) 51-58.

11. Caulton, M., MIC technology. IEEE Int Solid St Ccts Conf Dig (1970) 150-151.

12. Caulton, M., and Sobol, H., Microwave integrated circuit technology-a survey. IEEE J Solid St Ccts (Dec 1970) 5 (6) 292-303.

13. Chinchillo, A.R., Computer-aided design of practical microwave branchline couplers. SWIEECO Record of Tech Papers of 22 south western IEEE Conf (1970) 426-430.

14. Clar, P., Application of dielectric resonators to microwave integrated circuits. IEEE Int Microwave Symp (1970) 19-23.

15. Cohn, S.B., Microwave Integrated circuits - now. Microwave J (May 1970) 13 (5) 96. Review & editorial.

16. Collard, J.R., and Gobat, A.R., High dynamic-range low-noise microwave amplifier. Elecncs Lett (2 Apr 1970) 6 (7) 202-203.

17. Corr, D.G., Finite difference analysis of hybrid modes in microstrip structures. Ph D Thesis London Univ (Aug 1970).

18. Day, W.R., Dielectric resonators as microstrip circuit elements. IEEE Int Microwave Symp Dig (1970) 24-28 and IEEE Tr Microwave Theor Tech (Dec 1970) 19 (12) 1175-1176.

19. DeBrecht, R., and Caulton, M., Lumped-elements in MICs in the 1-12 GHz range. IEEE Int Microwave Symp Dig of Tech Pap (1970) 14-18.

20. Dell-Imagine, R.A., Parallel coupled microstrip filter design procedure. IEEE Int Microwave Symp Dig (1970) 29-32.

21. Deutsch, J., and Jung, H.J., Measurement of the effective dielectric constant of Microstrip lines in the frequency range from 2 GHz to 12 GHz. NTZ Commun J (DEC 1970) 23 (12) 620-624.

22. Earhart, C.E., Microwave integrated circuit fabrication. Proc Elecncs Comp Conf (1970) IEEE 562-569.

23. Edelman, H.C.M., New ferrite technology improves integrated circuits for microwaves. Microelecncs Rel (Sep 1970) 9 (5) 434.

24. Emery, F.E., Monolithic vs hybrid microwave integrated circuitry. IEEE Int Conv Dig (1970) 314-315.

25. Frigyes, I., Effect of packaging on microwave integrated circuits. Abs only. Proc Int Conf Adv Microelecncs. Paris (Apr 1970) 2p.

26. Furukawa, S., and Nakagami, T., Design theory of a self-excited frequency-multiplying oscillator. IEEE J Solid St Ccts (Jun 1970) 5 (3) 87-94.

27. Gall, F.M., Application of thin film techniques to microwave integrated circuits. Proc Elecncs Comp Conf (IEEE) (1970) 626-638.

28. Gall, F.M., Microwave integrated circuits. West Electr Engr (Jan 1970) 14 (1) 26-36.

29. Giorgis, R.O. et al., Computer design of stripline components. U.S. Govt R & D Rep AD-876732 (Jun 1970) 90p.

30. Grabowska, K. et al., On the technology of microwave integrated circuits components on the ceramic and ferrite substrate (in Polish). Elektronika (1970) (11) 449-455.

31. Grunberger, G.K. et al., Longitudinal field components and frequency-dependent phase velocity in the microstrip transmission line. Elecncs Lett (15 Oct 1970) 6 (21) 683-685.

32. Gunderson, L.C., Unique uses of ceramics in microwave engineering. Microwaves (Jan 1970) 9 (1) 38-42.

33. Hallford, B.R., Low noise microstrip mixer on a plastic substrate. IEEE Int Microwave Symp Dig (1970) 206-211.

34. Harris, D.H. et.al., Polycrystalline ferrite films for microwave applications, deposited by arc-plasma. J App Phys (Mar 1970) 41 (3) 1348-1349.

35. Hartwig, C.P., Basic parameters of microstrip transmission lines. IEEE Int Conv Dig (1970) 426-427.

36. Hartwig, C.P., and Ready, D.W., Ferrite film circulator. J App Phys (Mar 1970) 41 (3) 1351-1352.

37. Hunton, J.K. and Takeuchi, J.S., Recent developments in microwave slot line mixers and frequency multipliers. IEEE Int Microwave Symp (1970) 196-199.

38. Johnson, K.M., Recent advances in microwave integrated circuit solid-state source design. IEEE J Solid St Ccts (Jun 1970) 5 (3) 119-123 and IEEE Proc Microelecncs Symp (1969) p. b3.

39. Judd, S.V. et al., Analytical method for calculating microstrip transmission line parameters. IEEE Tr Microwave Theor Tech (Feb 1970) 18 (2) 78-87.

40. Kleppinger, R.E. et al., Thin film hybrid integrated power amplifiers. Proc Nat Elecncs Conf (1970) 26, 253-258.

41. Knox, R.M. and Toulios, P.O., Integrated circuits for the millimeter through optical frequency range. Symp on Submillimeter waves (1970) 2p.

42. Krzyzagorski, M., Some aspects of analysis of the microstrip junction circulators for MICs. Proc of 4th Colloq on Microwave Communc. Budapest (1970) 11p.

43. Kurpis, G.P. and Taub, J.J., Wideband X-band microstrip image rejection balanced mixer. IEEE Int Microwave Symp Dig (1970) 200-205.

44. Lemke, M. et al., Integrated subsystems on composite substrates. Advanced Radar Systems ed.by Voles R (U.S. Govt R & D Rep N-71-13901 523p.) Nov (1970) 11p.

45. Leppavuori, S. and Mannersalo, K., Finish integrated microwave circuits (in Finnish). Elektronikka (1970) 23 (3) 8-12.

46. Liu, S.G., Compact microstrip high power high efficiency avalanch-diode oscillator. IEEE Tr Microwave Theor Tech (Nov 1970) 19 (11) 982-983.

47. Lump, C.J., Design of microwave integrated circuits by optimal-seeking computer techniques. IEEE Int Conv Dig (1970) 320-321.

48. Maier, H.G., Ein verfahren zur messung der Komplexen dielectrizitatszahl von folien und dunnen platten im Mikrowellenbereich. **Frequenz (Oct 1970) 24 (10) 303-307.**

49. Malone, H. and Moore, R., Biphase modulator-Final tech report, May-Jul 1970. U.S. Govt R & D Rep AD-713003 (4 Sep 1970) 25p.

50. Mariani, E.A. and Agrios, J.P., Slot line filters. IEEE Int Microwave Symp Dig (1970) 190-195.

51. Mariani, E.A. and Agrios, J.P., Slot line filters and couplers. IEEE Tr Microwave Theor Tech (Dec 1970) 19 (12) 1089-1095.

52. Mehmet, K., Microwave integrated circuits preparation and measurement techniques for materials and components. Univ Warwick, Convetry, England, thesis, (29 Oct 1970) 111p.

53. Microwave integrated circuits. I & II. IEEE Int Microwave Symp Dig (1970) 2-37, 184-222.

54. Napoli, L.S. and Hughes, J.J., Characteristics of coupled microstrip line. RCA Rev (Sep 1970) 31 (3) 479-498.

55. Okoshi, T. and Miyoshi, T., Planar circuit - a novel approach to microwave circuitry. Proc Kyoto Int Conf on Circuit & Sys Theor (Sep 1970) B-5-1.

56. Oxley, T.H., X-band integrated circuit receiver. Microwave J (Apr 1970) 13 (4) 52-56.

57. Oxley, T.H. et al., Beam lead mixer and detector diodes for microwave integrated circuits applications. Abs only. Moga 70. 8th Int Conf Microwave & Optical Amp Gen (1970) 1p.

58. Patel, R.N., Microwave conductivity of thick-film conductors. Elecncs Lett (23rd Jul 1970) 6 (15) 455-456.

59. Podell, A., High directivity microstrip coupler technique. IEEE Int Microwave Symp Dig (1970) 33-36.

60. **Reuss,M.L. and Young, C.W., Interconnected 3 dB hybrids and** phase shifters; a combined switch, phase shifter and isolator. U.S. Govt R & D Rep AD-717341 (Nov 1970) 24p.

61. Riches, E.E. et al., Microstripline ferrite devices using surface field effects for MICs. Int Magnetics Conf Dig (1970) **1p. and IEEE Tr Magnetics (Sep 1970) 6 (3) 670-673.**

62. **Ronde de,F.C.,New class of microstrip directional couplers.** IEEE Int Microwave Symp Dig (1970) 184-189.

63. Schaub, W.E., S-band solid state amplifier miniaturization. Final Tech rep Aug 68 - Mar 70. U.S. Govt R & D Rep AD-871741 (May 1970) 147p.

64. Schilling, R.B., IC's for microwaves. Elecncs Wld (Jul 1970) 84 (1) 42-44.

65. Schmitt, H.J., Fundamentals of microwave integrated circuits. Proc Swedish Seminar on New Microwave Comp (Stockholm)(1970).

66. Schneider, M.V. and Snell, M.W., Hybrid integrated frequency multiplier from 10 to 30 GHz. Proc IEEE (Sep 1970) 58 (9) 1402-1404 and NEREM Rec (1970) 74-75.

67. Smith, W.R., Studies of microwave acoustic transducers and dispersive delay lines. Standford Univ California Ph D Thesis (1970) 158p. Abs. STAR (8 Dec 1971) 9 (23) 3734.

68. Sobol, H., Technology and design of hybrid microwave integrated circuits. Solid St Tech (Feb 1970) 13 (2) 49-57.

69. Solomon, A.H., Semiconductor devices in MICs. IEEE Int Conv Dig (1970) 430-431.

70. Special Issue, Solid state microwave circuits, IEEE J Solid St Ccts (DEC 1970) 5 (6) 291-367.

71. Stinehelfer, H.E., Microstrip circuit design. Microwave J (May 1970) 13 (5) 71-74.

72. Sutherland, A.D., Approach for heat sinks in monolithic microwave integrated circuits. Elecncs Lett (9 Jul 1970) 6 (14) 432.

73. Taft, D.R. and Robinson, G.H., MIC ferrite devices. IEEE Int Conv Dig (1970) 432-433.

74. Taft, D.R. and Robinson, G.H., MIC ferrite devices. Symp Microwave Integrated Circuits Digest (March 10, 1970).

75. Toulioa, P.P., Image line millimeter integrated circuits-directional coupler design. Proc Nat Elecncs Conf (1970) 26, 645-650.

76. Vergnolle, C., Modular conception of hybrid microwave IC subassemblies. Abs only. Proc Int Conf Adv Microelecncs Paris (Apr 1970) 2p.

77. Vincze, A.D., Design of a thick/thin film microwave receiver front end. NEREM Rec (1970) 76-77.

78. Ward, C.S. and Forest, R.G., Microstrip diode phase shifters at C and X-bands. IEEE Int Conv Dig (1970) 316-317.

79. Ward, C.S., X-band PIN diode phase shifter. Final tech rep.21 Feb 1968 - 20 Jul 1969. U.S. Govt R & D Rep AD-865280 (Jan 1970) 54p.

80. Welch, J.D., Beam lead tunnel diode amplifiers on microstrip. IEEE Int Microwave Symp (1970) 212-216 and IEEE Tr Microwave Theor Tech (Dec 1970) 18 (12) 1077-1083.

81. Wen, C.P., Coplanar-waveguide directional couplers. IEEE Tr Microwave Theor Tech (Jun 1970) 18 (6) 318-322.

82. Wharton, R.P., Dominant mode analysis of microwave hybrid integrated circuits transmission lines. Georgia Inst of Tech Thesis (1970) 208p.

83. Wolf, E.D. et al., Electron beam and ion beam fabricated microwave switch. IEEE Tr Electron Dev (Jun 1970) 17 (6) 446-449.

84. Yoder, D.F., Microwave integrated circuit fabrication and development using a laser. NEREM Rec (1970) 178-179.

1971

1. Afzelius, L. et al., Integrated stripline multiplier chain. Proc European Microwave Conf (1971) C10/5:1 - 5:4.

2. Aitchison, C.S. et al., Lumped-circuit elements at microwave frequencies. IEEE Tr Microwave Theor Tech (Dec 1971) 19 (12) 928-937.

3. Aitchison, C.S. and Newton, B.H.,Varactor-tuned X-band Gunn oscillator using lumped thin film circuits. Elecncs Lett (25 Feb 1971) 7 (4) 93-94.

4. Aitchison, C.S. et al., Lumped microwave circuits. Pt I & II. Des Elecncs (Sep 1971) 8 (11) 23-28 and (Oct 1971) 9 (1) 30-39.

5. Aitchison, C.S. et al., Lumped microwave circuits. III.Filters and tunnel diode amplifiers. Des Electron (Nov 1971) 9 (2) 42-51.

6. Antell, G.R., Monolithic mixers for millimeter wavelengths. Proc European Microwave Conf (1971) C2/3:1 - 3:4.

7. Araki, T. and Hirayama, M., 20 GHz integrated balanced mixer. IEEE Tr Microwave Theor Tech (Jul 1971) 19 (7) 638-643.

8. Aumiller, C.G. et al., Ferrite dielectric composite integrated microwave circuit development. IEEE Int Microwave Symp (1971) 66-69.

9. Baba, P.D. and Argentina, G.M., Development of lithium ferrites for array radar phasers. Final rep 1 Jul 1969 - 31 Aug 1971. U.S. Govt R & D Rep AD-736038 (Aug 1971) 56p.

10. Bady, I., Temperature stability of ferrite substrates in microwave integrated circuits. IEEE Tr Magnetics (Jun 1971) 7 (2) 276-280.

11. Bauer, R.J., Solid state array modules for use at L-, S-and X-band. IEEE Int Conv Dig (1971).

12. Beneking, H., Microwave circuitry on semiconductors - an approach to MIC's. Proc European Microwave Conf (1971) C2/1:1. - 1:4.

13. Blair, P.K. et al., 250 watt module using series operated Gunn diodes. Proc 3rd Biennial Cornell Electr Engng Conf (1971) 309 - 313.

14. Blanc, M. et al., Report of a supported study on microwave microelectronics with ferrite substrates. (in French). U.S. Govt R & D Rep N71-31696 (Jan 1971) 96p.

15. Booth, P. and Salmon, K., X-band microstrip mixer for television applications. (Abs only). IEEE Colloq on New Adv in Microstrip (Mar 1971) p.5.

16. Bosnell, J.R. and Lloyd, K.H., Comparison of manufacturing techniques for hybrid microwave circuits. Proc Tech Prog INTERNEPCON (1971) 287.

17. Bratschun, W.R. et al., Uses of ceramics in microelectronics survey. NASA Tech Surv (1971) SP-5097, 168p.

18. Cahalan, J.B. et al., Integrated X-band image and sum frequency enhanced mixer with 1 GHz IF. IEEE Int Microwave Symp (1971) 16-17.

19. Campbell, B.D. et al., Microwave solid state sources and integrated circuits. U.S. Govt R & D Rep AD-731756 (Jul 1971) 168p.

20. Caulton, M., Film technology in MICs. Proc IEEE (Oct 1971) 59 (10) 1481-1489.

21. Caulton, M. et al., Status of lumped elements in microwave integrated circuits. Present and future. IEEE Tr Microwave Theor Tech (Jul 1971) 19 (7) 588-599.

22. Cerniglia, N.P. et al., Multichip integrated circuit package. Final technical report 1 Jun 1968 - 31 Aug 1971. U.S. Govt R & D Rep AD-736790 (Aug 1971) 270p.

23. Chambers, D.R., MIC and the changing microwave industry. Proc IEEE Fall Elecncs Conf (Oct 1971) 492.

24. Charas, Ph., Integrated 6 GHz receiver front end for high capacity radio relay systems. IEEE Int Solid St Ccts Conf Dig (1971) 90-91.

25. Chrepta, M.M., Bulk semiconductor millimeter wave phase shifter. U.S. Govt R & D Rep AD-737183 (Nov 1971) 19p.

26. Chua, L.W., New broadband matched hybrids for MICs. Proc European Microwave Conf (1971) C4/5:1 - 5:4.

27. Cornbleet, S., Microwave stripline and lumped elements. Tech report Pt. 1 & 2. Microwave J (Jul 1971) 14 (7) 13B and Aug (1971) 14 (8) 10B-10D.

28. Dalle-Mura, P.H., Spacecraft integrated parametric amplifier development. Proc of the Space Shuttle Integrated Electronics Conf (1971) Vol 2 279-290 (U.S. Govt R & D Rep N71-35034 (1971) 365p.)

29. Day, W.R., MIC diode oscillator stabilized by a dielectric resonator. Proc 3rd Biennial Cornell Electr Engng Conf (1971) 257-264.

30. Denlinger, E.J., Frequency dependent solution for microstrip transmission lines. IEEE Tr Microwave Theor Tech (Jan 1971) 19 (1) 30-39.

31. Deutsch, J. and Jung, H.J., Measurement of the attenuation and effective dielectric constant of microstrip lines in the frequency range between 2 and 12 GHz. Proc European Microwave Conf (1971) C3/4:1 - 4:4.

32. Dietrich, N.R., TH-3 Microwave radio system; Microwave integrated circuits. Bell Syst Tech J (Sep 1971) 50 (7) 2175-2194.

33. Dobramysl, D., Integrated mixer for 18 and 26 GHz. IEEE Int Microwave Symp (1971) 18-19.

34. Dobramysl, D. and Bruntrup,H., Integrated receiver module for satellite transponders. Proc European Microwave Conf (1971) C3/1:1 - 1:3.

35. Domitz, R.W. et al., Microstrip tunnel diode amplifiers using matched beam lead diode pairs. IEEE Int Solid St Ccts Conf Dig (1971) 92-93.

36. Dwyer, W.J., Hermetically sealed broadband mixer assembly. IEEE Int Conv Dig (1971) 410-411.

37. Dwyer, W.J., and Howe, H., High reliability, high isolation microstrip receiver package. Proc European Microwave Conf (1971) C2/4:1-4:4

38. Eargle, J., How capacitor MICs produce cardiod patterns. D.B. Sound Engng (Apr 1971) 5 (4) 32-34.

39. Egami, S., 11 GHz band microstrip circuits. (In Japanese). ECL Tech J (Jul 1971) 20 (7) 1533-1546. Abs Rev Electrc Comm Lab (Nov-Dec 1971) 19 (11-12) 1275.

40. Ellis, D.J. and Gunn, M.W., Stripline Gunn oscillators are compatible with MICs. Elecnc Engng (May 1971) 43 (519) 50-53.

41. Eppinger, R.M., Integrated X-band microstrip transmit-receive switch on semi-insulating GaAs. Carnigie-Mellon Univ Ph D Thesis (1971) 165p.

42. Farinhelt, E.V., and Bandy, G.C., Design considerations for X-band phased array radar modules. IEEE Int Conv Dig (1971) 322-323.

43. Garcia, J.A., Wide band quadrature hybrid coupler. IEEE Tr Microwave Theor Tech (Jul 1971) 660-661.

44. Giannini, R.J. and others. L-band MIC front end for an IFF receiver. IEEE Tr Microwave Theor Tech (Jul 1971) 19 (7) 622-627.

45. Gutteberg, O. and Solbakken, K., New techniques for microwave systems (integration) gives large volume reductions. (in Norwegian). Elektrotek Tidsskr (Mar 1971) 84 (4) 31-37.

46. Haddad, G.I. et al., Microwave solid state device and circuit studies. Tech Rep 1 Dec 1970-1 Mar 1971 and 1 Mar - 1 Jun 1971. U.S. Govt R & D Rep AD-729711 (Jun 1971) 89p and AD-730636 (Aug 1971) 104p.

47. Hallford, B.R., 90 db microstrip switch on a plastic substrate. IEEE Int Microwave Symp (1971) 20-21.

48. Hamori, A. and Jansen, R.M., TH-3 Microwave radio system - Microwave transmitter and receiver. Bell Syst Tech J (Sep 1971) 50 (7) 2117-2135.

49. Harrison, G.R. et al., Ferrimagnetic parts for microwave integrated circuits. IEEE Tr Microwave Theor Tech (Jul 1971) 19 (7) 577-588.

50. Hasegawa, H. et al., Measurements on high frequency transmission characteristics of metallization patterns in monolithic IC's. Elecncs Comm Japan (Mar 1971) 54 (3) 52.

51. Hatsuda, T., Computation of coplanar stripline characteristics **by relaxation method. Elecncs Comm Japan (Nov 1971) 54-B (11)** 76-82.

52. Hayashi, Y. et al., Analytical method for microstrip transmission line with magnetized ferrite slab. Elecncs Comm Japan (Nov 1971) 54-B (11) 54-61.

53. Hempling, N.G. and Wayner, R.J., On array logic and phase shifter driver. Qly Prog Rep 1 Oct - 31 Dec 1970. U.S. Govt R & D Rep AD-725581 (May 1971) 20p.

54. Heng, T.M.S., Trimming of microstrip circuits utilizing microcantilever air gaps. IEEE Tr Microwave Theor Tech (Jul 1971) 19 (7) 652-654.

55. Herlich, G.A., Hybrid microelectronics S-band receiver. Proc Elecncs Comp Conf (1971) 21, 176.

56. Hines, M.E., Ferrite phase shifters and multiport circulators in microstrip and stripline. IEEE Int Microwave Symp Dig of Tech Pap (1971) 108-109.

57. Hines, M.E., Reciprocal and non-reciprocal modes of propagation in ferrite stripline and microstrip devices. IEEE Tr Microwave Theor Tech (May 1971) 19 (5) 442-451.

58. Hinte, A.F., Alumina sandwich line filters for high performance integrated circuit application. IEEE Int Microwave Symp Dig (1971) 26-27.

59. Hinte, A.F., Choosing a transmission line. Microwaves (Dec 1971) 10 (12) 46-50.

60. **Hopfer, S. et al., MIC techniques in the design of extremely** broadband reflectometers. IEEE Int Microwave Symp (1971) 40-41.

61. Horton, R. et al., Variation of microstrip losses with thickness of strip. Elecncs Lett (26 Aug 1971) 7 (17) 490-491.

62. Ito, Y. et al., MIC 6 GHz up converter and 4 GHz mixer. Fujitsu Sci Tech J (Mar 1971) 7 (1) 1-18.

63. Itoh, T., and Mittra, R., Dispersion characteristics of slotlines. Elecncs Lett (1 Jul 1971) 7 (13) 364-365.

64. Iveland, T.D., Dielectric resonator filters for application in microwave integrated circuits. IEEE Tr Microwave Theor Tech (Jul 1971) 19 (7) 643-652.

65. Jain, O.P. et al., Coupled-mode model of dispersion in microstrip. Elecncs Lett (15 Jul 1971) 7 (14) 405-407.

66. Jain, O.P. et al., Dispersive behaviour of microstrip transmission lines. Proc European Microwave Conf (1971) C3/5 :1-5 :4.

67. Jansen, R.M. and Prime, R.C., TH-3 Microwave radio system: System consideration. Bell Syst Tech J (Sep 1971) 50 (7) 2085-2116.

68. Johnson, D.R., Lead frame bonding: an interconnections development area for the first generation hybrid microcircuit technology development programme. U.S. Govt R & D Rep N72-22240 (Dec 1971) 35p.

69. Jones, R., MIC signal processor using YIG multiplexers. IEEE Int Conv Dig (1971).

70. Kaneki, T., Variational analysis for a ferrite-substrate microstrip. Elecncs Comm Jap (Nov 1971) 54A (11) 40-45.

71. Katoh, H., Capacitors in microwave integrated circuits. Elecncs Comm Japan (Mar 1971) 54B (3) 39-46.

72. Katoh, H., Capacitors in MICs. (in Japanese) Trans Inst Elec Comm Engrs Japan (Mar 1971) 54-B (3) 95.

73. Katoh, M. and Akaiwa, Y., 4 GHz integrated circuits mixer. IEEE Tr Microwave Theor Tech (Jul 1971) 19 (7) 634-637.

74. Krems, M. and Schmidt, W., Substrate material for integrated microwave circuits. Frequenz (Jan 1971) 25 (1) 2-9 (in German).

75. Lansdowne, K. and Kelly, A.J., Microwave logarithmic amplifiers using hybrid integrated circuits. IEEE Int Solid St Ccts Conf Dig of Tech Pap (1971) 94-95.

76. Lehrfeld, S.S., What if you don't use alumina? Microwaves (Mar 1971) 10 (3) 54-56.

77. Lemke, M., Low noise microslot and microstrip mixers on ferrite substrate for Ku band frequencies. Proc European Microwave Conf (1971) C3/3:1-3:3

78. Mariani, E.A., Evaluation of slot line bends at S-band frequencies. U.S. Govt R & D Rep AD-734814 (Oct 1971) 23p.

79. Maynard, E.D., Large scale integration of microwave circuits. Proc European Microwave Conf (1971) C1/1.

80. Meier, P.J. et al., Integrated X-band sweeping superheterodyne receiver. IEEE Tr Microwave Theor Tech (Jul 1971) 19 (7) 600-609 and IEEE Int Microwave Symp (1971) 24-25.

81. Microwave Semiconductor Corp. MIC 91,000 series S-band telemetry amplifiers. Microwave J (Aug 1971) 14 (8) 24.

82. Mittra, R. and Itoh, T., New technique for the analysis of the dispersion characteristics of microstrip lines. IEEE Tr Microwave Theor Tech (Jan 1971) 19 (1) 47-56.

83. Moschytz, G.S., Considerations in the design of manufacturable hybrid-integrated active fillers. Eurocon 71 Dig (Oct 1971) 1p.

84. Muckelroy, W.L., Thick film microstrip single-side band PIN diode modulator for X-band. U.S. Govt R & D Rep AD-735317 (Dec 1971) 24p.

85. Napoli, L.S. and Hughes, J.J., Simple technique for the accurate determination of the microwave dielectric constant for MIC substrates. IEEE Tr Microwave Theor Tech (Jul 1971) 19 (7) 664-665.

86. Ogasawara, N. and Kaji, M., Coplanar-guide and slot-guide junction circulators. Elecncs Lett (6 May 1971) 7 (9) 220-221.

87. Okean, H.C. and Meier, P.J., Microwave integrated tunnel diode amplifiers for broadband, high performance receivers. IEEE Int Microwave Symp (1971) 186-187.

88. Okean, H.C. and Pflieger, R., Octave band MIC electronically variable attenuators using PIN diodes. IEEE Int Microwave Symp (1971) 180-181.

89. Okean, H.C. et al., Integrated parametric amplifier module with self contained solid state pump source. IEEE Tr Microwave Theor Tech (May 1971) 19 (5) 491-493.

90. Okoshi, T. and Myoshi, T., Planar circuit-an approach to microwave IC. Proc European Microwave Conf (1971) C4/4:1-4:4.

91. Oliver, B., Hybrid ICs a major factor in microwave component development. Elecnc Engng (May 1971) 43 (519) 41-43.

92. Onno, P. and Plitkins, A., Miniature multi-kilowatt PIN diode MIC digital phase shifters. IEEE Int Microwave Symp (1971) 22-23,

93. Osterwalder, J.M., Ku-band electronically tunable monopulse receiver. IEEE Tr Microwave Theor Tech (Jul 1971) 19 (7) 627.

94. Oxley, T.H. and Ming, K.J., J (Ku)-band microwave integrated receiver. Proc European Microwave Conf (1971) C2/5:1-5:4.

95. Oxley, T.H. and Swallow, G.H., Beam lead mixer and detector diodes for MIC applications. Microwave J (Dec 1971) 14 (12) 31-34.

96. Poirier, A.L., Integrated microstrip circulator. IEEE Tr Microwave Theor Tech (Jul 1971) 19 (7) 661-662.

97. Ramadan, M. and Westgate, W.F., Impedance of coupled microstrip transmission lines. Microwave J (Jul 1971) 14 (7) 30-34, 56.

98. Renz, H., Integrated microwave switch and modulator for working in the range 0.5 to 18 GHz. Elektron Ind (20 Jun 1971) 2 (5/6) 108-110 (in German).

99. Roberts, R.J., Effect of tolerances on the performance of microstrip parallel-coupled band pass filters. Elecncs Lett (20 May 1971) 7 (10) 255-257.

100. Rowe, D.N.E., Design of 3 db branch line coupler centred on 1.25 GHz. Marconi Rev (1971) 34 (183) 253-258.

101. Sadler, W.G., FM/CW Gunn powered subminiature radar altimeter. Proc 3rd Biennial Cornell Electrical Engng Conf (1971) 265-272.

102. Savage, B.R. et al., MIC ferrite components and modules. IEEE Int Conv Dig (1971) 486-487.

103. Schelonka, E.P., Gunn effect microwave devices in microstrip. 21st Ann Southwestern IEEE Conf & Exhibition (1971) 160-161.

104. Schmitt, H.J. and Lemke, M., Microwave ICs on ferrite substrates. IEE Colloq on New Advances in Microstrip (Mar 1971) p6 Abs only.

105. Schneider, M.V. et al., Hybrid integrated circuits and integrated circuit technology at 30 GHz. Proc European Microwave Conf (1971) C1/2:1-2:4.

106. Schneider, M.V. et al., Hybrid integrated circuits for microwave systems at 30 GHz. Proc IEEE Fall Elecncs Conf (Oct 1971) 499-500.

107. Schneider, M.V. and Snell, W.W., Scaled hybrid integrated multiplier from 10 to 30 GHz. Bell Syst Tech J (Jul-Aug 1971) 50 (6) 1933-1942.

108. Schefer, J. and Kaplan, G.S., X-band vehicle location system. Proc IEEE Fall Elecncs Conf (Oct 1971) 253-257.

109. Snell, W.W., Low-loss microstrip filters developed by frequency scaling. Bell Syst Tech J (Jul-Aug 1971) 50 (6) 1919-1931.

110. Sobol, H., Applications of integrated circuit technology to microwave frequencies. Proc IEEE (Aug 1971) 59 (8) 1200-1211. **(Jun 1971) 60 (6) 756.**

111. Sobol, H., Microwave hybrid with impedance transforming properties. IEEE Tr Microwave Theor Tech (Sep 1971) 19 (9) 774-776.

112. Sobol, H., Radiation conductance of open-circuit microstrip. IEEE Tr Microwave Theor Tech (Nov 1971) 19 (11) 835-886.

113. Sobol, H., Review of the technology of microwave integrated circuits. Proc IEEE Fall Elecncs Conf (Oct 1971) 493-498.

114. Sobol, H. and Hilling, A.E., Application of integrated circuit technology to microwave frequencies. Proc European Microwave Conf (1971) Cl/S:1-S:9.

115. Special Issue , Microwave Integrated circuits. IEEE Tr Microwave Theor Tech (Jul 1971) 19 (7) 569-651.

116. Tomilinson, J., Integration with microwave circuits. IEE Colloq on New Adv. in Microstrip (Mar 1971) p 6.

117. **Toussaint, H.N. and Hoffman , R., Integrated microwave circuits** - status and direction of development (in German). Frequenz (Apr 1971) 25 (4) 100-108 Also see NLL. Trans. No. 2801 in English.

118. Trambarulo, R.F., 30 GHz inverted microstrip circulator. IEEE Tr Microwave Theor Tech (Jul 1971) 19 (7) 662-664.

119. Trougton, P., Design of complex microstrip circuits by measurement and computer modelling. Proc IEE (Mar-Apr 1971) 118 (3-4) 469-474.

120. Vaccaro, F.E. et al., Product design of a high power S-band MIC module for phased arrays. IEEE Int Conv Dig (1971) 326-327 and IEEE Tr Microwave Theor Tech (Jul 1971) 19 (7) 609-616.

121. Vogel, R., Properties of directional couplers employed in MIC's (in Polish). Rozpr Electrotech (1971) 17 (3) 517-540.

122. **(Vosburgh,M.C., Paradise phased array radars and divers** integrated semiconductor elements. IEEE Int Conv Dig (1971) 320-321.

123. Waugh, R.W., MIC converter for a 2-12.4 GHz receiver. IEEE Int Conv Dig (1971) 404-405.

124. Welch, J.D., Beam lead tunnel diode amplifiers. Int Microelecncs Symp (1971) 6-2-1- 6-2-5.

1972

1. Airborne Instruments Laboratory USA. Tunable C-band paramp **geared for communications work aboard NASA space shuttle.** Commun Des Equip Sys (Jun 1972) 12-14.

2. Andersen, E. and Olesen, S.T., Thick film microwave circuits; an evaluation of materials and processes for thick film striplines at microwave frequencies. U.S. Govt R & D Rep N72-20234 (Feb 1972) 66p.

3. Araki, T. and Hirayama, M., 20 GHz-band integrated balanced mixer. Elecncs Comm Japan (Oct 1971) 54-B(10) 56-60.

4. Araki, T. and Hirayama, M., Quasi-millimeter-wave integrated mixer, ECL-1384. Rev Electr Comm Lab (Sep-Oct 1972) 20 (9-10) 788-792.

5. Bardash, I. and Gayer, A.G., Slot line digital ferrite phase shifter. Final report 30 Oct 70 - 30 Jun 72. U.S. Govt R & D Rep AD-748431 (Aug 1972) 66p.

6. Blodgett, W.E., Ultrafine grain alumina substrates. Am Ceramic Soc Bull (Apr 1972) 51 (4) 361.

7. Brandle, R., Diode-phase-shifter using thin film technology (in German). Frequenz (Apr 1972) 26 (4) 84-88.

8. Caulton, M., Microwave integrated circuits. Topics in Solid State and Quantum Electronics Ed. by W.D. Hershberger. (1972) Wiley 420-468.

9. Charas, P., Integrated 6-GHz receiver front-end circuits for a radio relay system. IEEE J Solid St Ccts (Feb 1972) 7 (1) 61-63.

10. Cisco, T.C., Design of microstrip components by computer. U.S. Govt R & D Rep N72-22218 (Mar 1972) 266p.

11. Cohn, S.B., Slot-line field components. IEEE Tr Microwave Theor Tech (Feb 1972) 20 (2) 172-174.

12. Corr, D.G. and Davies, J.B., Computer analysis of the fundamental and higher order modes in single and coupled microstrip. IEEE Tr Microwave Theor Tech (Oct 1972) 20 (10) 669-678.

13. Cristal, E.G. and Frankel, S. , Hairpin-line and hybrid hairpin-line half-wave parallel-coupled-line filters. IEEE Tr Microwave Theor Tech (Nov 1972) 20 (11) 719-728.

14. Cristal, E.G. et al., "Microguide", a new mic transmission line. IEEE Int Microwave Symp Dig of Pap (1972) 212-214.

15. **Crowe, W., C-band integrated tunnel diode amplifier. Final** report Jan 67-May 68. U.S. Govt R & D Rep AD-842445 (Sep 1968) 55p. reported in Govt Repts Announcement (10 Jun 1972) 72 (11) 99.

16. De Santis, P. and Pucci, F., Novel type of MIC symmetrical 3-port circulator. Elecncs Lett (13 Jan 1972) 8 (1) 12-13.

17. Deutsch, J. and Jung, H.J., Measurement of the permittivity of ceramics especially of substrates for microstrips (in German). Nachrichtentechn Z (Oct 1972) 25 (10) 463-464.

18. Doyle, J., Mechanical design of microwave integrated circuit enclosures. Int Soc Hybrid Microelecnc Symp (1972).

19. Easson, R.M., Microwave integrated circuits. Electron (15 Jun 1972) (6) 19-20.

20. Ellis, D.J. and Gunn, M.W., YIG tuning of stripline Gunn oscillator. IEEE J Solid St Ccts (Feb 1972) 7 (1) 63-65.

21. From, W.H., Ceramic waveguide MICs. Int Soc Hybrid Microelecncs Symp (1972).

22. Gal'chenko, N.A. and Mikhalevsky, V.S., Use of schwartz method for computing electrical characteristics of strip transmission lines. Radio Engng Elecnc Phys (Feb 1972) 17 (2) 183-188.

23. Garcia, J.A. et al., Designing limiter/detectors for ECM receivers. Microwaves (Oct 1972) 11 (10) 52-55.

24. Glance, B., Low-Q microstrip impatt oscillator at 30 GHz. Proc IEEE (Sep 1972) 60 (9) 1105-1106.

25. Gorwara, A.K., Phase and amplitude balance: key to image rejection mixers. Microwaves (Oct 1972) 11 (10) 64-76.

26. Gould, J.W. and Talboys, E.C., Even-and odd-mode guide wavelengths of coupled lines in microstrip. Elecncs Lett (9 Mar 1972) 8 (5) 121-122.

27. Grace, M.I. et al., High efficiency solid state microwave amplifier. Final Technical report (4 Jun 1971-4 Mar 1972). U.S. Govt R & D Rep AD-747050 (Jun 1972) 82.

28. Grabowska, K. et al., On the technology of preconditioning treatment of teflon for an application to the realization of microwave integrated circuits. (in Polish). Elektronika (1972) (1) 15-19.

29. Groll, H. and Wiesbeck, W., Measurement of the dielectric constant of substrates and evaluation of the effective dielectric constant of microwave-striplines. NTZ Comm J (Jun 1972) 25 (6) 265-269.

30. Gunston, M.A.R., Microwave transmission line; impedance data. Van Nostrand (1972) 117p.

31. Hatsuda, T. et al., Wide band transistor amplifier used coplanar-type stripline. Trans IECE Japan (1972) 55-B (5) 267-268.

32. Herandez, L. and Smith, J.G., Space qualification and problems associated with hybrid MIC active devices. Int Soc Hybrid Microelecncs Symp (1972).

33. Hundewadt, J. and Stahlmann, R., Coupled microstrip transmission line parameters (English & German). VDE-Verlag GmbH, Berlin (1972) 116p.

34. Irish, R.T., Annular resonant structures and their uses as microwave filters. Radio Elecnc Engr (Feb 1972) 42 (2) 85-90.

35. Itoh, T. and Mittra, R., Analysis of microstrip transmission lines. U.S. Govt R & D Rep AD-746049 (Jun 1972) 124p.

36. Jain, O.P. et al., Open end and edge effect in microstrip transmission lines. IEEE Tr Microwave Theor Tech (Sep 1972) 20 (9) 626-628.

37. Jenners, J.A., Slot line investigations. M.Sc. Thesis Naval Postgraduate School, Calif. U.S. Govt R & D Rep AD-749070 (Jun 1972) 46p.

38. Joussemet, C., X-band Ku-band microelectric phase shifters (in French). Rev Tech Thomson CSF (Jun 1972) 4 (2) 407-431.

39. Kaneki, T., Phase constants of the slot line and coplanar waveguide loaded with the ferrite substrate. Abs Trans Inst Elecncs Commun Engrs Japan. (Jul 1972) 55 (7) 21-22.

40. Lacombe, D., Multioctave microstrip 50 Ω termination. IEEE Tr Microwave Theor & Tech (Apr 1972) 20 (4) 290-291.

41. Lacombe, D. and Cohen, J., Octave band microstrip DC blocks. IEEE Tr Microwave Theor Tech (Aug 1972) 20 (8) 555-556.

42. Lenzing, H.F., Measurement of dielectric constant of ceramic substrates at microwave frequencies (Abs only). Am Ceramic Soc Bull (Apr 1972) 51 (4) 361.

43. Maeda, M., Analysis of gap in microstrip transmission lines. IEEE Tr Microwave Theor Tech (Jun 1972) 20 (6) 390-396.

44. Michaelides, M., Interdigital filters and experimental microstrip filter design at 470 MHz. Mullard Tech Comm (Oct 1972) 12 (116) 188-192.

45. Michaelides, M., Microstrip band pass filter for 0.7 to 1.4 GHz. Mullard Tech Comm (Oct 1972) 12 (116) 193-196.

46. Michaelides, M., Microstrip directional couplers. Mullard Tech Comm (Oct 1972) 12 (116) 177-187.

47. Special Issue , Microstrip transmission lines: impedance matching, coupling and filtering. Mullard Tech Commun (Oct 1972) 12 (116) 170-196.

48. Michaelides, M., Tapered transmission lines for impedance matching sections. Mullard Tech Commun (Oct 1972) 12 (116) 170-176.

49. Microstrip technology, GTE Lenkurt Demodulator (Nov 1972) 7p.

50. Meier, P.J., Two new media with special advantages at mm wavelengths. IEEE G-MTT Symp Dig (1972) 221-224.

51. Neuf, D., Generate 1 to 12 GHz signals with this double balanced frequency converter. Microwaves (mar 1972) 11 (3) 54-59.

52. Okean, H.C. and Meier, P.J., Microwave integrated tunnel diode amplifiers for broadband high performance receivers. IEEE Tr Microwave Theor Tech (Feb 1972) 20 (2) 165-172.

53. Okean, H.C. et al., Microwave integrated oscillators for broad-band high performance receivers. IEEE Tr Microwave Theor Tech (Feb 1972) 20 (2) 155-164.

54. Okoshi, T. and Miyoshi, T., Planar circuit - an approach to microwave integrated circuitry. IEEE Tr Microwave Theor Tech (Apr 1972) 20 (4) 245-252.

55. Oliver, B., Thin film microcircuit technology-vhf to microwave. Elecncs & Power (Jan 1972) 18, 5-7.

56. Olyphant, M., Simple microwave permittivity measurements of microstrip and stripline substrates. Int Soc Hybrid Microelecncs Symp (1972).

57. Orr, W.H. and Robillard, T.R., Thin film hybrid integrated circuits for communications. IEEE Tr Pts Hybrids Pakag. (Jun 1972) 8 (2) 51-58.

58. Pal, S. and Babu, G.R., Characteristic impedance of microstrip transmission line. JAEU (1972) 5 (1) 21-27.

59. Parris, W.J., P-I-N variable attenuator with low phase shift. IEEE Tr Microwave Theor Tech (Sep 1972) 20 (9) 618-619.

60. Poole, W.E. and Renkowitz, D., M/W power transistors and MIC amplifiers: State-of-the-art. Microwave J (Oct 1972) 15 (10) 23-26, 28, 30.

61. Schmidt, D. and Wirbser, O., Functional integration of microwave integrated circuits into printed circuit-boards. NTZ Comm J (Aug 1972) 25 (8) 423-424.

62. Siemens, MIC in Europe. Microwave J (Apr 1972) 15 (4) 10B-10D.

63. Smolinski, A., Development in microwave solid state electronics (in Polish). Elektronika (1972) (3) 91-94.

64. Sobol, H., Review of the technological and electromagnetic limitations of hybrid circuits for microwave applications. IEEE Tr Parts Hybrid Packaging (Jun 1972) 8 (2) 59-66.

65. Takita, S. et al., Hybrid integrated 70 MHz band main IF amplifier. (Radio repeaters). Rev Electr Comm Lab (Jan-Feb 1972) 20 (1-2) 21-35.

66. Waugh, R. et al., Microwave integrated circuits using thick-film and thin-film technologies. Int Soc Hybrid Microelecncs Symp (1972).

67. Hughes, G.W. and White, R.M., MIS and Schottky barrier microstrip devices. Proc IEEE (Nov 1972) 60 (11) 1460-1461.

1973

1. Aitchison, C.S., Lumped components for microwave frequencies. Philips Tech Rev (1971) 32 (9-12) 305-314.

2. Allen, R.P.G. and Antell, G.R., Monolithic mixers for 60-80 GHz European Microwave Conf (1973) Pap A15.3.

3. Allen, R.P.G. and Antell, G.R., Monolithic mixers for 70 GHz (Abs only). Solid St Dev (1972) Proc 2nd European Conf.Inst of Phys (1973) 206.

4. Anon, Microwave integrated circuits. Pt.2.Inbelec Elecnc App News (Sep/Oct 1973) 10 (5) 39-47.

5. Armstrong, A.L. et al., Application of bulk semiconductor control components to MICs. IEEE Int Conv Dig (1972) 492-493.

6. Arnold, S., Single and dual Gate GaAs FET integrated amplifiers in C Band. IEEE-GMTT Int Microwave Symp (1972) 233-234.

7. Beachtold, W., X- and Ku-band amplifiers with GaAs Schottky-barriers field effect transistors. IEEE J Solid St CCTs (Feb 1973) 8 (1) 54-58.

8. Bennerstedt, T. and Henoch, B.T., Miniaturized active microwave filters. European Microwave Conf (1973) Pap B2.5.

9. Bestwick, P.R. et al., Direct frequency demodulation with CW Gunn and impatt oscillators. IEEE J Solid St Ccts (Feb 1973) 8 (1) 37-43.

10. Biethan, G. and Beneking, H., Hybrid conception for MICs on GaAs substrates. European Microwave Conf (1973) Pap A14.4.

11. Brundle, L.K. and Calemar, M.D., Wideband microwave isolator. Mullard Res Lab UK Ann Rev (1972) 4p.Abs.R & D Rep Announcement Elecncs Comm Edn (15 May 1973) p.l.

12. Bruning H.R., Packaging discrete variable capacitors in MICs. Microwave (Mar 1973) 12 (3) 56, 58, 60-61.

13. Brunton, L.T. et al., Noise performance of low-sensitivity active filters. IEEE J Solid St Ccts (Feb 1973) 8 (1) 85-91.

14. Bura, P., MIC Ku-band upconverter. IEEE Tr Microwave Theor Tech (Mar 1973) 21 (3) 136-137.

15. Cheek, T.M. et al., Wide-band low-shape-factor amplifier module using an acoustic surface wave bandpass filter. IEEE J Solid St Ccts (Feb 1973) 8 (1) 66-70.

16. Choi, S.D. and Boreham, J.F., High power microstrip RF switches. IEEE-GMTT Microwave Symp (1972) 52-54.

17. Chua, L.W., Recent advances in microwave integrated circuit frequency discriminators. Mullard Res Lab UK Ann Rev (1972) 5p. R & D Rep Announc Elecncs Commun edn (15 May 1973) p.1.

18. Climer, B. et al., MIC mixers and receivers for millimetric wave lengths. European Microwave Conf (1973) Pap A15.2.

19. Cohen, J. and Gilden, M., Temperature stability of an MIC Gunn-effect oscillator. IEEE Tr Microwave Theor Tech (Feb 1973) 21 (2) 115-116.

20. Cowpland, M.C.J., Microcircuit tone generator for pushbutton dialing. IEEE J Solid St Ccts (Feb 1973) 8 (1) 78-84.

21. Cristal, E.G. and Gysel, U.H., New compact multiplexer for stripline and MIC. European Microwave Conf (1973) Pap B3.1.

22. Cristal, E.G. et al., 'Microguide' a new microwave integrated circuit transmission line. IEEE-GMTT Int Microwave Symp (1972) 212-214.

23. Curran, J.E. and Lipscombe, R.A.H., Technology and mechanical design of thin film hybrid microwave circuits. Mullard Res Lab Ann Rev (1972) 6p. Abs R & D Rep Announc Elecncs Commun Edn (15 May 1973) p.5.

24. Curtis, J., Let us simplify MIC power amplifier design. Microwaves (Feb 1973) 12 (2) 46-50.

25. de Ronde, F.C., Recent developments in broadband directional couplers on microstrip. IEEE-GMTT Int Microwave Symp.(1972) 215-217.

26. de Santis, P. and Pucci, F., Experiments on the optimization of a novel MIC symmetrical three-port circulator. IEEE-GMTT Int Microwave Symp (1972) 238-240.

27. Digiondomenico, O.J. and Foster, T.M., Hybrids for microwave gear get boost from thick film technology. Electronics (16 Aug 1973) 46 (17) 104-109.

28. Emery, J.K., Surface characteristics of ceramic substrates for hybrid and microwave electronic circuits. Proc of symp on the Science of Ceramic Machinery and Surface Finishing, Gaithersburg Md. USA, 2nd Nov. 1970 (Washington, D.C. USA) N.B.S. p.293-300.

29. Fischer, W. and Wiesbeck, W., Wide band microstrip termination. NTZ Commu J (Feb 1973) 26 (2) 83-85.

30. Gelnovatch, V., Computer aided design, simulation and optimization for microwave integrated circuits. U.S. Govt R & D Rep AD-766677/9 (Aug 1973) 151p.

31. Glance, B. and Snell, W.W., Low-noise integrated millimeter wave receiver. European Microwave Conf (1973) Pap A15.1.

32. Hanson, D.C., Integrated X-band power amplifier using Gunn and impatt diodes. IEEE Int Solid St Ccts Conf Dig (1972) 38-39, 211-212.

33. Hanson,D.C. and Heinz, W.W.,Integrated electrically tuned X-band power amplifier utilizing Gunn and impatt diodes. IEEE J Solid St Ccts (Feb 1973) 8 (1) 3-14.

34. Hendricks, R.K., Space flight-qualified tunable C-band parametric amplifier systems. IEEE-GMTT Int Microwave Symp (1972) 11-13.

35. Holden, R.L. and Burns, R.W., High power UHF microstrip diode phase shifter. IEEE-GMTT Int Microwave Symp (1972) 47-48.

36. Howe, H., Why not use stripline. IEEE Int Conv Dig (1972) 488-489.

37. Howell, J.Q., Quick accurate method to measure the dielectric constant of MIC substrates. IEEE Tr Microwave Theo Tech (Mar 1973) 21 (3) 142-143.

38. Ito, Y. et al., K-band integrated microstrip modulator and mixer in waveguide for a high speed PCM radio repeater. IEEE Int Solid St Ccts Conf Dig (1972) 160-161.

39. Itoh, T., Method for analyzing the characteristics of some lumped elements in MICs. Abs. Tr Inst Elecncs Comm Engrs (Japan) (Aug 1973) 56 (8) 8-9.

40. Itoh, T. and Mittra, R., New method for calculating the capacitance of a circular disk for microwave integrated circuits. IEEE Tr Microwave Theor Tech (Jun 1973) 21 (6) 431-432.

41. Jones, R.R. and Moore, R.A., Elevated substrate ferrite film circulator. IEEE-GMTT Int Microwave Symp (1972) 241-242.

42. Kamnitsis,C. et al., Production techniques for MICs. IEEE Elecnc Comps Conf (1972) 32, 408-411.

43. Katoh, H., Lumped element approach to microwave integrated circuits. Abs Tr of the Inst of Elecncs & Comm Engrs Japan (Jun 1973) 56 (6) 10-11.

44. Kawamoto, H., Gigahertz-rate 100-V pulse generator. IEEE J Solid St Ccts (Feb 1973) 8 (1) 63-66.

45. Kitasawa, T. et al., Slot line on a magnetized ferrite substrate. Abs. Tr Instn Elecncs Comm Engrs Japan (Mar 1973) 56 (3) 9-10.

46. Knerr, R.H., 4-GHz lumped-element circulator. IEEE Tr Microwave Theo Tech (Mar 1973) 21 (3) 150-151.

47. Komizo, H. et al., A 0.5-W CW impatt diode amplifier for high capacity 11 GHz FM radio-relay equipment. IEEE J Solid St Ccts (Feb 1973) 8 (1) 14-20.

48. Koishi, Y., Microwave planar circuit mounted in wave guide. JEE (Aug 1973) 40-44.

49. Ladbrooke,P.H. et al., Comments on A quick accurate method to measure the dielectric constant of MIC substrates. IEEE Tr Microwave Theor Tech (Aug 1973) 21 (8) 570.

50. Ladbrooke, P.H. et al., Coupling errors in cavity resonance measurement on MIC dielectrics. IEEE Tr Microwave Theor Tech (Aug 1973) 21 (8) 560-562.

51. Lehrfeld, S.S., Hints and kinks for your MIC package design. Microwaves (Mar 1973) 12 (3) 62-64.

52. Lemke, M. et al., Integrated subsystems on composite substrates. AGARD Conf Proc No. 66 on Advanced Radar Systems (1970) Pap 11,10p.

53. Lenzing H.F., L-band high band-rate DC PSK detector/AFC discriminator in microstrip. IEEE-GMTT Int Microwave Symp (1972) 201-203.

54. Macdonald, I.H. et al., Pin limiter for microstrip circuits. Microwave J (Jan 1973) 16 (1) 52, 54, 56.

55. Mackey, W.R., Review of microwave integrated circuit technology. Bendix Tech J (1972) 5, 1-6.

56. Mackintosh, I.W., Circuit mode chart for high efficiency avalanche diode oscillators. IEEE J Solid St Ccts (Feb 1973) 8 (1) 44-53.

57. Meier, P.J., Two new integrated-circuit media with special advantages at millimeter wavelengths. IEEE-GMTT Int Microwave Symp (1972) 221-223.

58. MICs in instrumentation. IEEE Intercon Tech Pap (1973) Session 55 V. 8, 3 Papers.

59. Miyoshi, T. and Okoshi, T., Analysis of microwave planar circuits. Elecncs Comm Japan (Aug 1972) 55 (8) 24-31.

60. Newton, B.H. and William, J.C., Stabilization of microwave integrated circuit Gunn oscillators. Mullard Res Lab UK Ann Rev (1972) 3p. Abs. R & D Rep Announc Elecncs Commun edn (15 May 1973) p.1.

61. Okamura, S. and Nagai, T., Lumped-element circulator on ceramic substrates. IEEE-GMTT Int Microwave Symp (1972) 243-245.

62. Oxley, T.H., Application of hybrid integrated circuit techniques to microwave receivers. European Microwave Conf (1973) Pap A14.1 (1).

63. Oxley, T.H. et al., Hybrid microwave integrated circuits for millimeter wavelengths. IEEE-GMTT Int Microwave Symp.(1972) 224-226.

64. Payne, T.R. and Plumlee, H.R., Thermal printer. IEEE J Solid St Ccts (Feb 1973) 8 (1) 71-78.

65. Pflieger, R.H., Spaceborne hybrid MIC Pin diode radiometer switch. IEEE-GMTT Int Microwave Symp (1972) 8-9.

66. Pitzalis, O., Perspectives on lumped and distributed MICs. IEEE Int Conv Dig (1972) 490-491.

67. Raymond, R. and Gerard, O., Millimetric circulators and dielectric resonator filters in microelectronic technology. European Microwave Conf (1973) Pap A15.5.

68. Rippy, R.R. and Line, L.G., Discrete components microwave amplifiers. IEEE Intercon Tech Pap (1973) Sess 55 V.8 Pap3 4p.

69. Roschmann, P., Compact YIG bandpass filter with finite-pole frequencies for applications in microwave integrated circuits. IEEE Tr Microwave Theor Tech (Jan 1973) 21(1) 52-57.

70. Special Issue, Microwave Integrated Circuits. IEEE J Solid St Ccts (Feb 1973) 8(1) 1-100.

71. Sterzer, F., Microwave integrated circuits-an overview. IEEE Int Conv Dig (1972) 486.

72. Sullivan, F. and Perry, R., All Solid-state MIC transmit-receive module. IEEE-GMTT Int Microwave Symp (1972) 227-229.

73. Sweet, A.A., Multistage Gunn amplifiers for FM-CW systems. IEEE J Solid St Ccts (Feb 1973) 8 (1) 20-28.

74. Tajima, Y., Numerical analysis of MIC circular resonators. Elecncs Comm Japan (July 1972) 55-B (7) 46-53.

75. Tenenholtz, R., Broadband MIC multithrow PIN diode switches. Microwave J (Jul 1973) 16(7) 25-28.

76. Tsai, W.C. and Lee, C.W., C-band all ferrite integrated wideband high power GaAs avalanche diode amplifier. IEEE-GMTT Int Microwave Symp (1972) 179-181.

77. Upadhyayula, C.L. and Perlman, B.S., Design and performance of transferred electron amplifiers using distributed equalizer networks. IEEE J Solid St Ccts (Feb 1973) 8 (1) 29-36.

78. Ura, F., Thin film hybrid microwave integrated circuits. IEEE Intercon Tech Pap (1973) V.8 Sess 55 Pap 1, 8p.

79. Van Heuven, J.H.C. and Van Nie, A.G., Microwave integrated circuits. Philips Tech Rev (1971) 32(9-12) 292-304.

80. Weinreb, S. and Kerr, A.R., Cryogenic cooling of mixers for millimeter and centimeter wavelengths. IEEE J Solid St Ccts (Feb 1973) 8 (1) 58-63.

81. William, Ceramic waveguide microwave integrated circuits. IEEE-GMTT Int Microwave Symp (1972) 218-320.

82. Wittig, K.A., Microwave amplifier for L-band phased array. IEEE Reg Six (US Western Reg.) Conf Rec. (1972) 158-161.

83. Zavadil, J.L., Slot transmission line. Naval postgraduate School, Montersey Calif. Masters Thesis, US Govt R & D Rep AD-751657 (Dec 1971) 36p.

84. Lemke, M. and Schilz, W., Microwave integrated circuits on a ferrite substrate. Philips Tech Rev (1971) 32 (9-12) 315-32.

85. Tischer, F.J., Integrated circuitry for millimeter wave. European Microwave Conf (1973) Pap A15.4.

SUBJECT INDEX

In subject index (and also in author index) first two digits indicate the year of publication and the digits after the hyphen indicate the reference number in the bibliography for that year.

AUTHOR INDEX

ACKNOWLEDGEMENTS

It is pleasure to thank various authors and publishing houses who have kindly extended their permission to reproduce illustrations and tables from their publications. Thanks are due to:

Prof. Dr. H. Beneking for Figures 12.1a and 12.1b from his paper in Proc. European Microwave Conf., Paper No. C2/1, 1971.

Dr. George R. Brewer for Figures 13.7 and 13.8 from his paper in IEEE Spectrum, Vol.8, Jan. 1971.

Dr. Martin Caulton for Figures 3.6, 3.7, 3.15, 3.16 and 6.8 from his papers in RCA Review and IEEE Trans. Microwave Theory Tech.

Dr. Tom Hyltin for Figures 12.3 and 12.4 from his paper in IEEE Trans, MTT-13, Nov. 1965.

Dr. A. Schwarzmann for Figures 5.3 and 5.4 from his paper in Microwave Journal, Vol.12, May 1969.

Dr. Harold Sobol for Tables 11.3 and 11.4 from his paper in Proc. European Microwave Conf., Vol.II, C1/S:8, 1971.

Dr. Harold E. Stinehelfer for Fig. 3.10 from his paper in IEEE Trans., ED-13, July 1968.

Dr. Rudolf E. Thun for Tables 11.2 and 11.5 from his paper in IEEE Spectrum, Vol.6, Oct 1969.

Electronics (McGraw-Hill Publications) New York (USA) for Figure 12.2 from Electronics, June 10, 1968.

Horizon House, Dedham, Mass (USA) for Figures 5.3 and 5.4 from Microwave Journal, May 1969.

Institute of Electrical and Electronics Engineers, Inc. New York (USA) for Figures 3.11, 3.12, 4.4, 6.8, 6.11, 6.12, 6.13, 12.3, 12.4, 13.7 and 13.8 and Tables 11.1, 11.2, 11.5 and 11.6 from various IEEE publications.

The Institution of Telecommunication Engineers, New Delhi, for granting permission to publish Bibliography which appeared in various issues of Jour. Instn. Telecomm. Engrs (India).

McGraw-Hill Book Company New York (USA) for Figure 8.5 from 'Microwave Semiconductor Devices and their circuit Applications' by H.A. Watson, 1969, McGraw-Hill Book Company.

Microwave Journal for Figure 5.3 and 5.4 from Microwave Journal, Vol.12, May 1969.

Mullard Research Laboratories, Red Hill (England) for Figures 6.11, 6.12 and 6.13 from Dr. C.S. Aitchison's paper in IEEE Trans., MTT-19, Dec 1971.

RCA, Missile and Surface Radar Division Moores town N.J. (USA) for Figures 5.3 and 5.4 from Dr. A. Schwarzmanns paper in Microwave Journal Vol.12, May 1969.

Royal Swedish Academy of Engineering Sciences, Stockholm, Sweden, for Figures 3.13, 12.1a and 12.1b, Tables 11.3 and 11.4 from Proc. European Microwave Conference, 1971.

Thomson - CSF, Department Microelectronique Hyperfrequence (DMH) Paris (France) for Figures 2.2, 2.3, 2.6, 2.12, 2.13, 10.1, 10.2, 10.3, 10.5, 10.6, 10.8 from miscellaneous Thomson-CSF documents.

IBM RESEARCH LIBRARY
MAY 5 1975
YORKTOWN, HEIGHTS, N. Y.